Human–Computer Interaction Series

More information about this series at http://www.springer.com/series/6033

Tracy Hammond • Aaron Adler • Manoj Prasad
Editors

Frontiers in Pen and Touch

Impact of Pen and Touch Technology on Education

Springer

Editors
Tracy Hammond
Sketch Recognition Lab
Department of Computer
Science & Engineering
Texas A&M University
College Station, TX, USA

Aaron Adler
Information and Knowledge Technologies
Raytheon BBN Technologies
Columbia, MD, USA

Manoj Prasad
Microsoft
Redmond, WA, USA

ISSN 1571-5035
Human–Computer Interaction Series
ISBN 978-3-319-64238-3 ISBN 978-3-319-64239-0 (eBook)
DOI 10.1007/978-3-319-64239-0

Library of Congress Control Number: 2017954388

Printed on acid-free paper

This Springer imprint is published by Springer Nature
The registered company is Springer International Publishing AG
The registered company address is: Gewerbestrasse 11, 6330 Cham, Switzerland

Foreword

The 2016 Conference on Pen and Touch Technology in Education was the tenth annual iteration of this annual event. It was a special pleasure for me to chair the conference at Brown, where I also organized a workshop on pen computing in the spring of 2007 (see Fig. 1), the forerunner of WIPTTE (now renamed CPTTE). Some of this year's speakers and attendees participated in that first workshop. This year's conference combined both researchers and educators exchanging best practices. The conference featured the usual CPTTE blend of keynote speakers, research papers, presentations by practicing teachers about techniques they have found worthwhile in the classroom, and hands-on You-Try-It sessions. Participants also had ample opportunities to connect with one another to explore ideas, research applications, and practices, allowing for a stimulating cross-pollination between members of this diverse community.

Microsoft continues to dominate the field with its offering of hardware (Surface tablets and the large Surface Hubs); platform support for digital ink, pen, and touch interaction; and popular applications such as the OneNote family. Other manufacturers such as Samsung, Fujitsu, and Wacom continue to expand their product offerings, and even Apple, who eschewed the pen under Steve Job's leadership, has finally started offering the Apple Pencil for the iPad Pro models. Classroom adoption of tablets grows slowly but steadily but has not yet hit the knee of what we all still hope will be an exponential growth curve. As styli become more prevalent, there will be greater interest in expanding the minimalist interaction vocabulary popularized by the iPhone and its competitors (tap, swipe, pinch-zoom) by adding digital ink and character, gesture, and sketch recognition as first-class citizens. It is encouraging to see companies like Adobe putting increasing emphasis on tablets rather than just desktops. Our CPTTE keynoters and speakers, along with the resulting papers in this manuscript, provide us with inspiration and vision so we can continue building interest and community in this still young field.

We are grateful to the many sponsors who have so generously funded us this year and without whom this whole affair would have been impossible. Continuing their support from WIPTTE 2015 (and previous years) are Microsoft (Windows,

Fig. 1 In 2007, a kick-off pen-centric computing workshop was held at Brown. Many of the people in the initial photograph attended CPTTE 2017 and are still involved in the pen-computing community

OneNote, Surface, and Research) as platinum-level sponsor, Fujitsu and Wacom as gold-level sponsors, and PDF Annotator as bronze-level sponsor. All are vital to the success of this conference. Thanks are also owed to host sponsor Brown University, in particular the Department of Computer Science and the Office of University Event and Conference Services.

It's a pleasure to acknowledge here the hard work put in by all of the members of the organizing committee. I especially want to acknowledge last year's organizer/chair Jonathan Grudin; program co-chairs Aaron Adler, Mark Payton, and Manoj Prasad; program committee member Eric Hamilton; past chair Tracy Hammond; conference coordinator Lisa Manekofsky; and the rest of the CPTTE 2016 Organizing Committee. The amount of work done by these folks has been well beyond reasonable, and we simply would not be here without their efforts.

Thank you all for contributing—welcome to what we hope you will find to be a stimulating exchange of ideas.

CPTTE 2016 Chair Dr. Andy van Dam

Dr. Andy van Dam: Andries van Dam is the Thomas J. Watson Jr. university professor of technology and education and professor of computer science at Brown University. He has been a member of Brown's faculty since 1965, was a cofounder of Brown's Department of Computer Science and its first chairman from 1979 to 1985, and was also Brown's first vice president for research from 2002 to 2006. His research includes work on computer graphics; hypermedia systems; post-WIMP and natural user interfaces (NUI), including pen and touch computing; and educational software. He has been working for over four decades on systems for creating and reading electronic books with interactive illustrations for use in teaching and research. In 1967 Prof. van Dam cofounded ACM SICGRAPH (the precursor of SIGGRAPH) and from 1985 through 1987 was chairman of the Computing Research Association. He is a fellow of ACM, IEEE, and AAAS and a member of the National Academy of Engineering and the American Academy of Arts and Sciences. He has received the ACM Karl V. Karlstrom Outstanding Educator Award, the SIGGRAPH Steven A. Coons Award for Outstanding Creative Contributions to Computer Graphics, and the IEEE Centennial Medal and holds four honorary doctorates from Darmstadt Technical University in Germany, Swarthmore College, the University of Waterloo in Canada, and ETH Zurich. He has authored or coauthored over 100 papers and nine books, including *Fundamentals of Interactive Computer Graphics* and three editions of *Computer Graphics: Principles and Practice*.

Contents

Contributors

Aaron Adler Information and Knowledge Technologies, Raytheon BBN Technologies, Columbia, MD, USA

Phillip Z. Ai MIT, Cambridge, MA, USA

Matt Antantis South Fayette School District, McDonald, PA, USA

Calvin Armstrong Appleby College, Oakville, ON, Canada

Robert Baker Cincinnati Country Day School, Cincinnati, OH, USA

Randy Brooks Lovejoy Independent School District, Lucas, TX, USA

Aqib Niaz Bhat Sketch Recognition Lab, Department of Computer Science & Engineering, Texas A&M University, College Station, TX, USA

Bill Buxton Microsoft Research, Microsoft, One Microsoft Way, Redmond, WA, USA

Miranda Chao Graphics Lab, Computer Science Department, Brown University, Providence, RI, USA

Steven Chapman stevecode.com, Cambridge, MA, USA

Tiffany Citra Graphics Lab, Computer Science Department, Brown University, Providence, RI, USA

Samuel Lawrence Cohen South Fayette School District, McDonald, PA, USA

Philipp Eichmann Computer Science Department, Brown University, Providence, RI, USA

Jennifer French Department of Mathematics, MIT, Cambridge, MA, USA

Daniel W. Goldberg Department of Geography, Department of Computer Science & Engineering, Texas A&M University, College Station, TX, USA

Trent Green Graphics Lab, Computer Science Department, Brown University, Providence, RI, USA

Jonathan Grudin Microsoft, Microsoft Way, Redmond, WA, USA

Judith V. Gutierrez Cuba Department of Chemical and Food Engineering, Center for Science, Engineering, and Technology Education, Universidad de las Americas Puebla, Cholula, Puebla, Mexico

Eric Hamilton Graduate School of Education and Psychology, Pepperdine University, Malibu, CA, USA

Tracy Hammond Sketch Recognition Lab, Department of Computer Science & Engineering, Texas A&M University, College Station, TX, USA

Michael Helms IDREEM Lab, George W. Woodruff School of Mechanical Engineering, Georgia Institute of Technology, Atlanta, GA, USA

Nicholas Jackiw SRI International, Menlo Park, CA, USA

Prateek Jukalkar South Fayette School District, McDonald, PA, USA

Girish Kasiviswanathan Sketch Recognition Lab, Natural Language Processing Lab, Department of Computer Science & Engineering, Texas A&M University, College Station, TX, USA

Richard Kassissieh University Prep, Seattle, WA, USA

Hiroo Kato Graduate School of Education and Psychology, Pepperdine University, Malibu, CA, USA

Lily Ko TERC, Cambridge, MA, USA

Jung In Koh Sketch Recognition Lab, Department of Computer Science & Engineering, Texas A&M University, College Station, TX, USA

Kimberle Koile Computer Science and Artificial Intelligence Laboratory, MIT, Cambridge, MA, USA

Joseph J. LaViola Jr. Interactive Systems and User Experience Research Cluster of Excellence, Department of Computer Science, University of Central Florida, Orlando, FL, USA

Aurelio Lopez-Malo Department of Chemical and Food Engineering, Center for Science, Engineering, and Technology Education, Universidad de las Americas Puebla, Cholula, Puebla, Mexico

Pengchen Lyu Sketch Recognition Lab, Department of Computer Science & Engineering, Texas A&M University, College Station, TX, USA

Shuo Ma Sketch Recognition Lab, Department of Computer Science & Engineering, Texas A&M University, College Station, TX, USA

Sara Mata Whitfield School, Creve Coeur, MO, USA

Christy Maria Mathew Sketch Recognition Lab, Department of Computer Science & Engineering, Texas A&M University, College Station, TX, USA

Caio D.D. Monteiro Sketch Recognition Lab, Department of Computer Science & Engineering, Texas A&M University, College Station, TX, USA

Meenakshi Narayanan Sketch Recognition Lab, Department of Computer Science & Engineering, Texas A&M University, College Station, TX, USA

Carlene Niguidula Graphics Lab, Computer Science Department, Brown University, Providence, RI, USA

Aileen Owens South Fayette High School, McDonald, PA, USA

Michel Pahud Microsoft Research, Redmond, WA, USA

Enrique Palou Department of Chemical and Food Engineering, Center for Science, Engineering, and Technology Education, Universidad de las Americas Puebla, Cholula, Puebla, Mexico

Mark Payton Peraza Dermatology Group, Claremont, NH, USA Whitfield School, Saint Louis, MO, USA

Vinay Pedapati South Fayette School District, McDonald, PA, USA

Seth Polsley Sketch Recognition Lab, Department of Computer Science & Engineering, Texas A&M University, College Station, TX, USA

Manoj Prasad Microsoft, Redmond, WA, USA

Erik Prout Department of Geography, Texas A&M University, College Station, TX, USA

Nelly Ramirez-Corona Department of Chemical and Food Engineering, Center for Science, Engineering, and Technology Education, Universidad de las Americas Puebla, Cholula, Puebla, Mexico

Ashumi Rokadia South Fayette School District, McDonald, PA, USA

Andee Rubin TERC, Cambridge, MA, USA

Martin A. Segado Department of Mechanical Engineering, MIT, Cambridge, MA, USA

Parv Shrivastava South Fayette School District, McDonald, PA, USA

Nathalie Sinclair Simon Fraser University, Burnaby, BC, Canada

Thomas Stahovich Bourns College of Engineering, University of California, Riverside, CA, USA

Anna Stepanova Sketch Recognition Lab, Department of Computer Science & Engineering, Texas A&M University, College Station, TX, USA

Yongbin Sun Sketch Recognition Lab, Department of Computer Science & Engineering, Texas A&M University, College Station, TX, USA

Eugene M. Taranta II Interactive Systems and User Experience Research Cluster of Excellence, Department of Computer Science, University of Central Florida, Orlando, FL, USA

Andy van Dam Department of Computer Science, Brown University, Providence, RI, USA

Lucy Van Kleunen Graphics Lab, Computer Science Department, Brown University, Providence, RI, USA

Andrés N. Vargas González Interactive Systems and User Experience Research Cluster of Excellence, Department of Computer Science, University of Central Florida, Orlando, FL, USA

Nicholas Wilke South Fayette School District, McDonald, PA, USA

Michelle Hoda Wilkerson Graduate School of Education, University of California-Berkeley, Berkeley, CA, USA

Part I
Introductions and Welcome

Chapter 1
Introduction: Frontiers in Pen and Touch Research

Tracy Hammond, Aaron Adler, Manoj Prasad, and Anna Stepanova

Abstract Derived from contributions to the Conference on Pen and Touch Technology in Education (CPTTE) in 2016, this edited volume highlights recent developments for pen and tablet research within the domains of computer science, education, and outreach. This book will be of great interest for researchers and educators at every level. This book covers topics such as teaching math at various levels, from elementary to K16 and higher education classroom, as well as new technologies for K-12 classrooms, revealing novel tutoring systems in STEM education, and creating informal education, outreach, and games. Throughout this book, practitioners and theorists propose new ideas to advance accessibility to knowledge and outline future perspectives in digital teaching and learning.

1.1 Introduction

This monograph represents a collection of papers from the 2016 Conference on the Impact of Pen and Touch Technology on Education (CPTTE) held from March 31 through April 2 at the Brown University Campus in Providence, RI, USA. The conference was chaired by Dr. Andy van Dam, Professor of Computer Science at

T. Hammond (✉) • A. Stepanova
Sketch Recognition Lab, Department of Computer Science & Engineering, Texas A&M University, College Station, TX, 77843, USA
e-mail: hammond@tamu.edu; anna_stepanova@tamu.edu

A. Adler
Information and Knowledge Technologies, Raytheon BBN Technologies, 9861 Broken Land Parkway, Suite 400, Columbia, MD, 21046, USA
e-mail: aaronadler@alum.mit.edu

M. Prasad
Microsoft, 14820 NE 36th Street, Building 99, Redmond, WA, 98052, USA
e-mail: dasarpjonam@gmail.com

T. Hammond et al. (eds.), *Frontiers in Pen and Touch*, Human–Computer Interaction Series, DOI 10.1007/978-3-319-64239-0_1

Brown University, and co-chaired by Dr. Aaron Adler, Senior Scientist at BBN Technologies, Mark Payton, Director of Technology and Library Resources at Whitfield School, and Dr. Manoj Prasad, Software Engineer at Microsoft and Researcher at Sketch Recognition Lab.

The 2016 CPTTE marks the 10th annual instantiation of the conference, organized first in 2006. Each year the conference gets stronger and stronger, held in a different location each year. Success over the years has encouraged us to transform this conference from a workshop—WIPTTE, Workshop on the Impact of Pen and Touch Technology in Education—to a conference. In 2017, the conference will be held on the Northwestern University Campus in Evanston, IL, USA.

As highlighted in the Foreward by the workshop chair, Dr. Andy van Dam, this year marks a great amount of innovation for pen and tablet research as styli become more prevalent and many companies are suddenly putting a strong emphasis on pen-integrated technologies. The workshop and this volume define how pen and touch computing are positively impacting the whole of education.

1.2 CPTTE 2016: Venue

CPTTE 2016 was held at Brown University in Providence, Rhode Island (Figs. 1.1, 1.2, and 1.3). Brown University is a private institution that was founded in 1764 and the seventh-oldest college in the United States. It was ranked #14 in the National Universities category, #4 in best Undergraduate teaching, and #5 best Value Schools in the 2017 edition of Best Colleges. Brown is one of the leading research universities and is frequently recognized for its global outreach, its rich cultural events, its diverse campus groups and activities, its active community service programs, its highly competitive athletics, and its bucolic environment with state-of-the-art facilities located in a New England historic urban setting.

1.3 CPTTE 2016: Keynotes

In 2016, the conference had four keynote speakers Jeff Han, Tracy Hammond, Bill Buxton, and Rob Baker. Jeff Han started off on Thursday, the first day of the conference. Tracy Hammond's keynote was on Friday, and Bill Buxton and Rob Baker gave their keynotes on Saturday, the last day of the conference.

Jeff Han opened the conference with his keynote, in which he talked about "how much more touch can do than just swipe." He spoke about his personal journey that he took over the last 10 years. It started at Cornell University where he studied electrical engineering and computer science. His research interests were

Fig. 1.1 Brown University

Fig. 1.2 Brown University

Fig. 1.3 Brown University

in computer graphics. Among his projects were CU-SeeMe, the early internet multi-party videoconferencing application, which subsequently led to his first startup in the late 1990s; real-time rendering and realistic images of such complex images like fabric weave; robot navigation and social network mapping; and eventually the multi-touch sensing device that allowed finger scaling and image producing. In 2006 he gave his famous TED talk—dramatic live demonstration of multi-touch interaction techniques he published a year prior, launching the interest into the public mindset well before the introduction of Apple's iPhone and Microsoft Surface Table. Jeff talked about the foundation of the Perspective Pixel (PP), a successful venture-backed startup dedicated to the research and productization of advanced user interfaces for the knowledge worker, which was acquired by Microsoft in 2012. One of the main goals of the PP was to produce devices that even an untrained user could use for his job. In 2008, such touch-screen devices were used by CNN to monitor the presidential election and showing real-time voting maps by states. Later Jeff's research moved to the multi-point interaction in the 3D domain, as well as other domains such as oil exploration and art. In the recent years Jeff has been working with the Microsoft Surface Hub that is targeted for working groups, designed to accommodate multiple users and to provide a shared collaborative space to work.

In 2008 Jeff was named one of the *"Time 10"* most influential persons of the year. In 2009 Jeff received the Smithsonian National Design Award in the interaction design category (Fig. 1.4).

Fig. 1.4 Jeff Han asking a question

Tracy Hammond gave a talk entitled "Learning Through the Lens of Sketch" and spoke about how both her research and the field of Sketch Recognition evolved over the last decade. One of Tracy's motivations of her career was to develop algorithms that provide insights into human brain activity. Tracy explained that the best applications are those that allow us to communicate with each other and ourselves the best way. Tracy's initial work focused on domain-independent recognition methods through the LADDER Platform. Her current work focuses on harnessing the power of sketch recognition in education. The Mechanix system recognizes hand-drawn student answers and provides immediate feedback to Mechanical Engineering students taking a statics course. Two applications focused on using sketch recognition to teach drawing: iCanDraw (faces) and PerSketchTivity (3D perspective). In these projects she took advantage of a principle that she learned along the way that "people do not draw what they see, they draw what they perceive." Her work on corner finding and primitive recognition helped push forth several innovations. Because one's pen slows down when drawing a corner, and because sound is directly correlated to pen speed, the sound of the pen can be used to recognize what people are drawing. Additionally, when making corners, minuscule features, called NDDE and DCR, identify low level planning that goes on when drawing a corner; these features show up at particular times during child development. Hammond concluded her talk by showing how sketch recognition techniques can provide insights to other types of human motion, such as using eye-tracking to predict a diagnosis expertise, identify what one is looking at, or identify demographic information or even the user him/herself, as well as recognizing other

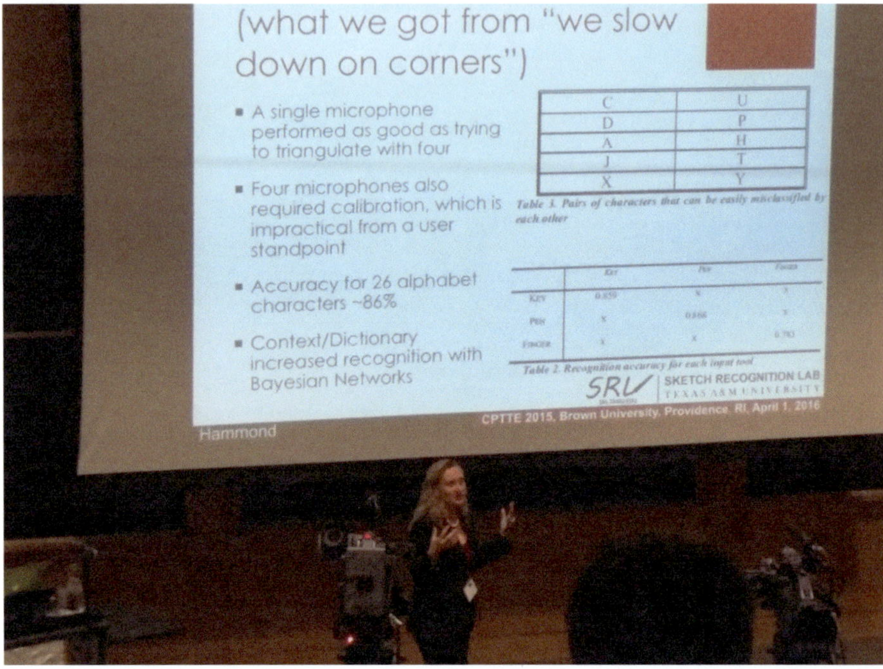

Fig. 1.5 Hammond giving her keynote

human motions and predicting their behaviors through accelerometer data gained through wearables.

Tracy Hammond is an international leader in activity recognition, focusing on eye, body and sketch motions, haptics, intelligent fabrics, smartphone development, and computer human interaction research. She is the Director of the Sketch Recognition Lab and Professor at the Department of Computer Science & Engineering at Texas A&M University. She has published over 150 scientific articles in a wide range of venues. Dr. Hammond is an editor of conference volumes and a contributor to previous WIPTTE/CPTTE conferences. Tracy holds a Ph.D. in Computer Science and FTO (Finance Technology Option) from MIT and four degrees from Columbia University: an M.S. in Anthropology, an M.S. in Computer Science, a B.A. in mathematics, and a B.S. in applied mathematics. Prior to joining Texas A&M University, Dr. Hammond taught for 5 years at Columbia University and was a telecom analyst for 4 years at Goldman Sachs (Fig. 1.5).

Bill Buxton made a presentation entitled, "Consistency and Ubiety in Interaction with Displays Large to Small" in which he talked about challenges in designing "Universal apps" and tools to support their development. In 2015, Jeff Han and Bill Buxton organized a workshop at the Microsoft Faculty Summit on large displays. Jeff had seeded a number of universities with 55″ Microsoft displays, and at the workshop, invited the recipients to present what they had done over the preceding

Fig. 1.6 Buxton giving his keynote

year. As a bonus, Han and Buxton offered them the opportunity to present on the new 84″ version of the display—a display that they had never seen before. Most accepted. The greatest lesson from this was how poorly things designed for the 55″ worked on the 84″—despite all other things besides size being equal. In his keynote, Bill shed some light on what happened and why. The main message of his talk was, that by pushing the extremes, insights emerge which enlighten not only how we design for the extremes, but also how we can improve design for what lies in-between, and in the process, raise the promise of attaining far more consistent interaction across form factors.

Bill Buxton is a singer, musician, lecturer, writer, teacher, critic, and researcher. His focus has always been at the human level—from motor-sensory to cultural—with a particular interest in the creative disciplines. His work has helped shape how we communicate both with technology, and each other through technology, with our human qualities enhanced (Fig. 1.6).

Rob Baker presented a talk entitled, "INK Everywhere—Digital Ink is a powerful Modality," where he talked about his experience in deploying devices and going 1:1 at the Cincinnati Country Day School, where he is a Director of Technology. Rob said that both, deploying and going 1:1 were not the finish line. These merely enable collaboration, engagement, alternate forms of assessment, creativity, personalized learning, and problem solving. Schools need to focus on functionality and deploy the right kind of hardware that allows users to choose the modality that fits the subject and task at hand rather than constrain learning, or cater only to certain types of learning tasks. In education, the device DOES matter! Baker's focus has always been on removing constraints and enabling great pedagogy. Unfortunately, most tech implementations in education only ever aspire to be great add-on programs. Technology can empower transformational pedagogy and learning environments. Digital Ink was a key driver for Baker in 2007 when he hacked OneNote and created their shared OneNote Class Notebooks that were the conception of the powerful framework the OneNote team has made available to every teacher and student in the world. Rob shared his insight and at the transformational effect Cincinnati Country Day School has experienced as a result of incorporating Tablet PCs with digital ink, touch and keyboard for the last 13 years.

Rob Baker is responsible for creating some of the most powerful teaching and learning environments in the world. He has been working on developing technology programs both in United States and internationally focused on embedding a range of new technologies into the learning process. Rob is an educator, which has allowed him to look at initiatives through the eyes of a classroom teacher focused on pedagogy, not technology. His work at Country Day shows what is possible when school deploys the right kind of hardware, which removes all constraints and allows teachers to use the modality that fits the task. Some of Rob's most important work has been with Microsoft on replicating Country Day's shared OneNote environment that he developed and implemented in 2007, and the fruit of that labor, the Class Notebook Creator App for Office 365 has made this a powerful framework for teaching and learning available to every teacher and student in the world (Fig. 1.7).

Fig. 1.7 Baker giving his keynote

1.4 CPTTE 2016: Sponsor Talks

While sponsors always play a large part in the CPTTE conference from year to year, this year the conference was especially blessed with a number of sponsor talks highlighting riveting new advances from the world of industry. Last year's conference at Microsoft kicked off many new technologies, and we saw more this year. We list some of the abstracts from those talks below.

Sponsor Talk: Microsoft OneNote: Designing OneNote Ink for Education (Ian Mikutel and Sarah Sykes). Ian Mikutel, Project Manager for OneNote Ink, and Sarah Sykes, Designer for OneNote Ink, talk about the process of building OneNote's ink technology for teachers and students worldwide. Topics include a look inside the new customer connected design process, illustrated by real examples from upcoming new inking features, and a discussion on where we are now and what's next for OneNote and digital ink.

Sponsor Talk: Microsoft Surface: Unlocking the Potential of Pen (Amy Sousa, Vineet Thurvara, and Daryl Wilson). At Microsoft, we have been committed to Pen from the very beginning with the Stylus. Today, Surface carries on this mission with a digital inking experience that blends powerful technology and elegance to make writing, drawing, and learning more intuitive and productive than ever before. Being customer obsessed, we also realize that users sometimes struggle to fully integrate digital inking into their everyday lives. The newest Surface Pen addresses these barriers head on with technologies to make digital inking as indispensable, intuitive, and satisfying as pen and paper. In this talk, we combine a discussion of the Surface Pen's feature set with rich ethnographic data to explore the value digital inking creates in the daily lives of students and professionals alike.

Sponsor Talk: Wacom: WILL Universal Ink for the Twenty-First Century (Sierra Modro). Amazing digital pens need amazing ink. Learn about updates to the WILL SDK to make it easier to integrate digital ink into any application. We'll also cover a case study of using WILL in the new Bamboo Spark smart folio. Wacom has new initiatives for 2016 that can keep you on the forefront of changes in the world of digital pen and ink. Learn how to get involved in the movement.

1.5 CPTTE 2016: You-Try-Its

One of the unique features of the CPTTE/WIPTTE conference series is the hands-on quality of many of the sessions. Beyond the traditional talk series, researchers and practitioners are encouraged to run interactive demo sessions where the users can interact with the various hardware and software solutions to see how they could work in their own classrooms, schools, or research. Other conferences may call these workshops or courses, but they are a little of both, and neither of those terms convey the significant amount of hands-on learning that goes on during these sessions.

Microsoft supplied tablets for the various demo experiences with all of the sample software pre-loaded so that every participant could take part in real-time no matter what their own personal device was.

You-Try-It: Experience Pen Input and Video Editing in Education Media Making (Hiroo Kato, Debra Sterling Chittur, and Eric Hamilton; Pepperdine University) This is an introduction to creating short educational videos, with an emphasis on pen input and video editing, which contribute to rapid video production and high learner engagement. In the limited time provided, participants will experience video making through the screen-capturing, pen-input, and also through editing multiple video/audio tracks. The activities will be heavily scaffolded using pre-made project templates, where much of the initial learning curve is suspended in order to experience the more creative and engaging aspects of educational media making. Various software will be utilized, and alternatives, including free options, will be demonstrated and/or displayed. This short session has proven popular in numerous conferences in the US and overseas. The overall effort is connected to a Fulbright research fellowship in sub-Saharan Africa and to a National Science Foundation Cyberlearning project in California and in Nevada. The workshop gave a meaningful initial experience of video making/editing, however, further resources will be provided so that willing participants can master other skills in their own time.

You-Try-It: Classroom Learning Partner: Pen-based Software for Creating and Sharing Visual Representation (Kimberle Koile, Andee Rubin, Steven Chapman, and Lily Ko; MIT CS & AI Lab, TERC, StevCode) This session provided hands-on experience with Classroom Learning Partner (CLP), a pen-based classroom interaction system whose goal is to improve STEM teaching and learning by supporting the creation and sharing of visual representation. Participants played the role of students in an elementary math classroom, using CLP's digital tools to create visual representation for multiplication and division problems. They also were able to view the teacher's version of the software, which provides information about students' use of the digital tools, with the goal of providing teachers with insights into student' mathematical thinking.

You-Try-It: Mechanix Vector Analysis Deployment in High School (Randy Brooks and Tracy Hammond; LoveJoy Highschool and TAMU Sketch Recognition Lab) In June of 2015, three high school teachers spent 2 weeks with the team at the Texas A&M Sketch Recognition Lab (TAMU SRL) as part of an NSF funded RET project (NSF EEC 1129525: Collaborative Research: Enabling Instructors to Teach Statics Actively) learned both the instructional and learning impacts of Mechanix, a software-based teaching tool addressing truss and free body diagram analysis. The high school teacher mission was to return to their schools in the Fall of 2015 and pilot the use of Mechanix in the high school environment. In this CPTTE 2016 you-try-it, attendees had the opportunity to work through the same problems in Mechanix that instructors created for their students. The problems begin with simple vector analyses with significant scaffolding to a truss analysis with just the base initial data that would be encountered in a first year engineering course.

You-Try-It: NuSys: Collaborative Insight Extraction on Shared Workspaces (Philipp Eichmann; Brown University) Gaining insights—the act of acquiring a deep and intuitive understanding of a thing or concept—is often achieved through discussing in small work groups. Students frequently conduct informal study group sessions where scholarly material is revisited, and questions and ideas are bounced back and forth as a way to gain and present new insights. NuSys is a software framework that enhances how groups develop ideas when a shared work space on and interactive whiteboard (IWB) and smaller devices driven by pen & touch gestures replace a physical whiteboard as the collaborative focal point. The core component of our system provides a fluid interface to an unbounded two-dimensional work space, where study material can be seamlessly imported, laid out, grouped, annotated, linked, and tagged collaboratively on multiple clients in real time and in a free-form fashion.

You-Try-It: Sketch Worksheets (Kenneth D. Forbus, Madeline Usher, and Maria Chang; Qualitative Reasoning Group, Northwestern University) Sketch worksheets are a new kind of sketch-based education software designed to facilitate spatial learning. Each worksheet represents a particular exercise, which the student does on a computer. Students get feedback, based on automatic comparison of their sketch with a hidden solution sketch. A software grade book, which uses scoring rubrics in the solution sketch, is intended to help instructors in grading. Sketch worksheets have been used in classroom experiments with college students and with middle-school students. They are domain-independent, requiring only that the exercise involves visual distinctions that the software can understand. This session provided hands-on experiences with sketch worksheets, and with the authoring environment that is used to make them. Participants will be guided through making a simple sketch worksheet themselves, using the authoring environment built into CogSketch, the underlying software. (CogSketch is being developed by the NSF-sponsored Spatial Intelligence and Learning Center, and is freely available online.)

You-Try-It: INKstitute (Rob Baker; Cincinnati Country Day School) Tablet PCs with touch, ink, and laptop functionality can remove all constraints for educators and allow IT departments to manage, maintain, support, upgrade and scale efficiently and effectively. In 1996, Cincinnati Country Day School (CCDS) was the first school in the United States to go 1:1 in grades 5–12. Robert Baker, the school's Director of Technology and Nate Johnson, MS science teacher and member of the Tech team, shared insights, experience and implementation strategies in this extended You-Try-It. It is an overview of the 3-day Tablet Conferences that they host three times per year at CCDS. The use of stylus/pen-based learning with tablets shows what is possible when you deploy hardware that removes constraints and allows teachers to focus on pedagogy, not technology. Hardware and software have just become widely available that make it easier and more affordable to reproduce the CCDS experience, empowering every student with a device that "can do it all." Find out what the educators who come to Country Day from around the world learn in the pursuit of capturing the educational power of its benchmark one-to-one tablet PC program. A mix of Surface Pro 3 and Surface 3 tablets running Win10 and Office 2016 will be provided to allow attendees to get a taste of the collaborative and creative power this type of environment provides.

1.6 CPTTE 2016: Socialization

One of the main benefits of the conference is the networking opportunities that encourage industry, university, and high school pen and touch researchers to intermingle. CPTTE is unique in that it blends perspectives from a wide variety of researchers, including high school teachers, technology administrators, college deans, university professors, and others interested in pen and touch technology and its impact in education.

The conference provided two socialization sessions. The first evening of the conference (Thursday) started off with a poster and dinner at Kasper Multipurpose Room in Faunce House. The evening agenda provided networking opportunities for all the participants, from schoolteachers to college deans and professors. About one hundred researchers and educators with various backgrounds were discussing innovative technologies for the classroom.

On Friday evening, CPTTE participants enjoyed Games Night and Dinner at Kasper Multipurpose Room in Faunce House (Fig. 1.8).

Fig. 1.8 Members of the TAMU Sketch Recognition Lab get together for a photo. Members from left to right are: Ph.D. alumnus Dr. Manoj Prasad, Ph.D. student Seth Polsley, M.S. student Christy Mathew, M.S. student Girish Kasiviswanathan, Ph.D. student Jung In Koh, M.S. student Aqib Bhat, M.S. student Swarna Keshavabhotla, and M.S. student Caio Monteiro. Missing attending members include lab director Dr. Tracy Hammond and RET HS teacher Randy Brooks

1.7 Monograph Organization

CPTTE 2016 included four keynotes, five emerging technology long papers and ten short papers describing new technology in development, one practical application of pen and touch technology in education describing research experiments in the classroom, five You-Try-It sessions, and three highlights of previously published research at other peer-reviewed conferences. All paper submissions were peer-reviewed by three to five expert researchers in the field. Dr. Aaron Adler (BBN Technologies), Mark Payton (Whitfield School), and Dr. Manoj Prasad (Microsoft) were the Program Co-Chairs of the conference, and handled the assignment of the paper reviews. Accepted papers were invited for re-submission as a chapter in this edited volume. Those papers are organized as follows:

1.7.1 Foreward and Part I: Introductions and Welcome

This manuscript starts off with a Foreward by Andy van Dam, CPTTE Conference Chairman and Computer Science Professor at Brown University. This introduction chapter provides a detailed overview of CPTTE 2016 and this book.

1.7.2 Part II: Math for All: Elementary Math

There are a total of six chapters on teaching math. The chapters explore elementary, K16, and higher education math teaching and learning. Three of these chapters highlight advances in Elementary math. In these three chapters, the authors describe new tools that help teachers get insight into students' mathematical thinking and students explore new applications and learn the basics of arithmetical operations.

In the chapter "Classroom Learning Partner: Pen-based Software For Creating and Sharing Visual Representations," the authors Kimberle Koile, Andee Rubin, Steven Chapman, and Lily Ko describe the session on hands-on experience with Classroom Learning Partner (CLP), a pen-based classroom interaction system whose goal is to improve STEM teaching and learning by supporting the creation and sharing of visual representations. In this chapter, one can learn how the Classroom Learning Partner system allows creating visual representations for multiplication and division problems. The great advantage of this tool is that the teachers are able to view the students' representations and gain insights into students' mathematical thinking.

In the chapter "Machine Analysis of Array Skip Counting in Elementary Math," the authors Kimberle Koile, Andee Rubin, and Steven Chapman describe the development and use of pen-based technology in elementary math classes as part of the "INK-12: Teaching and Learning Using Interactive Inscriptions in K-12" project.

The paper presents results of machine analysis of students' visual representations created using a combination of freehand drawing and a digital array tool that supports learning multiplication. The goal of the project was to provide insights into students' mathematical thinking by developing machine analysis routines that operate on students' ink annotations, providing accurate information about problem-solving strategies such as skip counting. The authors present evaluation results for their routines and two other versions of routines that do not use this knowledge and that consequently suffer from high error rates.

In the chapter "TouchCounts and Gesture Design," the authors Nicholas Jackiw and Nathalie Sinclair describe educational iPad application TouchCounts for early number learning. TouchCounts represent a gesture-based system for developing ordinal and cardinal facets of a young learner's number sense. The application allows exposing 3–7 year olds to flexible conceptions of numbers and to the basic mechanics of arithmetic. The application is designed as a game where a child produces numbers when tapping on the screen. Numbers can be merged into larger value by dragging them into each other. Two learning aspects are visualized in *two worlds*, the first of them only whole numbers beginning with 1 are produced, that encourages counting by 1, while in the *second world*, it is possible to push numbers together and pull them apart, thus introducing the basics of arithmetical operations.

1.7.3 Part III: Math for All: High School and College Math

The next section includes three chapters on K16 and higher education math. The authors introduce automatic sketch recognition systems for chart and graph recognition, as well as discuss how pen and touch technology in K16 mathematics classrooms can promote learner flow.

In the chapter "Sketch based interaction techniques for chart creation and manipulation," the authors Andrés N. Vargas González, Eugene M. Taranta II, and Joseph J. LaViola Jr. describe a new prototype application, SketChart, that uses a set of novel interaction techniques for the creation and manipulation of 12 chart types. Users can select a data-set from which a desirable type of a chart will be automatically generated. Several chart types are recognized with 100% accuracy, while others are very difficult to detect, with box plots, stacked bar graphs and group bar graphs being the most difficult with 48%, 67%, and 69% accuracy respectively. Nevertheless, the system has an overall recognition accuracy of 81% and received positive feedback. The participants in the study performed data analysis and provided answers to questions that mimic practical test example problems. All aspects of the system were well received including the recognition engine, visual feedback mechanisms, interaction techniques, and graph visualizations. Similarly, all of the interaction techniques scored high, participants especially enjoyed being able to select or hide specific data by interacting with the legends.

In the chapter "Digitizing Mathematical Symbolism through Pen and Touch Technology: A Beginning Taxonomy of Affordances for Learner Flow in Mathematics," the author Eric Hamilton discusses a rationale for pen and touch in K16 mathematics classrooms, arguing that digitizing mathematical cognition can promote learner flow. In this research, Eric Hamilton seeks means for learners to experience high engagement and flow levels of immersion in mathematics. The author identifies a three-tier initial taxonomy for classifying such applications. He suggests that development of a substantial research agenda around how and why pen and touch technologies can activate learning-sensitive mechanisms.

In the chapter "Sketching Graphs in a Calculus MOOC: Preliminary Results," the authors Jennifer French, Martin A. Segado, and Phillip Z. Ai describe a new free-form sketch-input and automatic grading tool designed for use with pen, touch, and mouse input for use in a calculus MOOC. The new tool allows students to construct graphs using mouse, touch, or pen input and provides immediate targeted feedback. Analysis of 8187 responses collected with the graphical input tool showed that the correct response is usually reached after two to three attempts, with the 90th percentile using between four to nine attempts depending on the difficulty of the problem. The authors also suggested that the use of such graphical tools in both MOOCs and classroom environments enhances learning gains in student ability to sketch functions and provides a rich source of data for future study.

1.7.4 Part IV: Personalized Feedback and Sketching in Science and Engineering

STEM subjects have always been a great place for sketch interfaces to help improve learning. While the first two sections highlighted advancements in mathematics, this chapter highlights advancements in engineering (including mechanical and electrical engineering) and science (geoscience and food chemistry).

In the chapter "Tablet PC Goal-Directed Practice Coupled with Real-Time Targeted Feedback Enhances Food Chemistry Learning," the authors Enrique Palou, Judith V. Gutierrez Cuba, Nelly Ramirez-Corona, and Aurelio Lopez-Malo describe the redesign of the Food Chemistry and Advanced Food Chemistry courses at UDLAP (Mexican private institution of higher learning) to improve undergraduate and graduate food chemistry teaching and learning. In the study they used Tablet PCs and InkSurvey and Classroom Presenter software to create learning environments that promote an interactive classroom. As a result they were able to enhance student learning by implementing goal-directed practice and instantaneous targeted feedback, making prompt pedagogical adjustments as needed. Formative assessment results showed improved student participation and higher test scores.

In the chapter "Score Improvement Distribution When Using Sketch Recognition Software (Mechanix) as a Tutor: Assessment of a High School Classroom Pilot," the authors Randy Brooks, Jung In Koh, Seth Polsley, and Tracy Hammond deployed and tested Mechanix, an interactive sketch-recognition software for mechanical and civil engineering courses, in a high school environment with the goal of improving the effectiveness of student practice regarding free body diagrams, vector analysis, and truss problem solving. Mechanix provides immediate, constructive feedback both to the learner as well as student-level assessment to the instructor. It was made available to students in a STEM-infused classroom as part of the Project Lead the Way (PLTW). Assessment of students' progress showed an average increase of 1.65 points on a 12 point scale between a pre- and post-test ($p < 0.005$). The present study found higher improvement scores for 'A-level' students, while previous studies suggested that progress may be achieved at all levels.

In the chapter "DCSR: A Digital Circuit Sketch Recognition System for Education," the authors Shuo Ma, Yongbin Sun, Pengchen Lyu, Seth Polsley, and Tracy Hammond present DCSR (Digital Circuit Sketch Recognition) system that recognizes hand-drawn digital logic circuits through a web interface and provides immediate feedback as the user draws. DCSR aims to provide an interactive, sketch-based approach for educators to assist students in learning digital logic. A user study with 15 students revealed recognition accuracy of 93.19%. User feedback included suggestions to improve the system's drawing ability, particularly ability to draw gates in one stroke and more freedom over drawing pins.

In the chapter "An Intelligent Sketching Interface for Education using Geographic Information Systems," the authors Aqib Niaz Bhat, Girish Kasiviswanathan, Christy Maria Mathew, Seth Polsley, Erik Prout, Daniel W. Goldberg, and Tracy Hammond present a new application to teach students to identify and draw rivers on maps. This new system is based on sketch recognition techniques and is aimed at providing help in learning geography. The new web application allows users to draw rivers on a map and uses a similarity measure to evaluate students' work. The user study included 10 users across multiple tests, and the findings showed that the application helps students gain geographic knowledge in an intuitive and effective way through sketching. The future work for this project will include adding more geographic features, such is river directions and automatic grading of students' sketches.

1.7.5 Part V: Increasing Liberal Arts Engagement and Learning with Sketch

Not just STEM subjects are affected by the use of the pen; this section focuses on how language learning can be improved through pen usage.

In the chapter "Insights from Exploration of Engaging Technologies to Teach Reading and Writing Story Baker," the author Michel Pahud describes a project that was terminated in 2007 due to issues around producing a successful device given the hardware available at the time. Modern technology progress and greatly reduced prices for tablet computers now give a second chance and a new perspective. The author proposes to go back and look at the experiments that were done and use them for future work inspiration.

From 2004 to 2007 the research team developed and tested over 30 prototype systems designed for children 5–7 years old who have challenges in learning to write. The intent was to create highly engaging interactive experiences for children who might have limited attention spans. Children were writing stories, sentence by sentence, with each completed sentence instantly converted to an animation. Several animation styles were used. The prototypes used animation, typing, a stylus for writing and sketching, machine-recognizable tangible objects for input, and simple robot activity for viewing the outcome of actions.

The author suggests that such an application can be used to contribute to teaching literacy, and/or to be used for story creating and interaction.

In the chapter "A Multilingual Sketch-based Sudoku Game with Real-time Recognition," the authors Caio D. D. Monteiro, Meenakshi Narayanan, Seth Polsley, and Tracy Hammond introduce a new multilingual, sketch-based version of the Sudoku game. The current version supports Chinese and Hindi languages and can serve as an educational tool for those learning a new language. The recognition algorithm for the application is based on the Hausdorff metric and allows extending the application to support other languages. The overall accuracy of sketch recognition algorithm is over 93% when recognizing Chinese and Hindi numbers at the same time. Hindi numbers seem to be harder to classify, as they are very curvy and share many similarities with one another, making template matching sensitive to noise.

1.7.6 Part VI: Let's Collaborate

Sketching also has a profound affect on collaboration. This section includes two papers who have shows novel methods for collaboration with sketching.

In the chapter "NuSys: Collaborative Insight Extraction on Shared Workspaces," the author Philipp Eichmann presents a new application NuSys that is aimed at facilitating group discussions and work during study group sessions by providing a shared work-space on an interactive whiteboard (IWB) and smaller devices driven by pen & touch gestures that replace a physical whiteboard. The new application provides a fluid interface where study material can be seamlessly imported, laid out, grouped, annotated, linked and tagged collaboratively on multiple clients in real-time and in a free-form fashion.

In the chapter "Collaborative Spaces: One Note in the Middle School Classroom," the author Sara Mata—English teacher and reading specialist—describes the process of developing physical and digital collaborative spaces in a middle school English classroom while using Microsoft OneNote software, which incorporates digital ink and tablet technology. Students participated in book clubs for 3 weeks as they shared a space in the classroom where they could read together and discuss their ideas. Students utilized OneNote's collaboration space to help them work together as they read their book. The author used the digital pen and tablet technology to provide them instant feedback as she evaluated their reading progress. As a result students gained a deeper appreciation for literature and engaged in rich discussions, while struggling readers were able to demonstrate their understanding of novels through conversations with their peers. The average number of books read per year increased significantly and students better understood the material they were reading.

1.7.7 Part VII: Domain Independent Educational Tools

In the chapter "Design Considerations for a Large Display Touch Application in Informal Education," the authors Lucy Van Kleunen, Trent Green, Miranda Chao, Tiffany Citra, and Carlene Niguidula investigated design challenges for large 82-in touch screen exhibit about the life of Alfred Nobel. The efficacy of large displays as informal educational tools was accessed and authors suggested some practical guidelines for designers working with large touch screens. The main obstacles related to use of large touch screens were related to distal access of touch targets and limited field of vision. The authors concluded that designers must create applications that are effective teaching tools for proximal visitors who only see part of the screen as well as attractive for distal visitors who see the screen from afar, by using such steps as adjusting size and placement of icons.

In the chapter "DataSketch: A Tool to Turn Student Sketches into Data-Driven Visualizations," the author Michelle Wilkerson analyzes how data visualizations are perceived by K-12 students and introduces a new tool—DataSketch. This tool is designed to allow young learners to develop data visualization literacy by creating their own ink-based data-driven visualizations. In the DataSketch tool, the drawing created by a student is decomposed into constituent objects in a way similar to the way graphs can be decomposed into points and axes, and each sub-component can be assigned as set of rules for how it works. The goal is to connect learners' existing commonsense knowledge about data representation to the types of rules and mappings that define formal representational systems.

In the chapter "A Survey Based Study on Personalizing Student Learning Using MyEduDecks Application," the authors Samuel Cohen, Nicholas Wilke, Parv Shrivastava, Aileen Owens, Matt Antanti, Prateek Jukalkar, Ashumi Rokadia, and Vinay Pedapati present new pen-based digital flashcards application that has been in development since 2013 at South Fayette High School. The MyEduDecks app was created to analyze how K-12 students use and benefit from pen-based technologies. The application can be used to create custom study decks and give them the ability to use their stylus to input answers and rely on automated grading (based on ink recognition) or self-grading to progress through their studies. To evaluate the use of the application, 77 students from three seventh grade classes participated in pre- and post-test user study. The post-survey results showed that students reacted positively to the use of MyEduDecks. Students suggested adding audio features for future use.

1.7.8 Part VIII: Panels and Keynotes

The last section includes edited transcriptions of the two panels and one of the keynotes at CPTTE.

The education panel included Dr. Eric Hamilton, Rob Baker, Dr. Jonathan Grudin, Mark Payton, and Cal Armstrong. Topics included the practice, the difficulties, and the added value of what leads to success in BYOD (Bring Your Own Device) to the educational process. A philosophical discussion emerged including arguments for and against domain-specific and domain-independent solutions, as well as future directions to better preparing students for success.

The research panel included Dr. Andy van Dam, Dr. Bill Buxton, Dr. Tracy Hammond, Dr. Kimberle Koile, and Dr. Tom Stahovich. Topics included sketch forensics applied to education, representation fidelity, ubiquity, domain-specific versus domain-independent, accessibility concerns, funding concerns, and future directions. Special attention was also spent discussing ink accessibility and how it might improve the lives of those facing disability.

Dr. Tracy Hammond's keynote serves as the final chapter of the book, covering various sketch recognition methods and algorithms to understand sketches she has created over the years. She also shows their abilities in both sketch forensics and activity recognition, providing inspiration as per how this can allow for surprisingly intelligent personalized feedback.

1.8 Organizing Committees

The chair of CPTTE 2016 was Dr. Andy van Dam (Brown University). The paper chairs were Dr. Aaron Adler (BBN), Dr. Manoj Prasad (Microsoft), and Mark Payton (Whitfield School). The publication chair was Dr. Joseph Tront (Virginia Tech). The sponsor chair was Donald Carney (Fluidity Software). The conference coordinator was Lisa Manekofsky (Brown University). Additional Ex Officio members of the program committee included Dr. Jonathan Grudin (Microsoft), Dr. Eric Hamilton (Pepperdine University and University of Namibia), and Dr. Tracy Hammond (Texas A&M University).

1.9 Acknowledgements

We greatly thank the many contributing authors and presenters, reviewers, students, and staff at Brown University for the truly outstanding work that was done to produce, publish, present, and demonstrate at the conference, and in providing the behind-the-scenes work necessary to make it all possible, especially Lisa Manekofsky.

The generous sponsorship of several corporations and organizations has been crucial to enabling CPTTE to provide a high quality program at a very low cost to attendees. We would particularly like to thank our sponsors who have provided invaluable resources, both financial and in-kind. The 2016 Host Sponsor was Brown University, with special thanks going to Dr. Andy van Dam and Lisa Manekofsky. Our 2016 Platinum Sponsor was Microsoft, with special thanks going to Dr. Jonathan Grudin and Dr. Bill Buxton. The 2016 Gold Sponsors were Wacom, with special thanks to Sierra Modro, and Fujitsu, with special thanks to Mitzi Miller, David Nee, and Alan Fulcher. The 2015 Bronze Sponsor was PDF Annotator, with special thanks to Oliver Grahl.

CPTTE 2016 would not have been successful with our dedicated volunteer peer reviewers including Aaron Adler (Raytheon BBN Technologies), Folami Alamudun (Oak Ridge National Laboratories), Carol Carruthers (Seneca College), Jonathan Cohen (Brehm Preparatory School), Martha Crosby (University of Hawaii), Kenneth Forbus (Northwestern University), Jonathan Grudin (Microsoft), Ken Hinkley (Microsoft), Richard Kassissieh (University Prep), Gergory Klee (Chaminade College Prep), Kimberle Koile (MIT), Joseph LaViola (University of Central Florida), Juiteng Li (Texas A&M University), Anne-Marie Mann (University of St. Andrews), Sara Mata (Whitfield School), Okumu Moses (Washington University in St Louis), Cassandra Oduola (Texas A&M University), Enrique Palou (Universidad de las Americas Puebla), Mark Payton (Whitfield School), Manoj Prasad (Microsoft), Vijay Rajanna (Texas A&M University), Marilyn Reba (Clemson University), Carla Romney (Boston University), Walter Schilling (Milwaukee School of Engineering), Paul Taele (Texas A&M Unviersity), Steven Tanimoto (University of Washington), Andy van Dam (Brown University), and David-Paul Zimmerman (Renton Prep),

And an extra special thanks goes to Anna Stepanova (Texas A&M University) who greatly helped with all parts of the production of this manuscript.

1.10 Conclusion

The 2016 Conference on Pen and Touch Technology in Education brought together diverse audience of researchers interested in new educational tools and pen and touch technology.

This collection of papers highlights research performed by the users, researchers, developers, decision makers, students, and teachers. Scientists shared their experiences on using the pen and touch interfaces across various educational settings.

We thank you, the reader, for your growing interest in making education better through the intelligent application of digital ink and touch.

Acknowledgements Thanks to Dr. Anna Stepanova and Dr. Jan Hammond with their help with this introduction. Photo credits to Randy Brooks and Aqib Bhat.

Your editors,
Tracy, Manoj, & Aaron

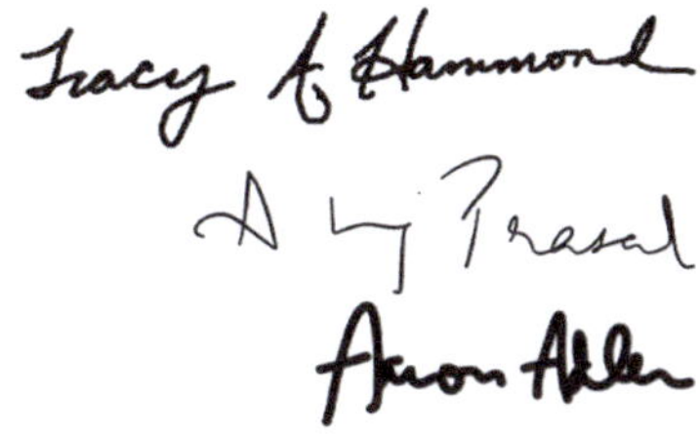

Dr. Tracy Hammond Director of the Sketch Recognition Lab and Professor in the Department of Computer Science and Engineering at Texas A&M University, Dr. Hammond is an international leader in sketch recognition and human-computer interaction research. Dr. Hammond's publications on the subjects are widely cited and have well over fifteen hundred citations, with Dr. Hammond having an h-index of 20, an h10-index of 38, and 4 papers with over 100 citations each. Her sketch recognition research has been funded by NSF, DARPA, Google, Microsoft, and many others, totaling over 3.6 million dollars in peer reviewed funding. She holds a Ph.D. in Computer Science and FTO (Finance Technology Option) from M.I.T., and four degrees from Columbia University: an M.S. in Anthropology, an M.S. in Computer Science, a B.A. in Mathematics, and a B.S. in Applied Mathematics. Prior to joining the TAMU CSE faculty Dr. Hammond taught for 5 years at Columbia University and was a telecom analyst for 4 years at Goldman Sachs. Dr. Hammond is the 2011–2012 recipient of the Charles H. Barclay, Jr. '45 Faculty Fellow Award. The Barclay Award is given to professors and associate professors who have been nominated for their overall contributions to the Engineering Program through classroom instruction, scholarly activities, and professional service.

Dr. Aaron Adler is a Senior Scientist at Raytheon BBN Technologies in Columbia, Maryland and Cambridge, Massachusetts. Dr. Adler has worked on variety of projects for DARPA and AFRL involving security, machine learning, robotics, artificial intelligence, and synthetic biology. Dr. Adler has a particular interest in creating intelligent user interfaces by automatically handling complexities to enable intuitive interfaces for users. He received his Ph.D. in Computer Science from M.I.T. where he also received his M.Eng. in Computer Science and Engineering and S.B. in Computer Science. His Ph.D. thesis centered on constructing multimodal interactive dialogues: combining speech recognition and sketch recognition for user input and generating speech and sketching for multimodal computer output. The system helps the user describe simple mechanical (Rube-Goldberg-like) devices by asking probing questions.

Dr. Manoj Prasad is a software developer in MSEG Customer Experiences R&D, Microsoft and researcher collaborating with Dr. Tracy Hammond, Sketch Recognition Lab. He holds a Ph.D. in Computer Science from Texas A&M University and a Bachelor's degree in Computer Science from Birla Institute of Technology and Science. His research interests include haptic user interfaces and sketch recognition. His dissertation involved creation of a haptic jacket for motorcyclists navigation (HaptiMoto) and pedestrian navigation (HaptiGo). This work was published in ACM CHI, a renowned conference in Interaction space. He is currently working on a sketching interface for an application in Windows. This application will let support agents remote into customer's box and annotate on the customers screen to guide them to their support issue.

College Station, TX, USA — Tracy Hammond
Columbia, MD, USA — Aaron Adler
Palo Alto, CA, USA — Manoj Prasad
January 2017

Part II
Math For All: Elementary Math

Chapter 2
Classroom Learning Partner: Pen-Based Software for Creating and Sharing Visual Representations

Kimberle Koile, Andee Rubin, Steven Chapman, and Lily Ko

Abstract A You-Try-It session at CPTTE 2016 provided hands-on experience with Classroom Learning Partner (CLP), a pen-based classroom interaction system whose goal is to improve STEM teaching and learning by supporting the creation and sharing of visual representations. Participants played the role of students in an elementary math classroom, using CLP's digital tools to create visual representations for multiplication and division problems. They also were able to view the teacher's version of the software, which provides information about students' use of the digital tools, with the goal of providing teachers with insights into students' mathematical thinking.

2.1 Context and Motivation for CLP

The goal of CLP is to improve STEM teaching and learning by supporting real-time use of student work created via a pen interface. It accomplishes this goal by means of a create, interpret, share interaction model: Students use a tablet pen to create representations; CLP employs AI methods that interpret the representations, when possible, in order to provide the teacher with feedback about student understanding and to enable the teacher to filter and sort the representations based on CLP's interpretation and on her own assessment; the teacher then can select student work to share as a basis for class discussion and to identify students who may need help.

K. Koile (✉)
Computer Science and Artificial Intelligence Laboratory, MIT,
77 Massachusette Avenue, Cambridge, MA, 02139, USA
e-mail: kkoile@mit.edu

A. Rubin • L. Ko
TERC, Cambridge, MA, 02140, USA
e-mail: andee_rubin@terc.edu; lily_ko@terc.edu

S. Chapman
stevcode.com, Cambridge, MA, 02139, USA
e-mail: stev.code@gmail.com

T. Hammond et al. (eds.), *Frontiers in Pen and Touch*, Human–Computer Interaction Series, DOI 10.1007/978-3-319-64239-0_2

In its early days, CLP employed Classroom Presenter [1] as its underlying infrastructure and was used in introductory undergraduate computer science and chemistry classes. Its digital ink interpretation routines focused on recognizing handwritten text strings and sequences, identifying shaded and enclosed regions, and recognizing sketched representations such as box-and-pointer (aka linked list) diagrams or chemical structures. CLP has been re-implemented several times, and for the past 6 years, new software tools have been developed for use in elementary school science and math classes. (See [2–10] for details of CLP's design, development, and use in classrooms.)

2.2 Current Work

Most recently, CLP has supported research projects aimed at improving upper elementary teaching and learning of multiplication and division, *INK–12: Teaching and Learning Using Interactive Ink Inscriptions in K–12*; and mathematical argumentation, *Technology For Mathematical Argumentation*. Videos describing CLP and the projects can be found on each of the project websites [11, 12] and. CLP's create, interpret, share interaction is illustrated in more detail on the INK–12 website's model of interaction page [11].

CLP is organized around the idea of an electronic notebook that contains lesson *pages*, along with digital tools for working through lesson problems on the pages. In the INK-12 project, current lessons focus on upper elementary math, and CLP's tools include not only digital ink but also tools that enable students to easily and quickly create visual representations using common mathematical constructs, such as arrays and number lines, and other novel tools designed to support new kinds of visual representations. In designing each tool, the goal was to balance freehand drawing with structure in order to support both students' creative expression and machine interpretation. Examples of student work created using CLP's array and stamp tools are shown in Fig. 2.1.

2.3 Session Overview

After a brief introduction to CLP, participants played the role of students, using CLP's tools to solve and explain multiplication and division problems by creating and manipulating visual representations. They also were able to view the teacher's version of the software, which provides information about students' use of the digital tools. Additionally, we discussed findings from our classroom trials.

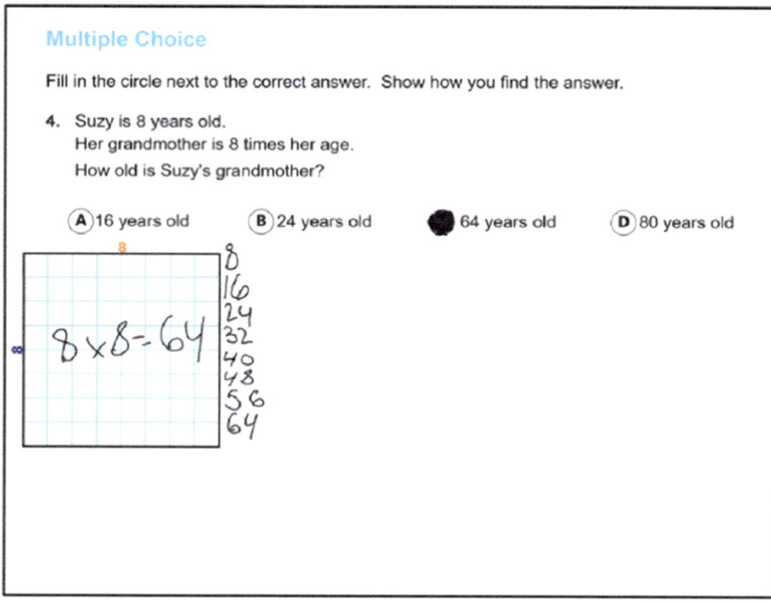

(a)

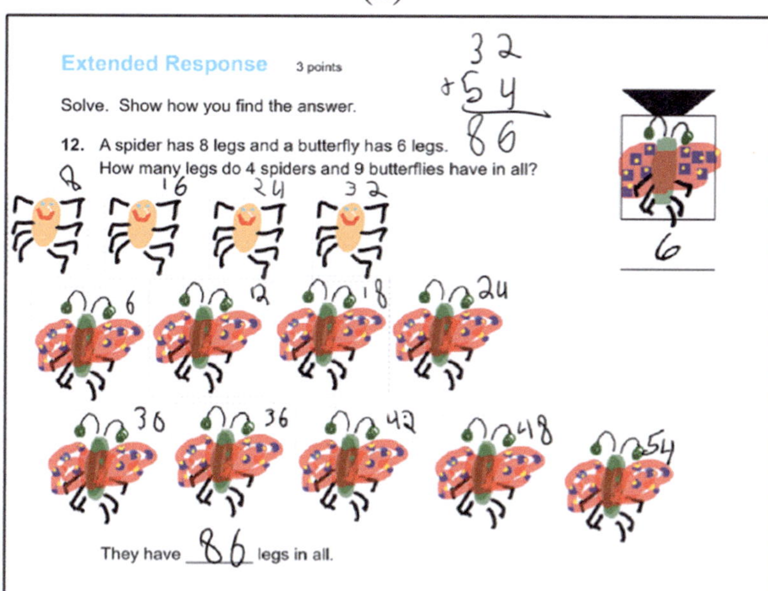

(b)

Fig. 2.1 Examples of student work created using CLP's array and stamp tools. (**a**) Using an array to solve 8×8. (**b**) Using a stamp to solve $8 \times 4 + 6 \times 9$

References

1. Anderson, R., Anderson, R., Simon, B., Wolfman, S.A., VanDeGrift, T., Yasuhara, K.: Experiences with a tablet PC based lecture presentation system in computer science courses. ACM SIGCSE Bull. **36**(1), 56–60 (2004)
2. Koile, K., Chevalier, K., Low, C., Pal, S., Rogal, A., Singer, D., Sorensen, J., Tay, K.S., Wu, K.: Supporting pen-based classroom interaction: new findings and functionality for classroom learning partner. In: First International Workshop on Pen-Based Learning Technologies, PLT 2007, pp. 1–7. IEEE (2007)
3. Koile, K., Chevalier, K., Rbeiz, M., Rogal, A., Singer, D., Sorensen, J., Smith, A., Tay, K.S., Wu, K.: Supporting feedback and assessment of digital ink answers to in-class exercises. In: Proceedings of the National Conference on Artificial Intelligence 1999, vol. 22, p. 1787. AAAI Press, Menlo Park; MIT Press, Cambridge/London (2007)
4. Koile, K., Reider, D., Rubin, A.: Ink-12: a pen-based wireless classroom interaction system for k-12 (2010)
5. Koile, K., Rubin, A.: Animated mathematical proofs in elementary education. In: Hammond, T., Valentine, S., Adler, A., Payton, M. (eds.) The Impact of Pen and Touch Technology on Education, pp. 67–79. Springer, Cham (2015)
6. Koile, K., Rubin, A.: Machine interpretation of students' hand-drawn mathematical representations. In: Hammond, T., Valentine, S., Adler, A., Payton, M. (eds.) The Impact of Pen and Touch Technology on Education, pp. 49–56. Springer, Cham (2015)
7. Koile, K., Rubin, A.: Tablet-based technology to support students' understanding of division. In: Hammond, T., Valentine, S., Adler, A., Payton, M. (eds.) Revolutionizing Education with Digital Ink, pp. 71–89. Springer, New York (2016)
8. Koile, K., Singer, D.: Development of a tablet-pc-based system to increase instructor-student classroom interactions and student learning. In: Proceedings of Workshop on the Impact of Pen-Based Technology on Education (WIPTE'06) (2006)
9. Koile, K., Singer, D.: Assessing the impact of a tablet-pc-based classroom interaction system. In: Reed, R.H., Berque, D.A., Prey, J.C. (eds.) The Impact of Tablet PCs and Pen-Based Technology on Education: Evidence and Outcomes, pp. 73–80. Purdue University Press, West Lafayette (2008)
10. Rubin, A., Storeygard, J., Koile, K.: Supporting special needs students in drawing mathematical representations. In: Hammond, T., Valentine, S., Adler, A., Payton, M. (eds.) The Impact of Pen and Touch Technology on Education, pp. 57–66. Springer, Cham (2015)
11. INK–12. http://ink-12.mit.edu
12. TMA. http://tma.mit.edu

Chapter 3
Machine Analysis of Array Skip Counting in Elementary Math

Kimberle Koile, Andee Rubin, and Steven Chapman

Abstract The *INK–12: Teaching and Learning Using Interactive Ink Inscriptions in K–12* project has been developing and investigating the use of pen-based technology in elementary math classes. This paper reports progress made on machine analysis of students' visual representations created using a combination of freehand drawing and a digital array tool that supports learning multiplication. The goal of the machine analysis is to provide insights into students' mathematical thinking as revealed through creation and manipulation of visual representations. For array representations, machine analysis involves interpretation of ink annotations that represent problem-solving strategies, one of which is counting by a number other than 1, aka skip counting. A subset of student work from a 5-week trial in a third grade class provides a corpus for development and evaluation of the machine analysis routines. This paper describes the routines and presents findings demonstrating that the routines are able to provide accurate information about students' skip-counting strategies. It discusses the key to the accuracy—using knowledge about the structure of arrays and the nature of skip counting to bias the machine analysis routines; and presents evaluation results for two versions of routines that do not use this knowledge and that consequently suffer from high error rates. The paper also discusses current work on extending the routines to analyze the process of creating representations and future work on using the routines on thousands of pieces of student work from the 5-week trial.

K. Koile (✉)
Computer Science and Artificial Intelligence Laboratory, MIT,
77 Massachusette Avenue, Cambridge, MA, 02139, USA
e-mail: kkoile@mit.edu

A. Rubin
TERC, Cambridge, MA, 02140, USA
e-mail: andee_rubin@terc.edu

S. Chapman
stevcode.com, Cambridge, MA, 02139, USA
e-mail: stev.code@gmail.com

T. Hammond et al. (eds.), *Frontiers in Pen and Touch*, Human–Computer Interaction Series, DOI 10.1007/978-3-319-64239-0_3

3.1 Introduction

Visual representations play a key role in mathematics education, helping students to gain insights into mathematical concepts and demonstrate their mathematical thinking, e.g. [2, 3, 19]. The National Council of Teachers of Mathematics (NCTM) in its Principles and Standards for School Mathematics devotes a strand to the importance of representations [13], and the Common Core State Standards for Mathematics, which incorporate the NCTM mathematical process standards, emphasize representations by specifying that students learn to "model with mathematics" [14]. Representations serve several functions for students of mathematics. Visual analogs of quantities and operations provide a way for students to explore mathematical relationships and can give students insight into the structure of the number system and the properties of operations, e.g., commutativity. Representations also serve a communicative function, giving students a non-verbal language for expressing their strategies and knowledge and a means for understanding others people's approaches.

To facilitate the use of mathematical representations, the *INK–12: Teaching and Learning Using Interactive Ink Inscriptions in K-12* project (ink-12.mit.edu) has been developing and testing pen-based digital tools in elementary math, focusing on multiplication and division [7, 8, 17]. With our tablet-computer-based software, called Classroom Learning Partner (CLP), students use a combination of digital tools and freehand drawing to create and manipulate mathematical representations in an electronic *notebook*. Using the tablet pen, students choose from among the digital tools, available via a command bar along the top of their tablet computer screens, to create representations that accurately reflect mathematical quantities; they use the tablet pen to annotate via drawing, writing, and highlighting in order to explore and record connections they see among representations, symbolic expressions, problem statements, and their own verbal explanations. They then wirelessly submit their work to their teacher, who can view the work on a tablet, e.g., to identify students who might need help or to choose for class discussion examples of alternate problem-solving strategies. Viewing student work in real time in a classroom, however, can be overwhelming—even in a class of 20 students, teachers can receive over a hundred submissions of student work in a single lesson. Our aim is to help teachers by developing machine analysis routines that can provide insights into students' use of visual representations and the mathematical thinking that is revealed through creation and manipulation of those representations. Such routines could, for example, alert a teacher about errors or apparent misunderstandings reflected in students' representations.

One of the visual representations important in teaching and learning multiplication is the array, which represents multiplication in terms of rows and columns [5]. Curricula often introduce the concept of multiplication with the idea of multiple copies of a group of a particular size. Using this idea, the multiplication problem 4×8 can be thought of as four groups of eight and can be represented by an array with four rows and eight columns. Each row represents a group of eight, and the product is the total number of cells in the array. Students can determine the product by employing a method called skip counting, in which they count by the size of the group, e.g., 8, 16, 24, 32. Over time, as students practice with and solidify understanding of relationships between skip counting and groups, they gain a foundation of meaning for the traditional multiplication facts and, with practice, begin committing the facts to memory.

Figure 3.1 shows an example of skip counting along the edge of an array in order to answer a problem that asks how many apples are in 4 bags of apples, with 8 apples in each bag. The student creates an array by tapping on an array tool icon on CLP's command bar and filling in numbers for rows and columns when prompted—4 rows and 8 columns in this example. She then counts by 8, and at the end of each row writes the total so far. The student's use of this representation suggests that she understands the idea of equal groups and has used each row of the array to represent a group, arriving at a correct final answer of 32. When a student's answer is incorrect, knowledge of the numbers in a skip-counting sequence can help a teacher distinguish between careless arithmetic mistakes and conceptual misunderstanding.

The goal of the work reported in this paper is to provide insights into students' mathematical thinking by developing machine analysis routines that will identify and interpret handwritten skip counting on arrays such as shown in Fig. 3.1. The task involves two steps: identifying digital ink strokes that are used in skip counting, while rejecting other strokes that are not; then interpreting the ink strokes in order to produce a sequence of numbers. Both steps are challenging. First, to identify likely skip-counting sequences, the routines must distinguish between ink that is part of a sequence and ink that is not. In the work shown in Fig. 3.2a–d, for example, the ink immediately outside the array edges should be considered part of the skip-counting sequences, except for the circles around the last numbers in the sequences in Fig. 3.2a. The ink inside the arrays in Fig. 3.2e should be considered part of a skip-counting sequence, but the ink inside all other arrays shown in Fig. 3.2a–h should not be. In Fig. 3.2f, one array should be identified as having skip counting while the other should not. Finally, none of the ink in close proximity to either inside or outside edges of arrays shown in Fig. 3.2g–h should be considered skip counting. Thus identifying ink that is potentially part of a skip-counting sequence is not simply a matter of grouping all ink that is outside an array edge, as might be surmised from the neatly written ink shown in Fig. 3.1.

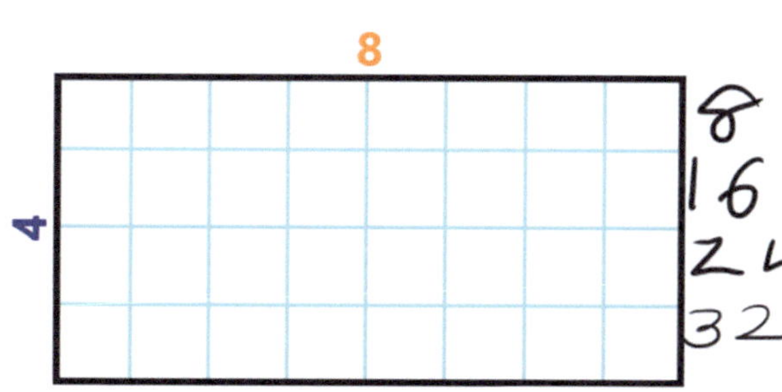

Fig. 3.1 Student's array for 4 bags of apples, 8 apples in each bag (4×8), with skip counting to find the product of 32

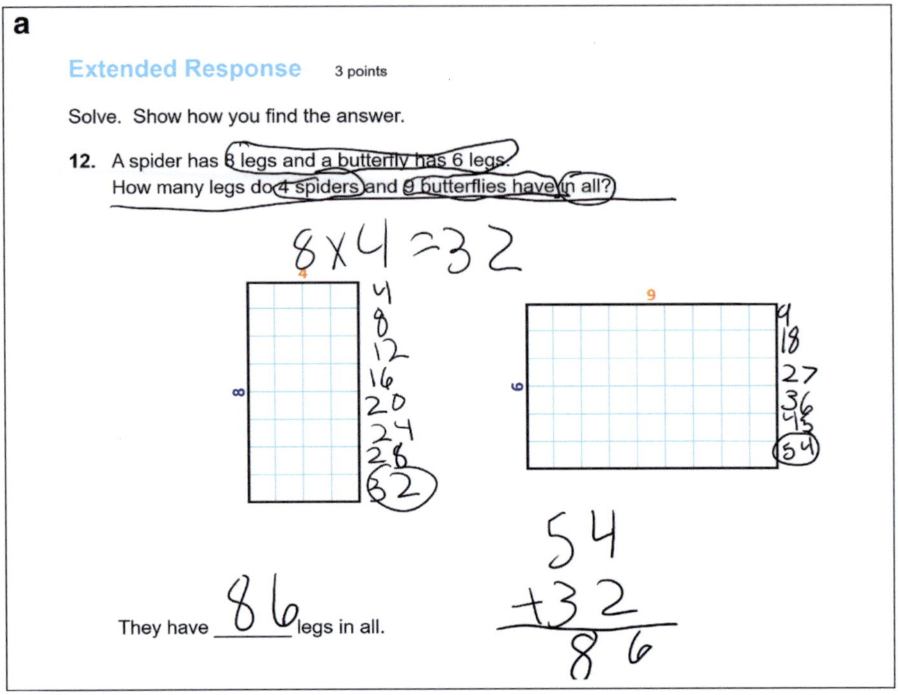

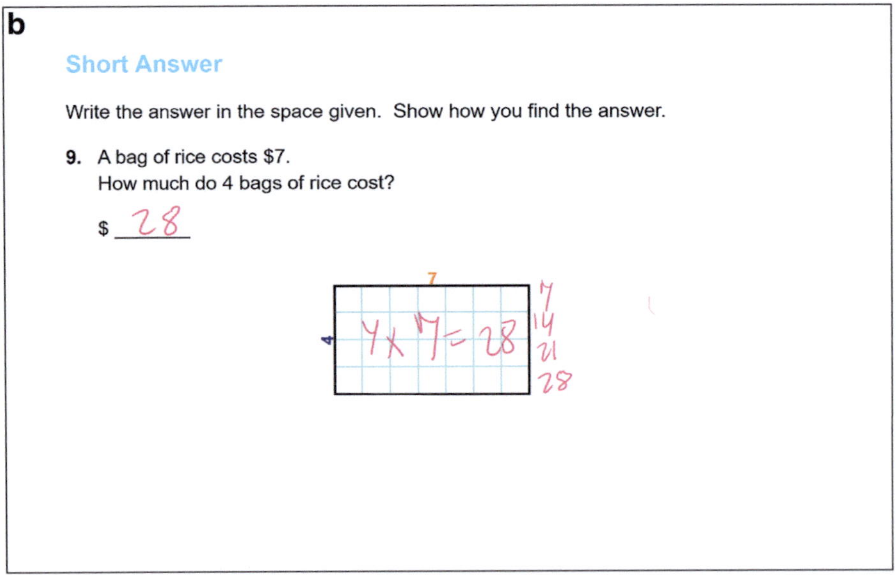

Fig. 3.2 Student work illustrating challenges in identifying and interpreting skip counting. (**a**) Extra ink encirling skip-count numbers. (**b**) Extra ink inside array. (**c**) Messy handwriting, extra ink inside. (**d**) Extra ink inside, crowded ink along edge. (**e**) Skip counting inside and outside arrays. (**f**) Extra ink inside and outside arrays. (**g**) Extra ink intersecting array edges. (**h**) Ink inside and outside arrays near edges

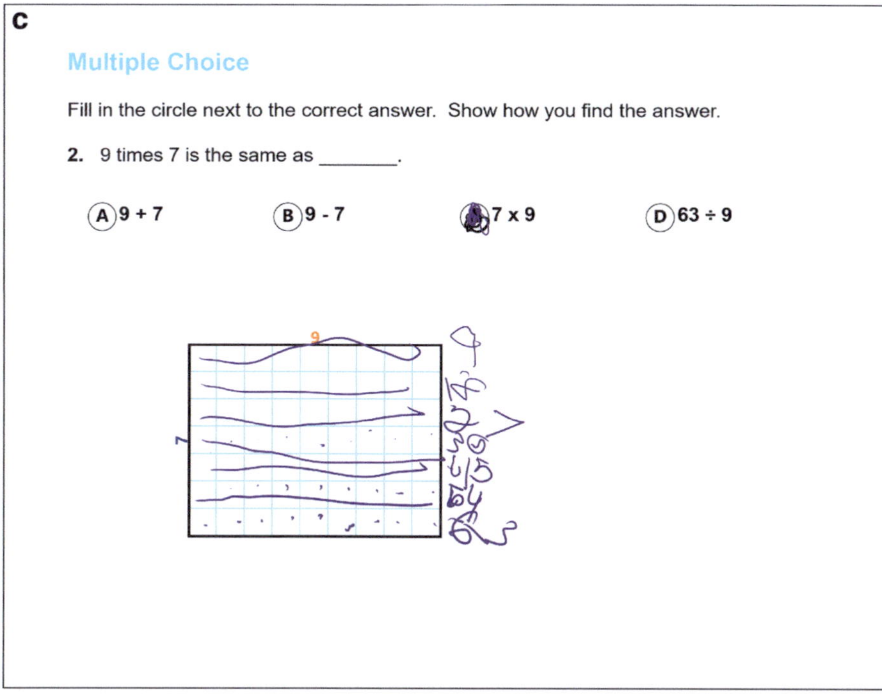

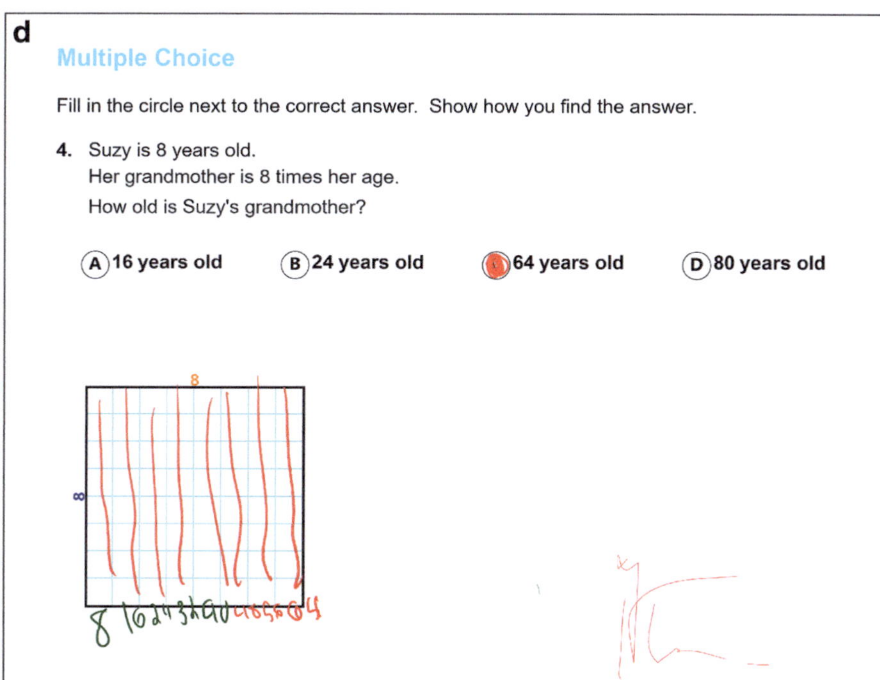

Fig. 3.2 (continued)

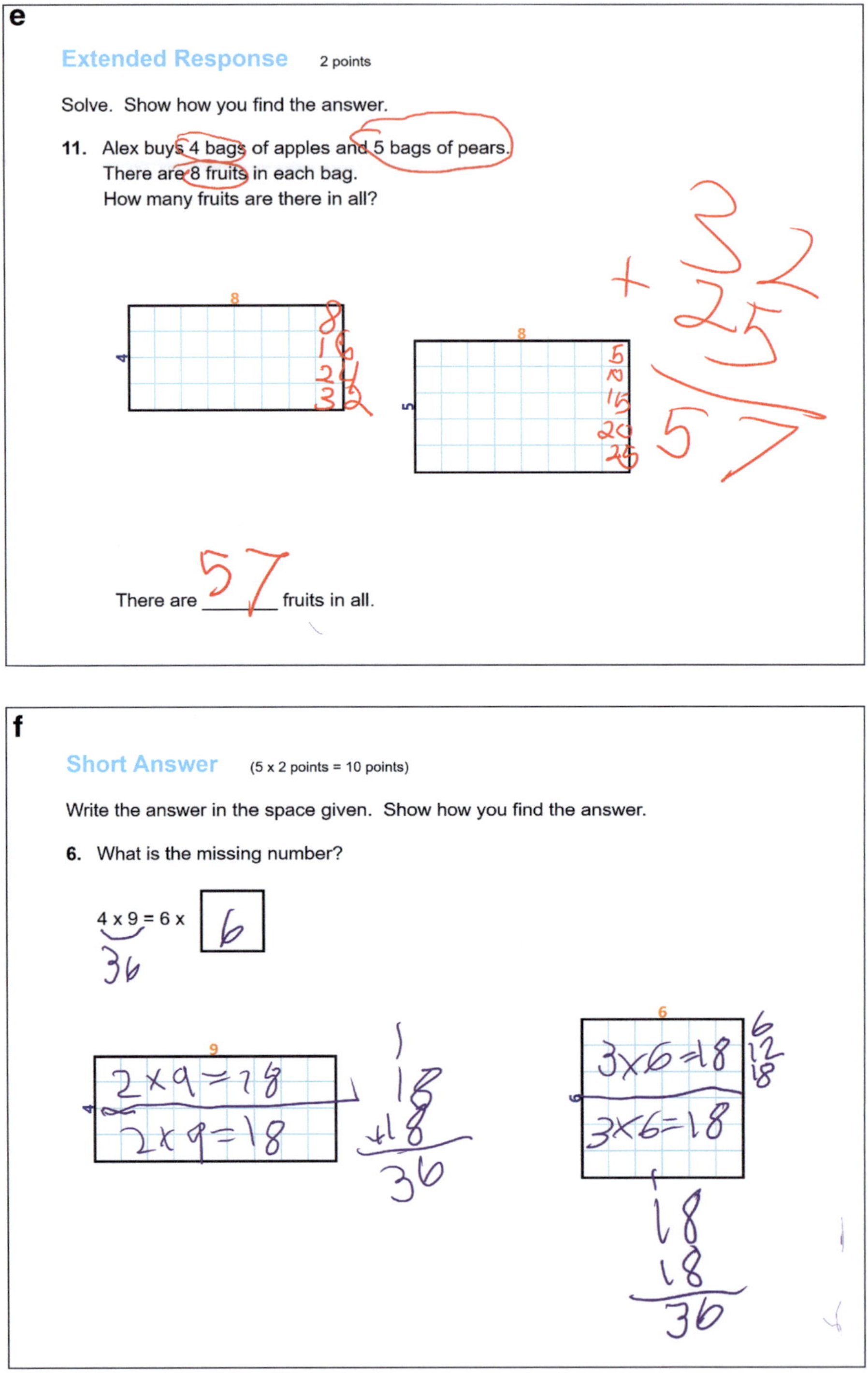

Fig. 3.2 (continued)

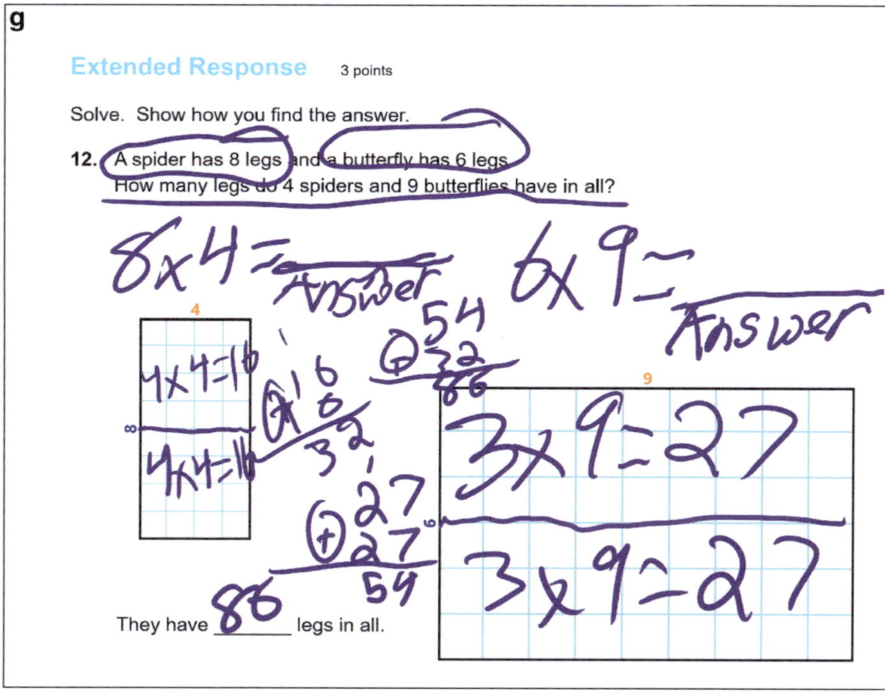

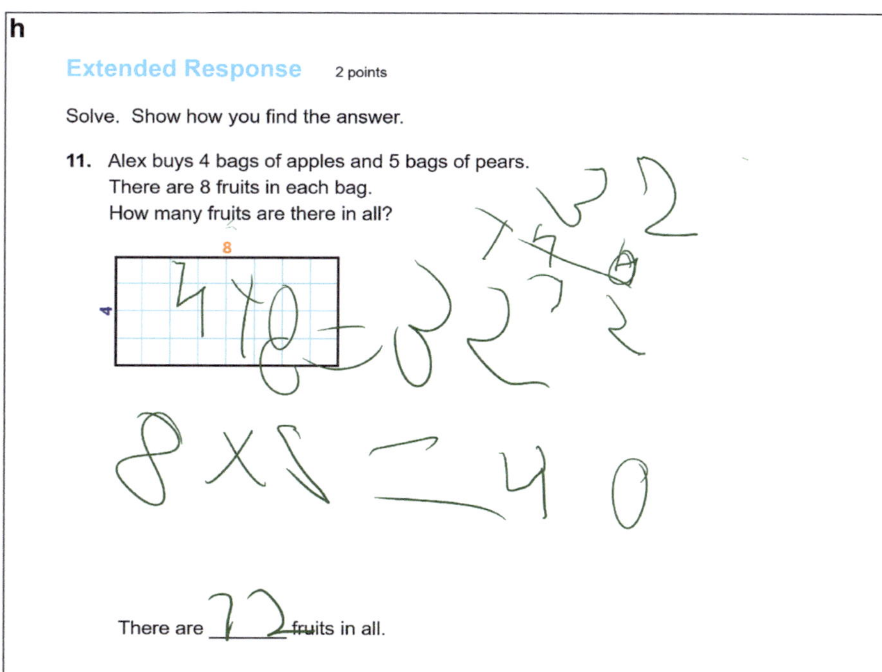

Fig. 3.2 (continued)

The second challenge is that of interpreting children's handwriting. Off-the-shelf handwriting recognition systems are trained on adult handwriting samples that are typically horizontal strings of text. Children's handwriting can be messy, but even when written neatly, it can differ significantly from adults' [6, 15, 16]. More importantly in our context, students write a combination of numbers, symbols, and random strokes in whatever locations they decide work best for their visual representations; the writing rarely appears in neat horizontal lines. The handwriting interpretation challenges are evident in Fig. 3.2a–h.

CLP's machine analysis routines meet the challenges illustrated in Fig. 3.2a–h, identifying 100% of all skip-counting sequences, with no ink sequences falsely identified as skip counting. The routines also are able to correctly interpret 94% of the numbers in skip-counting sequences to the right of an array. The key to the success is to use the structure of the array and knowledge about mathematics, in this case about skip counting, to bias the analysis routines. (Interpretation of sequences along the bottom of an array is more difficult and is discussed in the Current and Future Work section.) In the following sections we describe our data set of student work, the machine analysis routines, and results of running the analysis routines on the data set. We conclude with a discussion of contributions and implications for future development.

3.2 Methodology

Student work in the final assessment for a 5-week trial in a third grade class of 22 students is at the center of our current machine analysis efforts. The purpose of the trial was to investigate how pen-based technology that combines drawing with digital tool use can support elementary students in learning the concepts of multiplication and division. The assessment consisted of 12 multiplication and division problems—six word and six non-word—from Singapore Math's *Math In Focus* curriculum, which was being used in the school district.

Our data for developing and testing the machine analysis routines consists of the CLP electronic notebooks that contain the students' work on the assessment problems. Each student notebook contains final representations on each notebook page, along with re-playable interaction histories that capture the process of creating and interacting with the representations. This paper reports on machine analysis routines that operate on the final representations.

3.2.1 Representations Used

Students created representations using digital ink and three digital tools: a stamp, for drawing an image and creating multiple copies [8, 12]; a number line [1]; and an array [18]. Of the total of 264 pages of student work (22 students working 12 problems each), 74% of the pages include a final representation created using a digital tool. Approximately a third of those representations are arrays, for a total of 81 arrays, all of which are used in multiplication problems. In addition, 76 of the arrays (94%) have accompanying ink annotations, with 54 skip-counting sequences on 51 of the arrays—44 arrays with skip counting only on the right edge, 4 only on the bottom, 3 on both right and bottom. Other ink annotations include drawing a line across an array as part of a partial product strategy, e.g., as in Fig. 3.2f, g; equations inside an array, e.g., as in Fig. 3.2b, f–h; and dots or lines, e.g., as in Fig. 3.2c, d.

The large number of instances of skip counting on the right of an array is not surprising: The third grade curriculum being used represented a group as a row in an array, and the teacher modeled multiplication by writing a skip-counting number to the right of each row. Since the majority of skip counting in our data set was on the right of arrays—87% of the skip-counting sequences—we have focused our current machine analysis efforts on identifying and interpreting skip counting on the right.

3.2.2 Machine Analysis

The process of developing skip counting analysis routines started with the questions: How do math educators describe skip counting? How could a machine reproduce that description?

The action of writing the skip-counting sequence of 8, 16, 24, 32, shown in Fig. 3.1, would be described by math educators as skip counting by 8 from 8 to 32 along the right side of a 4 by 8 array. Written in terms of a coding scheme developed by the first two authors and two additional math education researchers, the expression for this action is ARR skip [4×8: 8, 8–32, right].[1] To reproduce this expression, machine analysis routines must output a sequence of numbers by identifying and interpreting the germane digital ink strokes, then summarize the sequence. Our current machine analysis routines perform the first of these tasks—outputting the sequence; the summarization step is discussed in the Current and Future Work section.

To meet the challenges in identifying and interpreting digital ink skip-counting sequences, CLP's machine analysis routines use knowledge about the structure of arrays and the nature of skip counting. Skip-counting numbers to the right of an array, for example, are likely to be written very close to an array, be aligned with consecutive rows of the array, and be about the same height as a row. Using these observations, the machine analysis routines can (1) use the structure of an array to group strokes associated with each row, (2) use the expectation of strokes aligned with consecutive rows to reject ink strokes that are not part of skip-counting sequences, and (3) use the expectation of a sequence of numbers and the values we would expect to see based on the array dimensions to improve handwriting

[1]The coding process and results of human analysis of the data set are described in a forthcoming paper.

recognition results. Employing these techniques, CLP is able to produce human-like coded expressions for skip-counting sequences. In the example discussed above, CLP produces the expression ARR skip [4×8: "8, 16, 24, 32", right].

CLP's first step in identifying and interpreting skip-counting sequences is, for each array on a page, to search all ink strokes on a page, using the array's location and size to reject strokes that are unlikely to be in a skip-counting sequence and returning a group of ink strokes that could be: It rejects any strokes that are much smaller than an array cell, e.g., stray dots, then uses the height of the array to set a bounding box, which extends from half a row height above to half a row height below the array, 80 pixels to the right of the right edge, and 40 pixels inside the right edge.[2] It rejects any strokes that have less than 50% of their height inside the bounding box or whose weighted center is not inside the bounding box. It prunes the remaining strokes by rejecting strokes that are too large (twice the row height), then prunes a second time by calculating an average stroke height for the remaining strokes and rejecting strokes that again are too large (twice the average stroke height). Using a factor of two when determining thresholds for stroke rejection works well with the students' ink annotations in our data set—no necessary strokes are ever rejected, and any unnecessary strokes are pruned in later steps. At this point, CLP has a set of candidate skip-counting strokes.

CLP's next step is to partition the candidate ink strokes into groups, where each group potentially corresponds to a skip-counting number. The partitioning is done by determining with which row ink strokes are aligned. This task would seem straightforward except for two challenges: Small strokes may be aligned incorrectly with an adjacent row, e.g., the small top stroke of a 5 may be drawn in the row above the body of the 5; and strokes may overlap more than one row, e.g., as with the "27" in Fig. 3.2c. The first of these challenges turned out to be easily met by our solution to the second challenge: The region tested for stroke overlap is not just a single row, but instead extends from half a row above to half a row below the row being checked. Each stroke then is grouped with the row whose extended region it overlaps the most. For cases in which a stroke overlaps two rows equally, the stroke is included in both rows.

Each row's group of ink strokes then is sent to a handwriting recognizer, which returns a list of possible interpretations.[3] CLP achieves high recognition rates by choosing number interpretations over non-numbers and, when presented with multi-digit options, choosing one that has the same number of digits as the value expected based on the array dimensions. In the student work shown in Fig. 3.2a, for example, for the last number in the sequence on the left, the handwriting recognizer returns choices of 32 and 320. CLP chooses 32 by using the dimensions of the array to

[2]Left and right bounds were determined empirically using the examples in the current data set. Further testing on a larger corpus may refine the dimensions.

[3]CLP is written in C# and currently runs in the Windows 8 and 10 operating systems; it uses the Microsoft English Handwriting Recognizer introduced in the Windows 8 (https://msdn.microsoft.com/en-us/library/ms840450.aspx).

determine that 32 is in the expected sequence of numbers. (Using the preceding interpreted numbers to predict expected sequence is discussed in the Current and Future Work section.) CLP also improves recognition rates by using heuristics to correct commonly misinterpreted strokes that occur in our data set: "&" is replaced with 8, ">" is replaced with 7, "Z" and "z" are replaced with 2, "|" and "l" (lowercase L) are replaced with the number 1.

CLP's final step is to use a set of rules based on knowledge of skip-counting patterns to reject any sequences that are likely to not be skip counting. Sequences are rejected by checking their interpreted values and finding that one or more of the following is true: the first two rows do not have values, fewer than two rows have values, more than two rows share the same value, the first row does not have a value and more than 50% of the rows do not have values (which handles arrays such as the one on the right in Fig. 3.2f), fewer than 34% of rows with values have numeric values, there is more than one gap of one row between values, or there is a gap of more than one row between values.[4]

3.3 Evaluation Results and Discussion

We evaluated CLP's machine analysis of skip-counting sequences in two ways: comparing CLP's results against human analysis results, and comparing CLP's results with results from two versions of machine analysis routines that use little or no knowledge of arrays or skip counting in identifying and interpreting sequences.

3.3.1 Comparison with Human Analysis Results

Using the analysis routines described in the previous section, CLP currently can identify skip counting along the right side of an array very accurately: It identified all 47 skip-counting sequences on the right of arrays in our data set and did not falsely identify any non-skip-counting ink as skip counting. It produced no interpretation errors in 37 of the sequences, thus reporting completely correct sequences in 79% of the sequences. It correctly identified 279 numbers out of a total of 296 numbers in all skip-counting sequences—94% of the numbers.[5]

The degree of accuracy achieved by the analysis routines means that CLP can present reliable information to a teacher about students' skip-counting strategies: It can identify which students used the strategy and present information about the accuracy of the students' sequences. It can identify, for example, that the student

[4]The threshold cutoffs were empirically determined.

[5]We mean number here, rather than digit, e.g., 16 is the number 16, not the "numbers" 1 and 6.

whose work is shown in Fig. 3.3a (and also in Fig. 3.2e) skip counted by the wrong dimension—5 instead of 8—which results in an answer of 25 for 5×8. This error is fairly common and may indicate some misunderstanding on the student's part about the difference between the size of a group and the number of groups or about how each quantity is represented in the array. Shown in Fig. 3.3b is an example of student work containing an arithmetic mistake: The student has added 4 to all elements except 16, which produces an incorrect answer of 26 rather than 28 for 7×4. This information suggests that the student has a relatively robust understanding of skip counting, but has made a careless arithmetic error, information that is not available just from knowing that the student's answer is 26.

3.3.2 *Effect of Using Array and Skip Counting Knowledge*

In order to evaluate the effect of using knowledge of array structure and skip-counting patterns, we compared the results of CLP's current machine analysis routines with two other versions of the analysis routines, each of which used less knowledge in attempting to identify and interpret skip-counting sequences. All versions prefer numerical to nonnumerical characters when the handwriting recognizer offers a choice. Version 1 employs only a bounding box along the right edge of each array, interpreting as a single string all strokes that fall within the box. Version 2 collects strokes using a bounding box and information about an array's row height, which allows it to improve results by rejecting strokes that are much shorter or much taller than the row height; it too interprets the collected

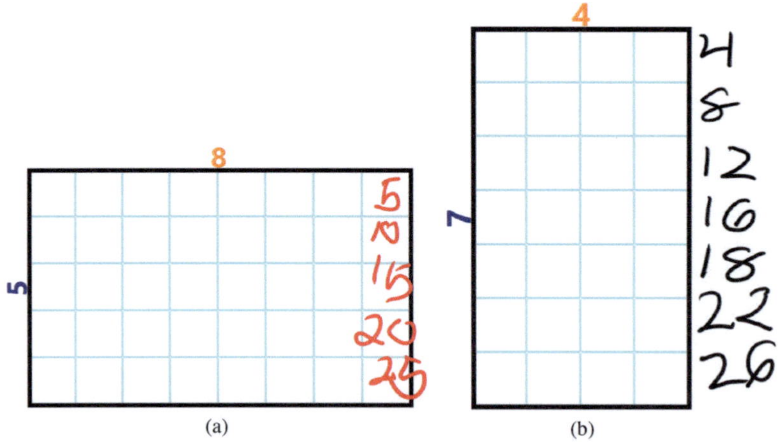

Fig. 3.3 Skip counting mistakes that CLP can identify. (**a**) Skip counting by the wrong dimension. (**b**) Arithmetic error from 16 to 18

Table 3.1 Results for different machine analysis routines

	Version 1 bounding box	Version 2 bounding box + row height	Version 3 all knowledge
# sequences interpreted correctly (of 47)	8 (17%)	10 (21%)	37 (79%)
# false positives	19	18	0
# false negatives	0	0	0

strokes as a single string. Version 3—our current version, described in Sect. 3.2.2—uses a bounding box and array row height to collect strokes, interprets groups of strokes aligned with rows, and uses heuristics to improve results. Version 3 returns a sequence of strings, each of which corresponds to a number, with one number per row of an array.

Table 3.1 presents results for how often each routine interpreted a sequence correctly, identified skip-counting sequences when it should not have (false positives), and failed to identify skip-counting sequences (false negatives). As is evident from the table, the more knowledge of arrays and skip counting used in the analysis routines, the better the results. Especially of note are the large counts of false positives for versions 1 and 2: These versions have no reliable way to distinguish skip-counting ink from non-skip-counting ink, so they consider all ink near the right edge of an array to be skip counting. In addition, versions 1 and 2 reject few or no ink strokes, so they will not miss any skip-counting sequences. Version 3, on the other hand, rejects quite a few ink strokes, but is still able to identify all skip-counting sequences.

As an additional measure of the interpretation accuracy for each of the versions, we calculated a *character error rate* (CER), which is a ratio of the number of errors in a string to the number of expected characters and is commonly used in handwriting recognition evaluation [15, 20]. To quantify the number of errors, we used *edit distance* (ED), which measures dissimilarity between two strings by counting the minimum number of steps needed to transform one string into the other [4]. We used Levenshtein edit distance [10], which sums insertions (I), deletions (D), and substitutions (S) required for the transformation: I + D + S. Dissimilar strings yield higher edit distances. CER is then ED ÷ N, where N is the number of expected characters in the string used as a reference. In our case, the reference string is created by concatenating the expected skip-counting sequence numbers, each separated by a single space, which is the delimiter used in the strings produced by versions 1 and 2.

Version 3 outperformed the other two versions, producing much lower CER values for all 47 skip-counting sequence on the right of arrays in our data set. Table 3.2 presents the average CER results for the 47 sequences and for the 10 sequences for which version 3 produced interpretation errors. Again, the use of knowledge of arrays and skip counting significantly improved version 3's performance over the other two versions.

Table 3.2 Character Error Rate (CER) values for different machine analysis routines

	Version 1 bounding box	Version 2 bounding box + row height	Version 3 all knowledge
Average CER	27	23	3
Average CER for v3 sequences with errors	43	39	18

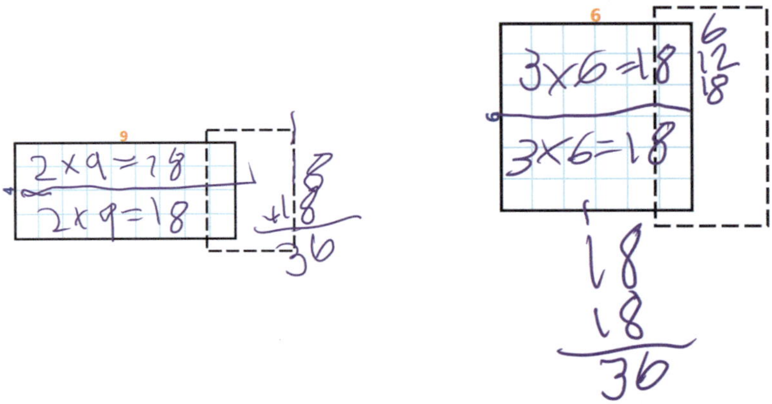

(a) 4x9 array
v1: "4"
v2: "1 -"
v3: no skip counting

(b) 6x9 array
v1: "in 88 8"
v2: "b 12 '8"
v3: "6" "12" "18"

Fig. 3.4 Two examples in which version 3's interpretations contain no errors

The examples shown below illustrate the differences in interpretations across the three versions of machine analysis routines. For the work shown in Fig. 3.4 (also shown in Fig. 3.2f), the bounding boxes for both arrays collect extraneous ink—the "+1" outside the 4×9 array and the "8" in each equation inside the 6×9 array—causing versions 1 and 2 to produce interpretations with very high error rates. Version 3, however, is able to reject the extraneous ink and produce correct interpretations for both arrays.

For the work in Fig. 3.4a,b, version 3's interpretation contains no errors. For the work shown in Fig. 3.5 (also Fig. 3.2c), version 3's interpretation contains three errors in the sequence of seven numbers for the skip counting by 9 from 9 to 63. Unlike the interpretations for versions 1 and 2 for this example, however, the majority of numbers are correct, and as discussed in next section, CLP can infer values for the misinterpreted numbers.

To further illustrate differences in interpretations across the three versions, Table 3.3 presents the bounding boxes, interpretations, and CER values for the student work examples shown in Fig. 3.2.

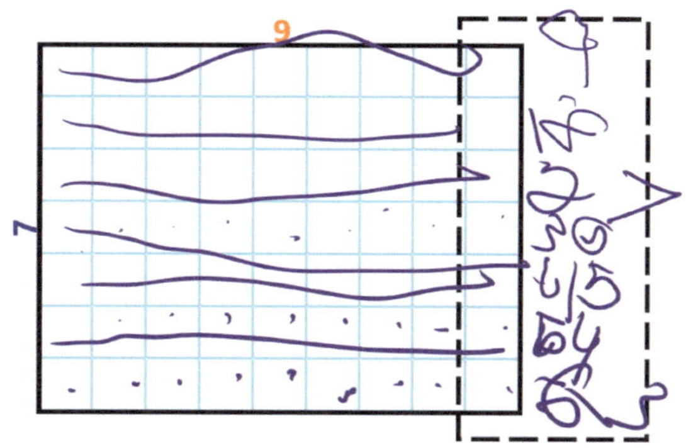

Fig. 3.5 Comparison of interpretations for different machine analysis routines
v1: "% 52 30s 05 tubby –863"
v2: "I 27 30s '6/5 tby 363",
v3: "9" "18" "27" "30" "05" "i4" "63"

3.4 Current and Future Work

We currently are focusing on several areas of development. To simplify our methods for aligning ink strokes with rows, we will investigate the idea of first grouping ink strokes that are aligned with each other, then aligning those groups with rows, as demonstrated by [21].

Our current machine analysis routines produce accurate results for identification and interpretation of skip counting along the right side of an array. They also reliably identify skip counting along the bottom of an array by using a bounding box and evaluating a single interpreted string with respect to an expected string determined by the array dimensions. The routines, however, do not yet attempt to produce a skip-counting sequence. We are working on interpretation techniques that will make use of ink alignment with columns, similar to CLP's alignment with rows. Interpretation of skip counting along the bottom of an array, however, poses the added challenges of lack of clear separation of sequence numbers, which is possible with one number written per row; and the limited horizontal space available with narrow arrays, which can cause numbers to not be aligned with columns.

We are implementing routines that can use knowledge of a skip–counting interval to infer a pattern when a skip-counting sequence contains handwriting recognition errors. Currently, skip-counting intervals are determined using the dimensions of an array. For the 7x9 array shown in Fig. 3.5, for example, a skip-counting sequence on the right would be expected to contain 7 numbers, start at 9, have an interval of 9, and end at 63. Using this information, the routines can infer that in the interpreted sequence 9, 18, 27, 30, 05, i4, 63, the 30 may be 36, the 05 may be 35, and the i4 may be 54. It then can report to the teacher both the original interpreted sequence and a summary: skip counting by 9 from 9 to 63 on the right, i.e., ARR skip [7x9: 9, 9–63, right]. We also are working on an alternate method for determining a skip-counting sequence interval—subtracting pairs of consecutive numbers in a student's sequence. This technique has the advantage of being able to report to a teacher when a skip-counting sequence is not related to an array, and it takes advantage of not just

Table 3.3 Interpretations and CER values for the student work shown in Fig. 3.2

Arrays + bounding boxes	Interpretations	CER
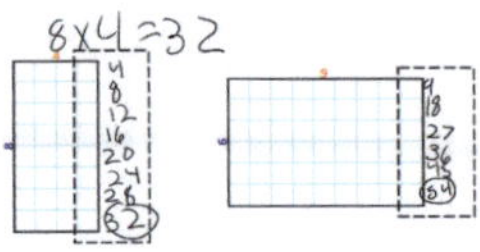 Fig. 3.2a	v1 4x8: "in's 160 2 24 28 320"	43
	v1 6x9: "98 27 36 45 540"	19
	v2 4x8: "4 F 12 16 20 24 28 32"	5
	v2 6x9: "918 2] 56 45 s☺"	31
	v3 4x8: "4" "8" "12" "16" "20" "24" "28" "32"	0
	v3 6x9: "9" "18" "27" "36" "45" "54"	0
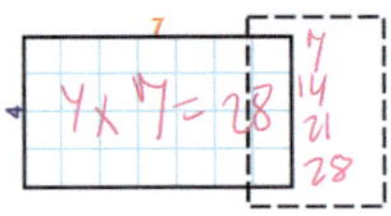 Fig. 3.2b	v1: "17% 8211 d8"	70
	v2: "17% 11 d8"	60
	v3: "7" "14" "21" "28"	0
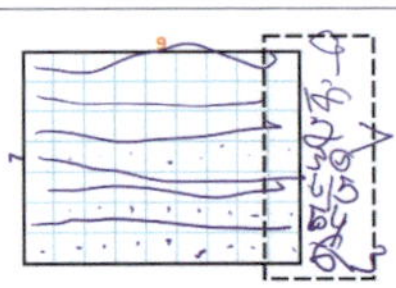 Fig. 3.2c	v1: "% 52 30s 05 tubby -863"	79
	v2: "I 27 30s '6/5 tby 363"	68
	v3: "9" "18" "27" "30" "05" "i4" "63"	16
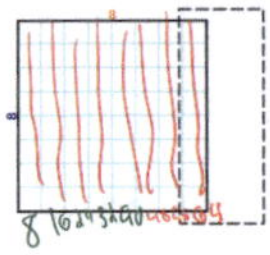 Fig. 3.2d	v1: "…."	n/a
	v2: "to"	n/a
	v3: "not skip counting"	n/a
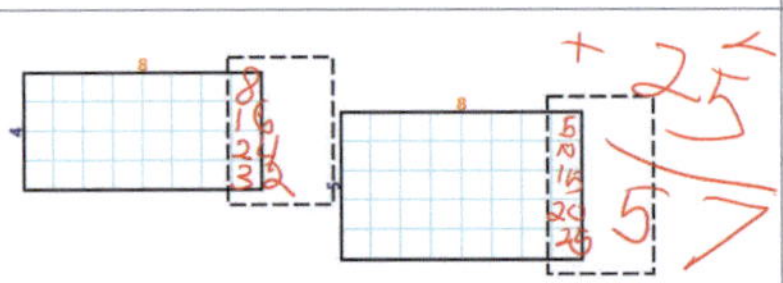 Fig. 3.2e	v1 4x8: "8. 6 24 32"	20
	v1 5x8: "is 205 25"	62
	v2 4x8: "£ 6, 22/ 39"	60
	v2 5x8: "15 15 20 25"	23
	v3 4x8: "8" "16" "24" "32"	0
	v3 5x8: "5" "10" "15" "20" "25"	0
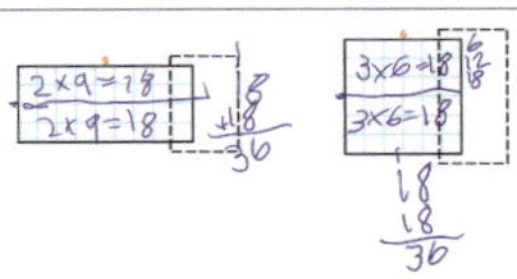 Fig. 3.2f	v1 4x9: "4"	n/a
	v1 6x6: "in 88 8"	71
	v2 4x9: "1 -"	n/a
	v2 6x6: "b 12 '8"	14
	v3 4x9: "not skip counting"	n/a
	v3 6x6: "6" "12" "18"	0

(continued)

Table 3.3 (continued)

Arrays + bounding boxes	Interpretations	CER
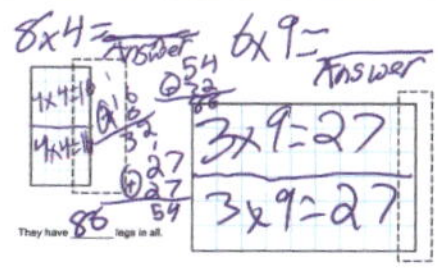	v1 8x4: "\Eigen"	n/a
	v1 6x9: "no strokes in boundary"	n/a
	v2 8x4: "-161 K f"	n/a
	v2 6x9: "no strokes in boundary"	14
Fig. 3.2g	v3 8x4: "not skip counting"	n/a
	v3 6x9: "not skip counting"	0
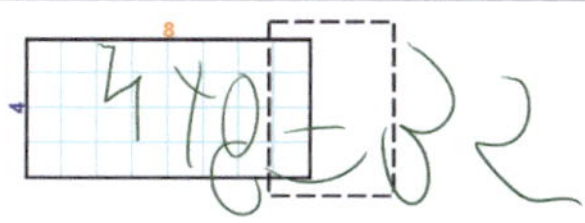	v1: "_-"	n/a
	v2: "5"	n/a
Fig. 3.2h	v3: "not skip counting"	n/a

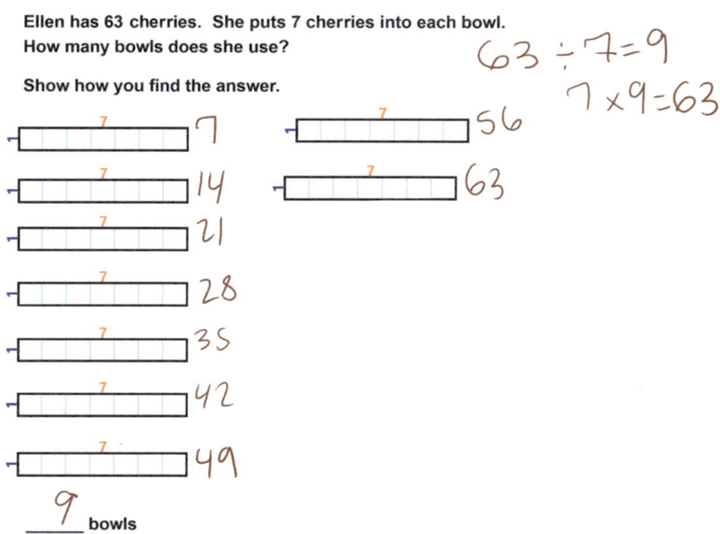

Fig. 3.6 Student's use of multiple arrays and skip counting

general information about array structure, but also specific information about what a student actually wrote.

An interesting variation on the array skip-counting representations in our current data set is shown in Fig. 3.6. In this work, created during the 5-week trial, a student has used a skip-counting strategy for division, building up to the dividend by creating single-row arrays and skip counting along the side of each array until she reaches the dividend. We are working to extend CLP to handle this case, as our current machine analysis routines assume skip counting occurs only on multi-dimensional arrays.

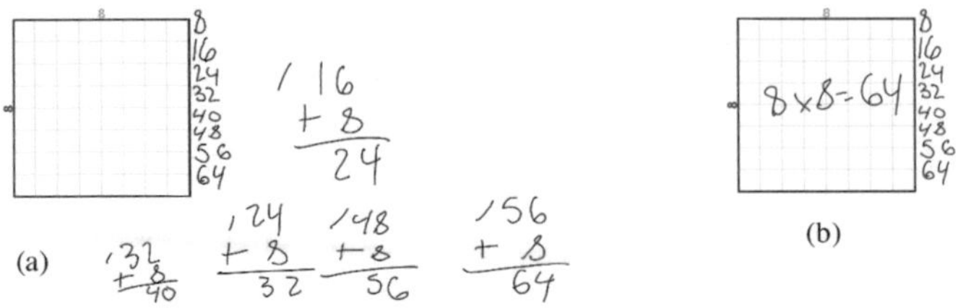

Fig. 3.7 Interaction history reveals student's use of arithmetic to complete sequence. (**a**) Arithmetic used to compute sequence. (**b**) Arithmetic erased in final representation

As mentioned earlier, CLP stores both a final representation and a re-playable interaction history, which captures a student's process of creating and interacting with the representation. We are working on routines that will analyze an interaction history in order to provide additional insights into a student's mathematical thinking—insights often not possible when looking at only a final representation. We have found, for example, that students often try several representations before settling on a final one to submit to their teacher. We also have found that students often do arithmetic, which they erase before submitting their work. In the example shown in Fig. 3.7, for example, the final representation does not give any indication that the student wrote 8 and 16 alongside the array, then did arithmetic for 24, 32, and 40, wrote 48 alongside the array, then did arithmetic for 56 and 64. This interaction indicates that the student is comfortable with her 8 times table up to 16, then must use arithmetic for larger numbers, except for 40 + 8.

Finally, we are working on machine analysis routines for other types of array ink annotations, such as the use of a line to indicate a partial product strategy as shown in Fig. 3.2f, g. We also are working on machine analysis routines for other types of representations such as a number line [1]; stamp, which enables students to draw an image and make multiple copies [8, 12]; division template, which is an interactive visual representation for the process of division [9, 18]; and bin, which supports what are called dealing out strategies [11, 22]. With an extended set of machine analysis routines, we will be able to increase our data set to include 8000 additional pieces of student work—work from the full 5-week trial in the third grade class whose assessment we are currently using, plus work from a previous 5-week trial in a fourth grade class.

3.5 Conclusion

This paper reports progress made on the development of machine analysis routines that accurately interpret elementary math students' visual representations created using a combination of freehand drawing and a digital array tool. Using the final assessment from a 5-week unit on multiplication and division in a class of 22 third grade students, the routines, which are part of our Classroom Learning Partner (CLP) software, are able to identify all final array representations that are accompanied by a handwritten sequence of numbers that exhibit a common problem-solving strategy called skip counting. The majority of the skip-counting sequences occur along the right side of arrays, and CLP is able to correctly interpret 94% of the handwritten numbers in the sequences in our data set, with 79% of the sequences interpreted completely correctly. The key to this success is to use knowledge about the structure of arrays and the nature of skip counting to bias the machine analysis routines. The effect of using such knowledge is evident from our evaluation of two versions of routines that do not use this knowledge—both suffer from high error rates in the interpretation of the handwritten skip-counting sequences and from the inability to distinguish skip-counting ink from non-skip-counting ink. With our analysis routines' current accurate identification and interpretation of skip counting, CLP will be able to present reliable information about students' skip-counting strategies. Such information can give teachers and others valuable insights into students' use of visual representations and the mathematical thinking that is revealed. Our current work focuses on expanding the machine analysis routines to operate on additional types of representations and on not just a final representation, but on an interaction history that captures a student's process of creating and manipulating representations. With these extensions, we hope to expand the usefulness of information we can provide to teachers about their students' mathematical reasoning, to math education researchers about the ways in which visual representations support student learning, and to software designers about ways in which design interacts with machine analysis.

Acknowledgements This research is funded by the NSF *INK–12: Teaching and Learning Using Interactive Ink Inscriptions* project, DRL–1020152 (Koile), DRL-1019841 (Rubin). We gratefully acknowledgment contributions from MIT's CLP research group members past and present, and from math education researchers Lily Ko and Marlene Kliman at TERC. We also thank Randall Davis, Jonathan Grudin, Tracy Hammond, YJ Kim, and Marlene Kliman for valuable feedback on drafts of this paper.

References

1. Benz, E.: Design and implementation of a number line tool for classroom learning partner. Unpublished UROP project (2014)
2. Brizuela, B.M., Gravel, B.E.: Show Me What You Know: Exploring Student Representations Across STEM Disciplines. Teachers College Press, New York (2013)
3. Cuoco, A.: The Roles of Representation in School Mathematics. Reston, NCTM (2001)
4. Damerau, F.J.: A technique for computer detection and correction of spelling errors. Commun. ACM **7**(3), 171–176 (1964)
5. Fosnot, C.T., Dolk, M.L.A.M.: Young Mathematicians at Work. Heinemann, Portsmouth (2001)
6. Kim, H.H.: Analysis of children's sketches to improve recognition accuracy in sketch-based applications. Ph.D. thesis, Texas A&M University (2012)

7. Koile, K., Rubin, A.: Animated mathematical proofs in elementary education. In: Hammond, T., Valentine, S., Adler, A.D., Payton, M. (eds.) The Impact of Pen and Touch Technology on Education, pp. 67–79. Springer, Cham (2015)
8. Koile, K., Rubin, A.: Machine interpretation of students' hand-drawn mathematical representations. In: Hammond, T., Valentine, S., Adler, A.D., Payton, M. (eds.) The Impact of Pen and Touch Technology on Education, pp. 49–56. Springer, Cham (2015)
9. Koile, K., Rubin, A.: Tablet–based technology to support students' understanding of division. In: Hammond, T., Valentine, S., Adler, A.D. (eds.) Revolutionizing Education with Digital Ink, pp. 71–89. Springer, Cham (2016)
10. Levenshtein, V.I.: Binary codes capable of correcting deletions, insertions, and reversals. In: Soviet Physics Doklady, vol. 10, pp. 707–710 (1966)
11. Markovits, A.: Machine interpretation of elementary math students' hand–drawn bin representations. Bachelor of Science Thesis. Electrical Engineering and Computer Science Department, MIT, Cambridge (2016). http://ink-12.mit.edu/documents/AMarkovits_UAP.pdf
12. Maynard, E.D.: Using a structured vocabulary to support machine understanding of student work. Master of Engineering Thesis, Electrical Engineering and Computer Science Department, MIT, Cambridge (2013). http://ink-12.mit.edu/documents/2013.02_MaynardE_MEng.pdf
13. NCTM: National council of teachers of mathematics: Principles and Standards for School Mathematics. National Council of Teachers of Mathematics, Reston (2000)
14. NGA: National Governors Association Center for Best Practices (NGA center). Common Core State Standards for English Language Arts & Literacy in History/Social Studies, Science, and Technical Subjects. National Governors Association Center for Best Practices and the Council of Chief State School Officers, Washington, DC (2010)
15. Read, J.C.: A study of the usability of handwriting recognition for text entry by children. Interact. Comput. **19**(1), 57–69 (2007)
16. Read, J.C., MacFarlane, S., Horton, M.: The usability of handwriting recognition for writing in the primary classroom. In: Fincher, S. (ed.) People and Computers XVIII—Design for Life, pp. 135–150. Springer, New York (2005)
17. Rubin, A., Storeygard, J., Koile, K.: Supporting special needs students in drawing mathematical representations. In: Hammond, T., Valentine, S., Adler, A.D., Payton, M. (eds.) The Impact of Pen and Touch Technology on Education, pp. 57–66. Springer, Cham (2015)
18. Simon, E.: Design and implementation of a division tool for elementary math education. Bachelor of Engineering Thesis, Electrical Engineering and Computer Science Department, MIT, Cambridge (2014)
19. Smith, S.P.: Representation in school mathematics: children's representations of problems. In: Kilpatrick, J., Gary Martin, W., Schifter, D. (eds.) A Research Companion to Principles and Standards for School Mathematics, pp. 263–274. National Council of Teachers of Mathematics, Reston (2003)
20. Soukoreff, R.W., MacKenzie, I.S.: Metrics for text entry research: an evaluation of MSD and KSPC, and a new unified error metric. In: Proceedings of the SIGCHI Conference on Human Factors in Computing Systems, pp. 113–120. ACM (2003)
21. Stahovich, T.F., Lin, H.: Enabling data mining of handwritten coursework. Comput. Graph. **57**, 31–45 (2016)
22. Yang, D.: Identifying bins-dealing strategies in Classroom Learning Partner. Bachelor of Science Thesis, Electrical Engineering and Computer Science Department, MIT, Cambridge (2015). http://ink-12.mit.edu/documents/DYang_UAP.pdf

Chapter 4
TouchCounts and Gesture Design

Nicholas Jackiw and Nathalie Sinclair

Abstract TouchCounts is an iPad app for early number learning that illustrates how cognitive facets of that domain can be mapped on to many of the affordances of the touchscreen tablet environment. Looking at the design of the *mathematical gestures* that learners use in the software's interface to count and perform basic arithmetic operations, we find that theories of embodied cognition and gesture analysis (which focus on only sometimes deliberate hand and body motions as supporting acts of communication and reasoning) usefully expand formulations of gestures common in the software design discourse (which constitute gesture as highly deliberate, and unambiguous physical imperatives articulated within a specific software application's command syntax). We develop this potential contribution in the context of educational software design, where the application purpose is to introduce users to novel conceptual domains, rather than to extend competence or enable expertise within already familiar ones.

4.1 Motivation

Informally, the TouchCounts app[1] seeks to answer the question, "What might you learn about numbers if you had an unlimited supply of fingers?" More formally, TouchCounts uses software and touchscreen technology to extend young learners' foundational experience of finger counting to more flexible and more generalized mathematical situations, and in so doing to expose three- to seven-year-

[1]http://www.touchcounts.ca. Developed by the authors with the support of a Canada Foundation for Innovation grant to Sinclair at Simon Fraser University, TouchCounts [12] can be freely downloaded at Apple's App Store.

N. Jackiw (✉)
SRI International, Menlo Park, CA, 94025, USA
e-mail: nicholas.jackiw@sri.com

N. Sinclair
Simon Fraser University, Burnaby, BC, Canada
e-mail: nathsinc@sfu.ca

T. Hammond et al. (eds.), *Frontiers in Pen and Touch*, Human–Computer Interaction Series, DOI 10.1007/978-3-319-64239-0_4

olds to flexible conceptions of numbers and to the basic mechanics of arithmetic. Recent neuroscientific research demonstrates considerable linkage between children's "basic finger sense" (which has a long clinical history as an indicator of various types of brain damage) and their mathematical ability in Grade 1 and later [10]. In this direction, TouchCounts' design uses the finger-centric interface of the touchscreen to augment the already powerful cognitive connection between physical digits and mathematical ones.

4.2 Touchcounts' Application, Functionality, and Use

TouchCounts[2] offers access to two open-ended exploration environments: a world of numbers and a world of operations. Each world begins as a (mostly) blank screen, in which "touching" produces numbers by world-specific mechanisms described below. Both worlds reset offer and home buttons, for clearing previous work or switching worlds. In the spirit of Papert's [8] constructionist micro-worlds, TouchCounts' worlds neither offer explicit instruction for their "proper use" nor score or otherwise assess student performance within them. Supplemental resources accessible outside the software (through the app's website) aim at parents or teachers to suggest various broad types of activity and conversation that might be collaboratively encouraged with the software, but ultimately advise adults to recognize "children benefit from making choices and being in control. TouchCounts encourages exploration and choice-making in an environment that is structured by mathematical rules, but that is not channeled into narrow opportunities or preconceived outcomes."

4.2.1 Mechanics of the Numbers World: Creating New Numbers by Counting

The first touch in the "Numbers World" produces, at one's finger tip, a small disk labeled with the symbol "1", and causes the software to announce—in the voice of a 5 year old girl—the number name "1". This "number object" remains under the user's finger as long as that finger remains in contact with the glass, and can be dragged or flicked around the screen. But when the user "lets go," the number falls rapidly off-screen under a downwards gravitational pull, and is gone

[2]This paper focuses on TouchCounts' gestural design and its theoretical foundation. Analysis of TouchCounts from the perspective of topics in mathematics education and research findings stemming from student use can be found in [13]. While educational activities using TouchCounts fall beyond the scope of this paper, interested readers can consult the app's website or a YouTube channel of video demos: http://bit.ly/1ZWwuYc.

forever. The next finger-touch produces a similar object named and numbered "2". Progressive finger taps produce successively higher number values. Multiple finger taps produce multiple numbers, one per finger, although when multiple numbers are produced simultaneously, only the highest value is vocally named in the counting voice. Thus, when a user repeatedly taps a pair of fingers in an initially empty world, she produces a rapid succession of falling numbers pairs—1 and 2, 3 and 4, 5 and 6—while the software counts by twos: "two... four... six...." Repeatedly pressing all of one's fingers on the screens counts off by 10s.

In addition to the endless succession of increasing values, the "Numbers World" features a discrete "shelf." Unlike other values, numbers put onto the shelf do not fall off the bottom of the screen; they are caught by the shelf and remain visible. The shelf offers a mechanism for holding values the user might consider "special," without defining that idea further. Users can empty the shelf by shaking it (sending its cached values to gravitational oblivion) or by resetting the world.

Figure 4.1 shows a Numbers world screen in the midst of activity. The user has just created four new values by pressing four fingers to the screen, and then these numbers are "falling away" after the user has lifted fingers. A few early values have been parked on the shelf. During several years observing activities like this, the largest value we have seen a child create in the "Numbers World" is three thousand and two hundred (3,200).

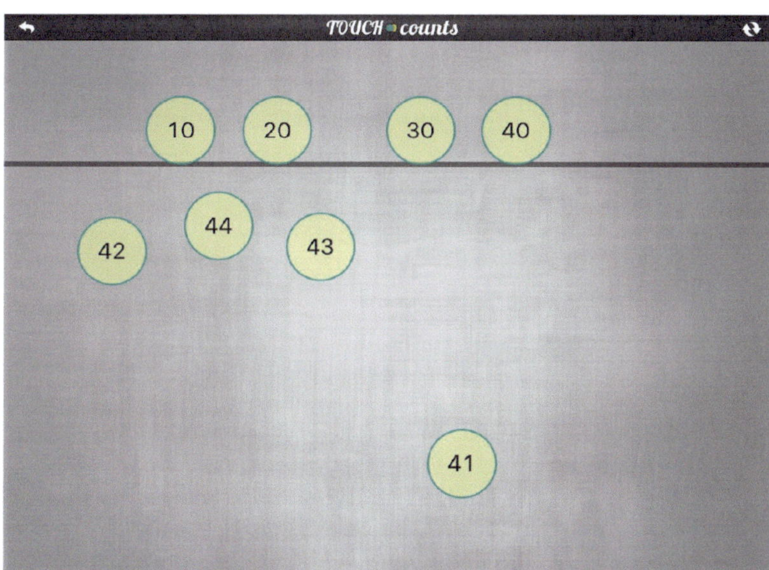

Fig. 4.1 Numbers World: four values on the shelf; four more (falling) beneath it

4.3 Mechanics of the Operations World

In the "Operations World," touch—and the number values it creates—behave differently. A single finger tap produces a circular disk named and numbered one (1), but upon release, the number remains on the screen, gently hovering its surface rather than being whisked quickly offstage by gravity. But two finger taps together—either simultaneously or in overlapping sequence—create two separate "unit chips" beneath the fingers that are circumscribed into a single number value object, numbered and named two (2), rather than two separate successive numbers. Upon release, the constituent unit chips contract into a slowly stately dance inside their "2", rather than vanish offscreen. Ten fingers "more or less" at once produce the number ten (10), which appears as a visually "larger" number than one (1) or two (2).

Experimentation rapidly leads to a generalized rule: pressing a first finger down begins composing a number, additional "overlapping" fingers increase the value of this new number until no fingers are touching the screen, at which point it is "completed" and its constituent unit chips begin their stately respiratory dance. By strumming fingers repeatedly and sequentially on the screen in the manner of overlapping rhythms or piano scales, one can create numbers much larger than ten. Touching an already-created number merely "pokes" it without further changing its value, but dragging an existing value relocates it on the screen. Figure 4.2 demonstrates several freshly created numbers in the "Operations World".

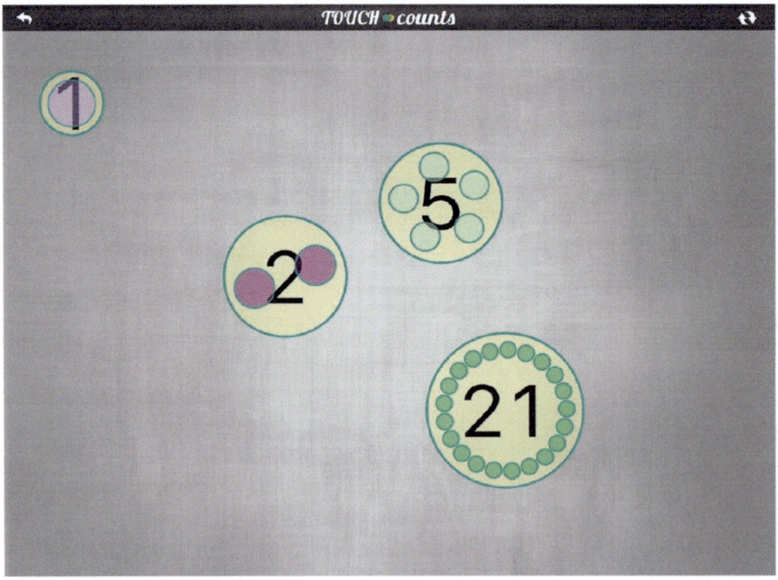

Fig. 4.2 Operations World: four newly minted numbers

Fig. 4.3 Adding 4 and 10 by pushing them together

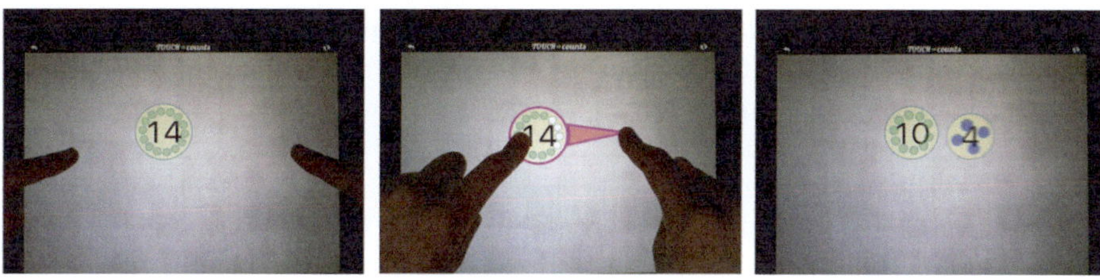

Fig. 4.4 Pulling 4 from 14, leaving 10 behind

In addition, numbers can be created by putting together and pulling apart other numbers. Since any existing number can be dragged, two or more numbers can be dragged simultaneously. When two (or more) number disks are dragged into each other, they are temporarily circumscribed by a new disk that encompasses both (or all) of the dragged numbers. Continuing to drag the constituent values out of overlap erases this temporary new disk, but finalizing the drag in this "combined" value causes the constituent dragged numbers to dissolve into the new larger value, with their constituent number chips reassembling themselves into a single new sum. Figure 4.3 illustrates the evolution of two numbers being "pinched together" to form a third.

A number larger than 1 can be broken into smaller numbers either by "pulling it apart" with two fingers, or equivalently by holding it in place with one finger while "pulling away" with another. When the dragging fingers are released, the original number splits into two parts: the amount pulled away (the subtrahend) and the amount remaining left over (the difference). The amount pulled away is determined by how "dramatically" the originally number is pulled apart, or more technically by the length of the vector in which one finger pulls away from the other. (The shortest gesture pulls only 1 unit chip out of the original minuend; the largest possible pulls a ll of the chips out of the original number.) Figure 4.4 demonstrates an example in which 14 is partitioned into 10 and 4, with the magnitude of the subtrahend, 4, ultimately determined by the length of the pink vector "pulling away" from the minuend, 14. Even before the fingers are released, a recoloration of the original 14 green unit chips provides a dynamic preview of the amount currently targeted for "pulling out."

4.4 TouchCounts' Mathematical Learning Design

The most fundamental facilitative learning goal of TouchCounts' design is to help students navigate four separate number representations and put them each into one-to-one correspondence with the others, verbal number names to objects in a set to deictic finger pointing to the Arabic numerals. The software interface encourages not only correspondence of these representations (by birthing each new number simultaneously in all of them), but also the learner's agency in co-producing or constructing them. While a conventional HCI interpretation of the software's dynamics might separate these representational equivalences, contrasting the role of the "pointing representation" (finger tapping) as formal input to the responsive output of sound and graphical image, we argue elsewhere [4] that when finger gesture interactions with tablet/touchscreen interfaces give rise to simultaneous, continuous virtual representations, such assemblages tremendously weaken the traditional perceptual divide between input and output. Thus in TouchCounts, when virtual yellow disks track your fingertips as you slide them across the glass, they feel more part of your fingertip than of the glass. The voice instantly announcing the number name at the second a child taps is a child's own voice, not the computer's. Children routinely claim they make numbers in TouchCounts, that this number—in all its semiotic dress forms—is not just 10, but "my ten."

The software's two worlds correspond to two fundamental, complementary aspects of number sense. In the first world, the whole numbers are produced in never ending sequence beginning with 1; the fundamental activity is counting or iterating by one; and there is a strict correspondence between successive finger taps and successive number values. This world emphases ordinal aspects of number. By contrast, the second world takes a more cardinal approach to number, allowing arbitrary values to be summoned into existence by act of will, and explored in terms of their equivalence to arrangements of other numbers. Operationally, the "pushing together" and "pulling one from another" gestures of the second world are not quite addition and subtraction, even if their numerical results are identical to those arithmetical operations. Rather they are conceptual building blocks leading toward those operations, and they can be gainfully explored by children who have yet to encounter more formal arithmetic. Importantly, their gestural forms embody specific mathematical features of the operations they invoke. In TouchCounts, two or more numbers can be pushed together simultaneously, just as addition is an n-ary operator "$(1 + 2 + 3)$", but numbers can be pulled one from another only serially ("5 3 2" can mean either "$5 - (3 - 2)$" or "$(5 - 3) - 2$", but not both!) And the pushing together gesture is physically commutative in a way that pulling apart gesture is not: where two fingers play symmetric roles in pushing two numbers together, (and can be swiveled into interchangeable positions without changing the net intermediate or terminal results—see Fig. 4.3, center or right), when pulling apart, the role of fingers is asymmetric (one touch anchors the original number to the mat while the other finger teases away from it).

Finally, TouchCounts' design attempts to layer various supporting pedagogical features on top of the mathematical features mapped by its basic gestural command syntax and corresponding visual response, as—for example—when the unit chips composing a "pushed together" number retain the colors they had in their original addend numbers, leaving a trace in this 14 (Fig. 4.3 right) that it came from a combination of some earlier 4 and some earlier 10 (Fig. 4.3 left).

4.5 Where TouchCounts' Gestures Come From

TouchCounts' gesture-based commands for counting and arithmetic do not derive from any standard palette of common iPad gestures but rather from a design process informed by two bodies of prior work. In one direction, the human computer interaction (HCI) community focuses on gesture based software command grammars for software interfaces, and on gesture design as well as automated gesture detection to define and recognize "verbs" or action imperatives gestural command languages. In another direction, researchers in gesture analysis focus on the usually spontaneous, and only sometimes intentional, gestures we make with our hands as a part of communicating, both with others and with ourselves. We develop both of these contributions here.

4.5.1 HCI Gesture Design and Analysis

Following the sudden and widespread availability of consumer grade touchscreen devices beginning a decade ago (e.g. with the Apple iPhone, released in 2007), the HCI community rapidly realized that "proper attention [to] how touch is developed and utilized in different interfaces" was still wanting [15]. Very broadly speaking, touch has been seen by researchers and technology evangelists alike as continuing a long evolution of "direct manipulation" interface technologies, running from the trackball (in the 1940s) through the light pen and mouse (of the 1950s and 1960s, but popularized only in the late 1980s with the GUI revolutionary). Some have even considered the touchscreen an evolutionary discontinuity: since there is no longer any mechanical prosthesis—no ball, pen, stick or mouse—intervening between "hand" and "eye," touch interfaces promise a breakthrough in intuitive accessibility, a "more natural" form of interaction. As Steve Ballmer, then CEO of Microsoft, stated: "We will look back on 2010 as the year we... started incorporating more natural forms of interaction such as touch, speech, gestures [into] what computer scientists call the 'NUI' or natural user interface." (quoted in [7], p. 6). This promise—of inherent naturalness or, equally, intuitiveness—has gone on to become an essential criterion in HCI recommendations and rubrics for considering the effectiveness of potential touchscreen gestures. Thus, the authors of Brave NUI World: Designing Natural User Interfaces for Touch and Gesture claim prospective gestures "must [...] create an experience that feels just as natural to a novice as to an expert user" (p. 13 [14])—and this is one of only two imperative guidelines for such designs. Similarly, Nielsen et al. [6] find that to offer an effective "human based gesture approach," gestures must be "easy [...]; intuitive; [...] logical; [and] ergonomic" (p. 12). Even in the popular press—as for example when in 2013 the blogosphere erupted with concern over Twitter's claiming patent ownership of a popular "intuitive" touchscreen gesture[3]—the equation is clear: effective designs are natural and intuitive; and natural and intuitive designs are effective.

4.5.2 Embodied Cognition

Psycholinguistic and cognitive treatments of gesture have approached them from a substantially different starting point, which in light of the previous conversation we can imagine to have been a question: "since gestures"—human gestures, primarily hand gestures—"appear to occur quite naturally"—i.e. spontaneously, "in the wild"—"then: what do they mean and how do they function?" Early work in gesture analysis [5] identified gestures as communication acts engaged in the same system of communication as language, with similarly recognizable facets of externally directed intentionality (our own gestures help us make meaning, to others) and internally directed orchestration or comprehension (our own gestures help us make meaning—of others, and of "things"—to ourselves). These readings are straightforward from the embodied cognition perspective, which locates meaning making and production in the physical body (i.e., in the perceptual motor system and in the body's situatedness in its environment) as much as in the brain. Thus, gesture is an embodied language in both its expressive and impressive aspects. Cognitive researchers have then looked at how gestures arise or are associated with particular knowledge domains, as when—in mathematics [10] show a correlation between finger gnosis and school mathematics performance, or when Carlson et al. [1] demonstrate that even adults count arrays of objects substantially faster when they are permitted to gesture point or nod their heads while counting. From here educational research has asked whether these associations can be used explicitly and productively and in support of learning. Can a teacher more effectively communicate an understanding, or can a student more effectively construct one, by gesturing? And can a gesture be found, or intentionally designed, to carry specific pedagogic meanings? Cook, Mitchell, and Goldin Meadow [2] show experimentally when students' exposure to new mathematical concepts is supported by the enactment of new, cognitively intentional, explicitly designed gestures, they demonstrate improved retention of those concepts both immediately and longterm. The authors conclude

[3]See e.g. www.webpronews.com/twitterhaspatentapplicationforpulltorefresh201203/, www.cnet.com/news/twittergrantedrefreshpatentvowstouseitonlydefensively/, and many similar others, which base arguments on the gesture's "intuitive" nature.

that "gesturing can...play a causal role in learning[...] by giving learnings an alternative, embodied way of representing new ideas. We may be able to improve children's learning just by encouraging them to move their hands." (p. 1047) Moving most recently into the realm of educational touchscreen software, Dubé and McEwan [3] measure adults using competing touchgestures—one discrete and one continuous in form—to estimate values on number lines, and note consistently better task performance with the continuous gesture, which they hypothesize reflects the continuous nature of the mathematical number domain. These findings "suggest that gesture use during touchscreen interaction may affect knowledge acquisition and may facilitate generalizable learning," and conversely, that "using a gesture that is incongruent with an underlying mathematics concept could decrease performance during the acquisition phase and could [adversely] affect learning" (p. 96).

4.5.3 A Synthetic Critique

Thus where the cognitive gesture analyst investigates how a gesture can be designed to create, reflect, or express an understanding (even if loosely), the HCI gesture designer is concerned with how a gesture specifically directs a concrete action. At first these seem not so far apart: both seek to harmonize the physical form of a hand gesture with some more virtual domain of operation, regardless of whether that domain is a broadly conceptual abstraction (such as surrounds a generalized mathematical understanding) or a specific virtual reality (such an individual contact page in a digital address book). Yet, considered from the perspective of learning theory, HCI's insistence on "natural" and "intuitive" forms works against the premise that learning is a (usually effortful) state transformation: if something is intuitive, it doesn't need to be learned. Thus, Wigdor and Wixon's [14] dictum—that a gesture's experience should be "just as natural to a novice as [...] to an expert"—can be read as categorically stating that gesture design should avoid any differential separating expert from novice. This further implies that gestures should never participate in shaping users' learning or domain experience, since it is precisely learning and experience that separate novice from expert! It's tempting to reconcile this impasse, by seeing such aspirations—to naturalism and intuition—as intended only for the form of the physical gesture rather than for the conceptualization of the deeper topic or idea to be learned. But the premise of embodied cognition is that these are two aspects of the same thing. Rather than deem this a fundamental opposition—intuition as opposed to learning—we suspect the two perspectives are better understood as concerned with very different phases of a user's experience of a gesture. Where the learning scientist is interested in the gesture's authentic expression in the context of grasping for meaning, perhaps the CHI designer is more concerned with its habituated form, even (or only) long after its potential for novel meaning making has been exhausted.

Thus when asked whether a gesture is "natural," we are not being asked to consider that pushing, stroking, and sliding one's fingers in specific choreographies over unresponsive sheets of plain glass might be an act found "in nature," nor to imagine such acts as "obvious" in the sense of corresponding to a-priori and predictably determinate actions or meanings. Instead, "natural"—like "intuitive"—aims to describe a user's ideal reaction to the gesture only after having already acquired it, and after having instrumentalized its meaning through some practiced repetition. "Now that I understand what it does," they encourage designers to imagine users asking, "Does it feel transparent and obvious, or awkward and irritating?" Thus a claim that Twitter's "pull to refresh" gesture is "very intuitive" does not imply that anyone has been expected to intuit that patiently pulling a scrollable list will suddenly alter that list's contents. Instead, once the gesture has been learned, its physical form and associated action seem reasonably convenient, recollectable, and, if not conceptually sensible to the domain, at least not non-sensible. It is natural and intuitive post-hoc. From this perspective, any conceptual or structural relationship between the form and meaning (or action) of a gesture—any degree to which the gesture physically acts as presentiment or echo to its effect—is largely irrelevant to the "natural" HCI designer, since the potential presence of such a link does not feature critically in a definition of natural, intuitive forms, and since the moment of post-hoc evaluation of the success of those forms happens long after the association between them and their actions has been cemented. These two concerns may of course be very different for the designer of learning experiences, whose primary emphases may be (following Cook, Mitchell, and GoldinMeadow [2]) on the usage moment where the association between gestures and their virtual meanings or actions is being first forged, and (following Dubé and McEwen [3]) how the form of a gesture supports or invokes a productive conceptualization of its meaning.

To help operationalize these differences as further principles of learning design, Peirce's semiotic framework for considering the relationship between a sign and its object [9] usefully informs how various touchscreen gestures relate to their meanings and consequences within a software environment. Following Peirce, we call gestures that directly mimic the form of their corresponding meaning iconic. Gestures that correlate to, "point to," or are conceptually congruent with their software consequence are considered indexical. Where the iconic relationship involves duplication, the indexical one involves implication. Finally, gestures that denote their object only through convention, law, or habit rather than through intrinsic attributes of form are what Peirce calls symbols. Thus a gesture of rhythmically nodding one's head might appear iconic to a software consequence in which one's virtual avatar nods its head in parallel; indexical to a consequence in which each head nod produces another drum beat in an ongoing song sequence, and symbolic to a consequence in which the active document is spell-checked. Of these types of gesture, we hypothesize that indexical gestures are uniquely situated to promote learning in environments where software consequences occur in a learning domain—that is, in the subject/concept/action arena to which as educational designers we aim to introduce learners and within which, to cultivate their facility. Both the iconic and the symbolic gesture lack inherent conceptual trajectory to their articulation. The icon involves strict duplication of form without particular invocation of that form's consequence or causality, and the symbol involves only arbitrary, rather than "meaningful," association between form and function. Each suggests a form of learning by rote, and of encouraging associations that are superficial on one hand or behaviorally engineered on the other. By contrast, indexes might promote learning by offering referential footholds to broader conceptual consequences. Where Twitter's pull gesture only arbitrarily relates to a list replacement operation, Dubé and McEwen's [3] continuous linear drag indexes the linear continuum of real numbers. Unlike a symbol, indices orient and indicate and leave a trace [11]; but—unlike an icon—indices do not simply duplicate or entirely assimilate. These distinctions are largely inessential to the CHI designer's metric of "naturalness," which we have shown is meaningful only in a phase where the differences in conceptual heft across gesture types has become immaterial.

Boiled down to a simple design insight, this argument amounts to the claim that, when promoting learning is the central goal of a touchscreen experience, concerns of natural or intuitive gestural forms are far less relevant than consideration of the conceptual relationship between gestures and their enactive consequences within the learning domain. In TouchCounts, we have applied this insight—as well as our hypothesis that indexical gestures are uniquely functional in a domain specific learning discourse—by designing physical forms for TouchCounts software command gestures that attempt to refer outside the software environment both to concrete physical activities (of fingers and hands, enumerating and manipulating countable objects), as well as to mathematical ideas and abstractions (such as the limitlessness of numbers). Specifically, the Number world's number creation gesture ("one by one") points from the serial physical act of counting on one's fingers—as when a young child is asked how many coins are on the table—and points to a more generalized 1:1 correspondence between (natural) numbers and objects. By contrast, the Operation World's number creation gesture ("all at once") points from the subitizing gesture of holding up fingers simultaneously, as when a young child is asked how old they are—and points to the more general cardinal identity to sets of objects. Similarly, TouchCounts' arithmetical gestures resituate "pushing things together" and "pulling something apart" into an explicitly quantized context, incrementally formalizing them into "adding together" and "taking away" while pointing not only to generalized addition and subtraction, but through serial addition ("adding four threes together") and equipartition ("splitting 12 into three groups, fairly") on to multiplication and division as well. In each case, the indexical gesture refers bidirectionally to a physical action known outside—and in some sense "before"—the software encounter, and to a mathematical abstraction or generalization laying "beyond" the software's (still finite, still temporalized) virtual worlds.

4.6 Conclusion

This paper has introduced TouchCounts, a gesture based system for developing ordinal and cardinal facets of young learner's number sense, and for introducing them to arithmetical operations through a set of virtualized affordances that occupy an intermediate or transitional zone between the physicality, and finitude, of their actual fingers and the abstract and unlimited conceptual realm of mathematics. We have outlined relevant contributions to TouchCounts' gesture design from learning science and software interface design communities, and proposed that indexical gestures that refer in their form to embodied characteristics of the learning domain offer unique opportunities to promote learning.

References

1. Carlson, R.A., Avraamides, M.N., Cary, M., Strasberg, S.: What do the hands externalize in simple arithmetic? J. Exp. Psychol.: Learn. Mem. Cognit. **33**(4), 747 (2007)
2. Cook, S.W., Mitchell, Z., Goldin-Meadow, S.: Gesturing makes learning last. Cognition **106**(2), 1047–1058 (2008)
3. Dubé, A.K., McEwen, R.N.: Do gestures matter? The implications of using touchscreen devices in mathematics instruction. Learn. Instruct. **40**, 89–98 (2015)
4. Jackiw, N.: Touch and multitouch in dynamic geometry: sketchpad explorer and "digital"İ mathematics. In: Proceedings of the 11th International Conference on Technology in Mathematics Teaching, pp. 149–155 (2013)
5. McNeill, D.: Hand and Mind: What Gestures Reveal About Thought. University of Chicago Press, Chicago/London (1992)
6. Nielsen, M., Störring, M., Moeslund, T.B., Granum, E.: A procedure for developing intuitive and ergonomic gesture interfaces for HCI. In: International Gesture Workshop, pp. 409–420. Springer (2003)
7. Norman, D.A.: Natural user interfaces are not natural. Interactions **17**(3), 6–10 (2010)
8. Papert, S.: Mindstorms: Children, Computers, and Powerful Ideas. Basic Books, Inc., New York, NY, USA (1980)
9. Peirce, C.S., Welby, V.: Semiotic and Significs the Correspondence Between Charles S. Peirce and Lady Victoria Welby. Indiana University Press, Bloomington (1977)
10. Penner-Wilger, M., Anderson, M.L.: The relation between finger gnosis and mathematical ability: why redeployment of neural circuits best explains the finding. Front. Psychol. **4**, 877 (2013)
11. Sinclair, N., Freitas, E.D.: The haptic nature of gesture: rethinking gesture with new multitouch digital technologies. Gesture **14**(3), 351–374 (2014)
12. Sinclair, N., Jackiw, N.: Touchcounts [software application for the ipad]. Tangible Learning Project, SFU (2011)
13. Sinclair, N., Pimm, D.: Whatever be their number: counting on the visible, the audible, and the tangible. In: Meletiou-Mavrotheris, M., Mavrou, K., Paparistodemou, E. (eds.) Integrating Touch-Enabled and Mobile Devices into Contemporary Mathematics Education, pp. 49–78. Information Science Reference, Hershey (2015)
14. Wigdor, D., Wixon, D.: Brave NUI World: Designing Natural User Interfaces for Touch and Gesture. Elsevier, Burlington (2011)
15. Wigdor, D., Fletcher, J., Morrison, G.: Designing user interfaces for multi-touch and gesture devices. In: CHI'09 Extended Abstracts on Human Factors in Computing Systems, pp. 2755–2758. ACM (2009)

Part III
Math For All: High School and College Math

Chapter 5
Sketch Based Interaction Techniques for Chart Creation and Manipulation

Andrés N. Vargas González, Eugene M. Taranta II, and Joseph J. LaViola Jr.

Abstract In a classroom environment, students are often required to extract descriptive statistics from charts, or find an appropriate graphical representation of the data they are given. On paper these drawings are static, but in an interactive learning environment, the exploration of data can be improved. To address this, we present a set of novel interaction techniques for the creation and manipulation of 12 chart types that leverage pen and touch in a prototype application, SketChart. We allow a user to sketch an appropriate chart type to visualize their data, after which, he or she can remove, merge, and filter data through direct manipulations. Data series and points are also managed through a pair of interactive legends. In a formative user study based on a set of typical statistics questions, we find that even with a chart recognition accuracy of only 81%, the overall system was still well received and can be used as a learning tool for descriptive statistics.

5.1 Introduction

Charts as a graphical representation of data are an integral part of education, data exploration, and scientific analysis. As a visualization of quantitative and qualitative information, charts are especially useful in helping people understand trends, find implications, and discover the meaning of their data. However, current techniques for visualizing and interacting with data are limited to the WIMP[1] interface metaphor [23]. Although WIMP is functional, it does not facilitate the natural, expressive, or familiar pen and paper workflow [7] that enables one to seamlessly transition between data interaction, annotation, handwritten mathematics, and free form sketching, all of which are common in an educational or data exploratory

[1]Windows, icons, menus, and point and click.

A.N. Vargas González (✉) • E.M. Taranta II • J.J. LaViola Jr.
Interactive Systems and User Experience Research Cluster of Excellence,
Department of Computer Science, University of Central Florida, Orlando, FL, 32816, USA
e-mail: andres.vargas@knights.ucf.edu; etaranta@gmail.com; jjl@cs.ucf.edu

T. Hammond et al. (eds.), *Frontiers in Pen and Touch*, Human–Computer Interaction Series, DOI 10.1007/978-3-319-64239-0_5

environment, and all of which many sketch based scientific interfaces aspire to support.

In accordance with the aforementioned, data visualization software must support a variety of additional interactions specific to dealing with charts. For example, such software should allow one to easily select and group data, interpret a sketch that depicts how the selected data should be visualized, and provide intuitive methods for switching between various views of data. Real-time combinatorial techniques for constructing complex charts are needed, such as to provide the ability to combine a bar chart with a line graph to create a Pareto chart. Techniques for data filtering and manipulation are also required. However, more importantly, the interface should lean towards a what you draw is what you get (WYDIWYG) philosophy [24], where interactions are direct, intuitive, and memorable. Therefore, in an initial effort to design the desired apparatus, we present as a foundation, a set of novel pen and touch interaction techniques that embrace the intuitiveness of pen and paper while incorporating modern but familiar touch gestures, such as pinch-to-zoom and swipe to pan. Further, in order to evaluate their effectiveness, we implemented these techniques in a prototype application SketChart (see Fig. 5.1) that, to the best of our knowledge, is the first WIMP-free data visualization software to allow free form sketch-based chart generation and data manipulation. Because recognition of hand drawn charts is a critical component of our interface, we also discuss our recognition strategy and report on its accuracy and effect on the acceptance of our proposed techniques.

Specifically, SketChart allows a user to select data and sketch a graph in order to generate a visualization of the data. Presently, 12 chart types based on Harris's categorizations [12] are supported. These types, in part, were selected in order to explore quantitative patterns, e.g., bar, line, and pie charts over varying distributions, and 100% stacked chart, box plot, and histogram. Once a chart is created, users are then able to modify values along different axes, change minimum and maximum values, add or remove data series, and so on, so that the visualizations and data are always in sync. Using SketChart, we conducted a formative user study to verify whether our design goals were met, and to determine if the interaction techniques indeed provide an intuitive, easy to use interface. In the evaluation, we pose a number of typical descriptive statistics questions that require participants to create and manipulate charts. Our findings show that despite recognition errors with certain chart types, participants enjoy the expressiveness and usability of our prototype.

5.2 Previous Work

5.2.1 Visualizations via Pen and Touch

In recent years, Microsoft researchers have been investigating user interfaces as well as interaction techniques for chart generation and data manipulation. An early

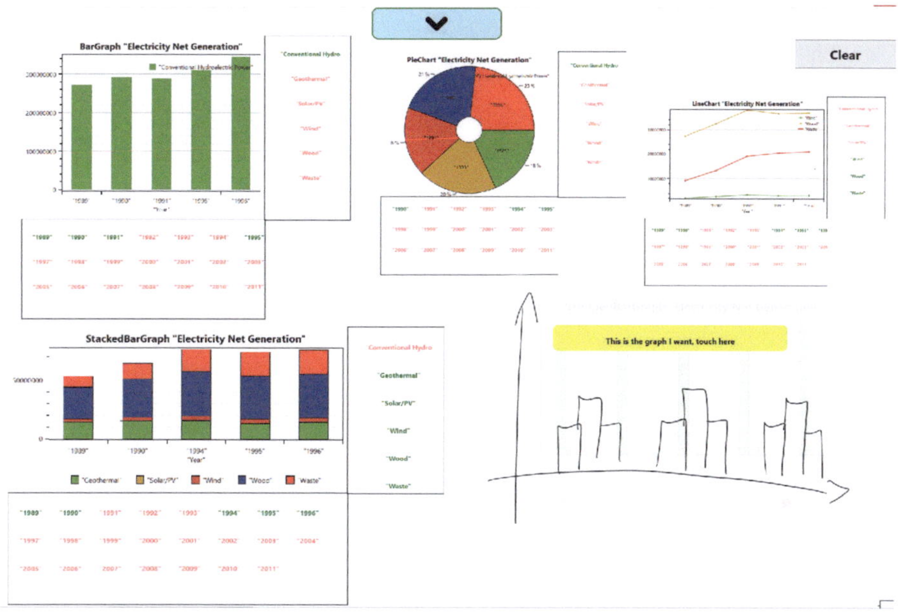

Fig. 5.1 SketChart in action. Several previously sketched graphs were recognized and their interactive charts are shown, including a bar graph, pie chart, and line chart (*top*), as well as a stacked bar graph (*bottom left*). In an empty space (*bottom right*), the user is sketching a grouped bar graph, which is correctly recognized and is shown behind the ink. The raw data from which the charts are populated is located in a floating spreadsheet window accessible by the drop *down arrow* (*top*)

effort was SketchVis [6]. In this work, gestures are used to create and modify chart types, as well as to execute functions over data, e.g., min, max, and average; however, SketchVis is rather limited in scope and only supports a small number of interactions. SketChart also uses gestures, though (unlike SketchVis) these are based on common knowledge representations of well-known operations, such as '+' to sum data and '−' to subtract data, among others. We believe that the use of gestures in this way reduces the cognitive load on users, as compared to a full gesture-based interface in which every interaction requires additional learning and recall. Following SketchVis, Walny et al. [24] conducted a Wizard of Oz study to further analyze and understand how people can combine sketch and touch interaction while exploring data. They found that although sketch interaction does not feel natural at first, after a short learning curve, such becomes routine. Touch interactions, on the other hand, were immediately natural and performed successfully by study participants, especially with those already common in other applications, such as pinch-to-zoom and drag-to-pan. For this reason, SketChart also leverages touch to perform these familiar actions, but not to directly manipulate data or generate charts. Lee et al. soon after presented SketchStory [15], where they explore how to "tell a story" with data visualizations. This work can be thought of as a sketch interface for creating PowerPoint like presentations with support for chart generation and data manipulation. However, they are similarly restricted to a purely gestural interface. More recently Lee et al. described SketchInsight [16], which is most similar to our work. However, their interface is, again, purely gestural. In contrast, we do not rely on gestures, but instead allow individuals to specify their desire through sketching, where a user only needs to draw an example of the chart type he or she wants to see, which falls in a space somewhere between gestures and WYDIWYG [24]. Therefore, a user is not required to learn or memorize a specific gesture in order to generate a particular chart; and, in this way, we can scale our system to support numerous chart types while simultaneously reducing the number of commands a user has to remember.

With respect to interacting with visual information, Baur et al. [3] as well as Sadana and Stasko [20] explore a set of multi-touch gestures such as drag and drop, zoom in and out, and others. Note that these works only consider touch interaction and are limited to stack charts and scatter plots. A similar approach is presented in Kinetica [19] where various multi-touch gestures are considered in the context of a physics application, and where multivariate data is visualized through a scatter plot. Furthermore, TouchViz [18], proposed a small set of touch gestures to interact with a single visualization. They compare their approach to a button-based WIMP interface and find their methods to be faster and have higher accuracy, although their interface requires a learning curve. Similar to these works, we adopt familiar touch gestures for chart manipulation including scale, translation and selection, although all other gestures that directly manipulate data utilize pen interactions. Finally, [26] present PanoramicData which aims to represent the relationship of different datasets through the use of visualizations. Gestures are performed to coordinate multiple visualizations rather than interactions to invoke and manipulate charts.

5.2.2 Sketch Understanding in Education Applications

Sketch understanding is a widely researched problem, crossing numerous domains. Specifically designed for learning environments we have for example, MathPad^2 which was proposed by LaViola and Zeleznik [14] to explore mathematical sketches, using handwritten mathematics to animate free-form diagrams. A real time recognition engine for chemical drawings was implemented in ChemInk [17]. PhysicsBook [8] was created for solving physics problems using a sketch-based interface; and in artificial intelligence, OntoSketch [5] was created to generate ontologies from free-form sketches. Schmieder et al. [21] using InkKit [9] developed a recognizer to support entity relation diagrams for software development. Even though these are only a few examples, little work has actually focused on recognizing chart diagrams for solving descriptive statistics problems. However, leveraging prior work in multi-domain sketch recognition, we modified SketchRead [1], a hierarchical based recognition system, to understand sketched chart diagrams. In summary, SketChart extends the state of the art by enabling users to generate charts from rough sketches and further, a new set of gesture interaction techniques are introduced that facilitate data exploration.

5.3 Methodology

SketChart's design was motivated by the Wizard of Oz from [24]: we sought to develop an interface that minimized the number of gestures one must remember in order to utilize the software without sacrificing interactive or exploratory functionality. A strong recognizer, good charts graphical representations, and a flexible dataset handler are required to achieve our goal. The prototype involved a set of unique pen and touch interactions to generate charts and work with data. Its interface comprises a floating spreadsheet window and an ink canvas. The spreadsheet view is used to select a working dataset, and once a file is loaded, the raw data is displayed in a typical grid of cells arrangement, similar to Microsoft Excel. A subset of data is selected using a stylus by stroking through the relevant data, so that the cells contained in the bounding box of the first and last stroke point define the data range. This technique was chosen for its ease of use and similarity to mouse based selection strategies used in other spreadsheet applications. Note that only a small portion of time is spent in this view whereas the majority of work actually occurs on the ink canvas; although a user can return to this window at any time to modify their working dataset.

With a dataset selected, the user can then create a new chart by sketching a representation of the desired visualization in an empty area of the canvas. Ink is parsed in real-time and using eager recognition, a proposed chart type is rendered semi-transparently behind the user's drawing. Although most sketches are quickly recognized, for chart types that require a greater level of detail such as box plots, a user can continue to sketch until the appropriate choice emerges. We chose to use this approach because of the expressibility of sketching as compared to gestural or menu based interactions for chart generation, as well as to help alleviate the need for users to learn several new commands—one for each unique chart type supported. Once a user is satisfied with the recognition result, the model is selected by tapping in a confirmation area above the ink, after which an interactive chart based on the working dataset is formally added to the canvas. Two interactive legends are also generated based on the raw data, where the first comprises column labels and the second comprises row labels. These are displayed to the right of and below the chart, respectively.

Presently, SketChart supports the 12 unique chart types listed in Fig. 5.2. Those most frequently used in practice are supported, such as the line, bar, and pie charts, and in addition, we also selected several very similar looking chart types, all of which utilize bar shapes. This was done in order to evaluate whether users are able to adequately express themselves through sketch when the desired visualization they need to draw is potentially quite similar to another chart type, and to test whether the interaction required to obtain the correct chart remained intuitive as well as fluid.

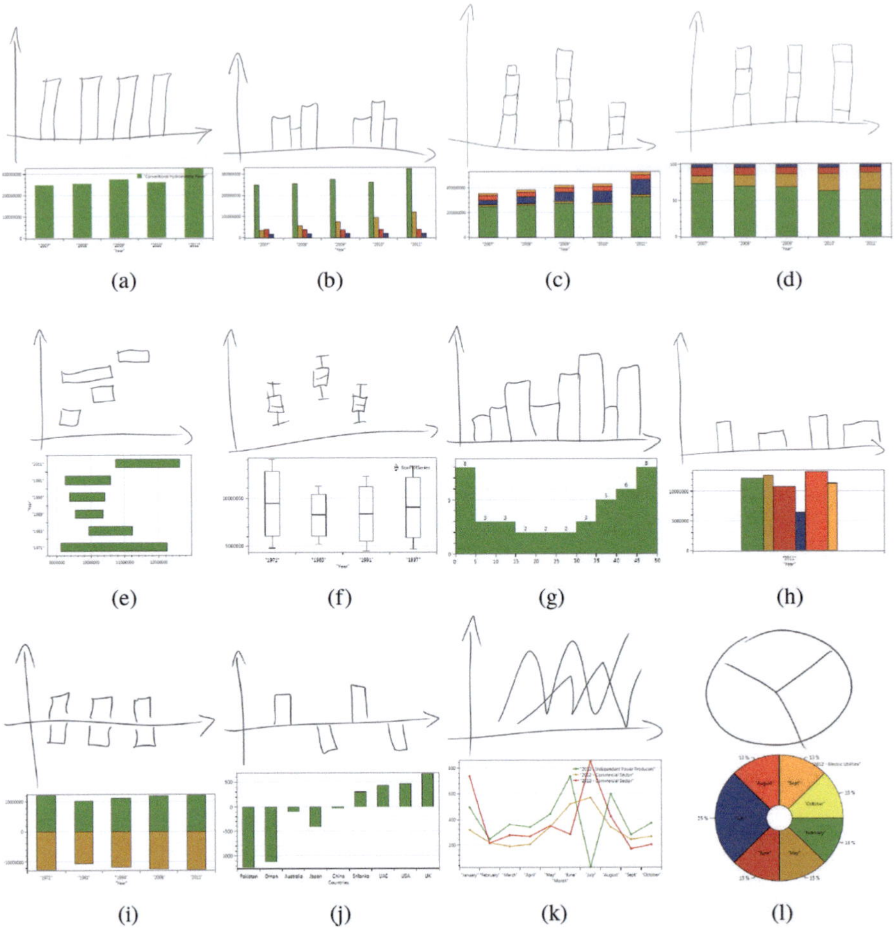

Fig. 5.2 The 12 chart types supported in SketChart. For each type, an example drawing is shown that when sketched in the ink canvas will generate the chart displayed underneath. (**a**) Bar graph (BG). (**b**) Grouped bar graph. (**c**) Stacked bar graph. (**d**) 100% stacked BG. (**e**) Range bar graph. (**f**) Box plot. (**g**) Histogram. (**h**) Area bar graph. (**i**) Paired bar graph. (**j**) Deviation bar graph. (**k**) Line chart. (**l**) Pie Chart

By interacting with a chart and its legends, a user can further manipulate the visualization and data through the set of gestures shown in Table 5.1. These gestures were developed through an iterative design process and, in some cases, chosen because of their ubiquity. Those functions that are currently implemented in SketChart, such as for basic arithmetic and filtering, were included because they represent frequently used operations in data analysis that enable users to explore a wide variety of questions. One gesture that is available while drawing a chart, or while working with data, is the scribble-erase gesture.

Table 5.1 Full set of 12 gestures supported by SketChart. The left six actions and scribble-erase are pen-based gestures whereas scale, translate, and scroll are touch-based gestures. The remaining two are pen or touch gestures

Gesture	Operation	Gesture	Operation
+	Element-wise summation of data series	Scribble-erase	Erase ink
−	Element-wise subtraction of data series	Pinch-to-zoom	Scale chart
X	Removal of two or more data series	Drag-to-pan	Translate chart
<	Filter data less than specified value	Swipe	Scroll legend
>	Filter data greater than specified value	Tap on chart	Show value at specified point
Lasso label	Select multiple data series	Tap on label	Show or hide a data series or point

5.3.1 *Data Manipulation*

The manipulation techniques were motivated by different common operations a user working with data can do in Microsoft Excel such as summations, subtractions, data selection and filtering based on [4]. Tasks are communicated to SketChart through pen-based gestural interactions that map to common knowledge arithmetic concepts, e.g., plus sign for data summation and minus sign for subtraction; or they are intuitive, contextually-based actions. For example, a data series can be hidden or again shown by tapping on its associated label in the right side legend. The color of the label provides visual feedback about the data's inclusion state, where red indicates that the data is hidden and green indicates that the data is shown. Similarly, points across the data series can be hidden or included by tapping on the associated labels in the legend below the chart.

Element-wise summation of multiple data series is supported using the summation gesture, a plus sign symbol ('+') followed by a tap. Those data series appearing on the chart that overlap with the bounding box of the gesture are combined together to form a new data series, which is subsequently affixed to the chart and legend as shown in Fig. 5.3 (right). Similarly the subtraction gesture ('−' followed by a tap) performs an element-wise subtraction, however in this case, the order of the data series is important. Therefore, the first element selected by the hyphen represents the initial data series from which the remaining series are subtracted. To completely remove a data series from the chart, one can use the removal gesture, 'X' and tap. Note that the exact meaning of this gesture will depend on the visualization. An X-mark over a bar chart element will remove the associated bar, whereas the same gesture over this data in a pie chart will remove the associated slice. The lasso gesture is used to select a group of data series from the legend. The selected group is highlighted and with repeated application of the lasso, multiple disjoint data series

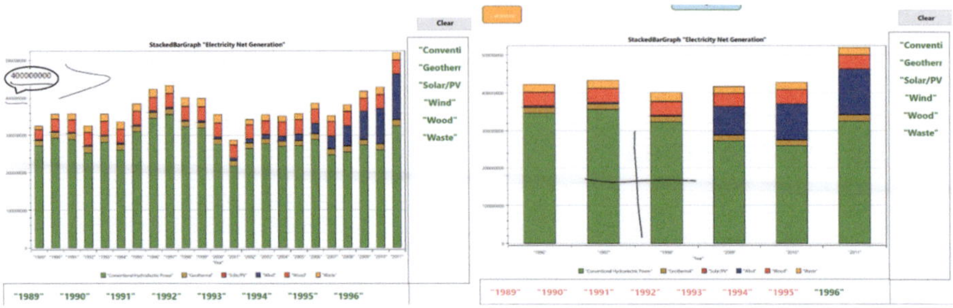

Fig. 5.3 Filtering and summation example. A student is asked for the electricity generation in high kilowatt-hour (kWh) years. First she filters the data (*left*) by underlining 400,000,000 along the y-axis label, and then draws a greater-than sign. What remains are the six highest kWh years. Now curious about the total energy generated between 1997 and 1998, she uses the summation gesture (*right*) to simultaneously select the data from those years and sum them together

can be combined into a group. In this context, the summation, subtraction, and removal gestures work on the group rather than on chart elements directly. Both methods are supported based on early feedback, where users prefer to operate on data elements directly when possible, but conversely prefer to work with the legend in situations where data is not collocated in the chart area.

Another common data operation supported by SketChart is to filter data based on a threshold value. For this, the greater-than and less-than filter gestures are provided. First a threshold is selected by underling a value in the y-axis of the chart and then an inequality is chosen by drawing the appropriate symbol (>or <), followed by a tap to complete the operation. An example of this is depicted in Fig. 5.3 (left). Initially, we allowed users to specify the threshold by writing a numerical value on the chart directly, but found that for large thresholds, this was impractical—users do not enjoy writing 1,000,000 for example. However, because many data exploratory tasks do not require an exact threshold, users instead prefer to provide an approximate value simply by selecting this from the axis as discussed.

5.3.2 *Touch Interface*

A small number of touch interactions are also available. A user can touch a data point on the chart to generate a tooltip showing the value at that specified location. Dragging with two fingers allows one to move the chart through the ink canvas, whereas the familiar two finger pinch-to-zoom gesture is used to scale a chart. Legends and series labels are likewise sensitive to touch and behave identically to their pen-based tap gesture counterpart used to show and hide data series. Finally, for situations where legends have more values than available screen real estate, a user can use the swipe (or flick) gesture to pan the legend data left, right, up, or down.

5.3.3 *Sketch Recognition*

Given that free form sketching is an integral part of our design for specifying data visualizations, we require a robust recognizer for such diagrams. However, our goal here is not to develop new recognition methods, but rather to enable SketChart. We therefore turn to well-known techniques and tailor these to our purpose. An overview of our system is shown in Fig. 5.4 (top), which is based on SketchRead [1]. The basic unit of processing is an ink stroke. Each stroke is filtered [13], corners are identified [25], and combined strokes are segmented using MEGIS [22]. The basic components of charts are primitives, namely lines and circles, which lends itself well to the hierarchical description language of LADDER [11] to define complex shapes which in turn define chart types. Using a Bayes network, concepts such as graphs are proposed as high level hypotheses and are validated using lower level hypotheses based on said primitives and shapes; the exact hierarchy is shown in Fig. 5.4 (bottom). The recognizer then yields a probability for each chart type and the highest probability chart is shown to the user as previously described, as he or she is drawing. After a chart has been confirmed by the user and is committed to the ink canvas. Finally, gesture interactions are handled using the $N recognizer [2].

5.4 Evaluation

In order to evaluate the effectiveness of our interaction techniques in generating graphs and manipulating data using SketChart, we conducted an informal user study. The study started with a pre-questionnaire to gather demographic information, followed by the experiment and ended with a post-questionnaire. Aspects of the system were rated using a Likert scale (1=Negative to 7=Positive) as were the interaction techniques (1=Very difficult to 7=Very easy). Finally, qualitative feedback was also requested. In the experiment, we asked participants to perform data analysis and provide answers to questions that mimic practical test example problems. From the University of Central Florida, we recruited 10 volunteer students (four male, six female) having at least some prior knowledge of chart generation in Microsoft Excel. Participants were all enrolled in graduate or undergraduate programs in computer science, computer engineering, electrical engineering, industrial engineering, literature and communications. The experiment took approximately 60 min to complete, and each participant was paid $10 for their time. The apparatus used in this experiment was a multi-touch, active stylus tablet PC (Surface Pro 3) running Windows 8.1.

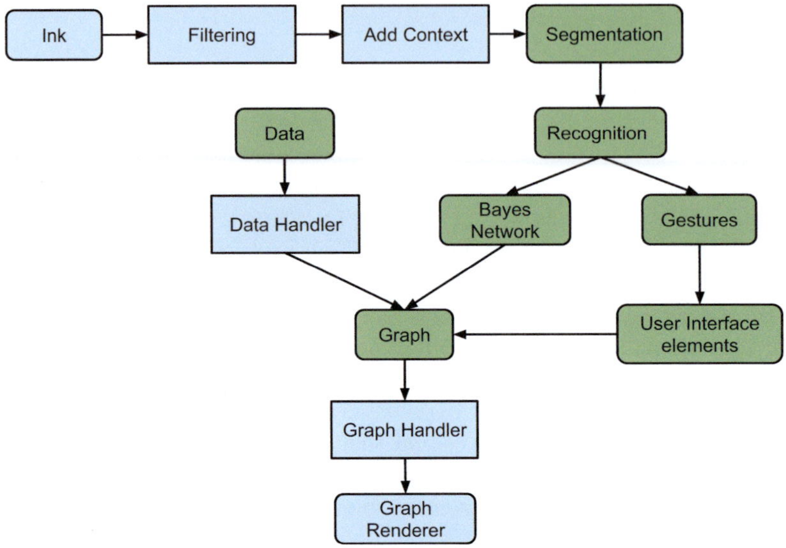

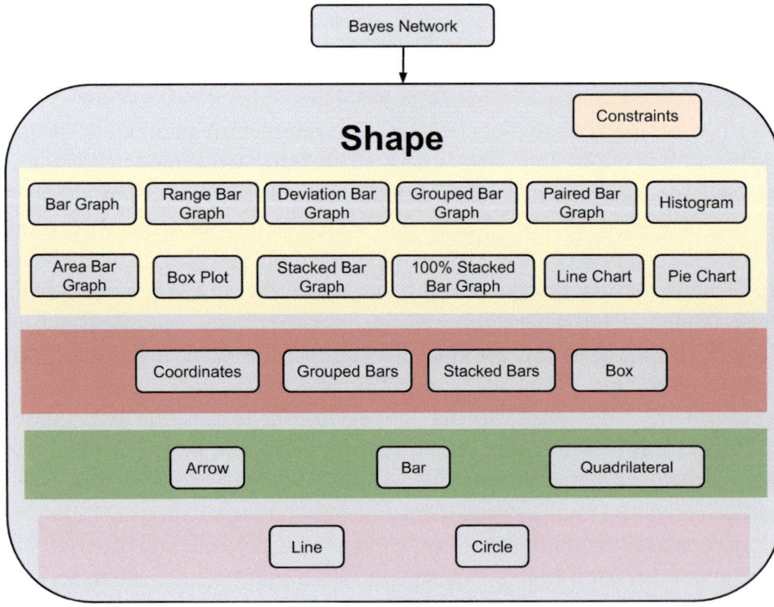

Fig. 5.4 Flowchart of SketChart's overall system architecture (*top*). The Bayes network is based on a hierarchy of ink stroke primitives that are combined together into more complex shapes and concepts (*bottom*)

5.4.1 Experimental Design

We developed a set of 10 questions, shown in Table 5.2, based on four different publicly available datasets. The first[2] and second[3] dataset contain timeline information. The third and fourth datasets involve petroleum consumption and SAT scores, respectively. Further, the design of the questions are close to ones found in descriptive statistics exams and they ensure that participants utilize a wide variety of charts as well as all gestural interactions. Note that only ten of the 12 chart types were included in this evaluation to ensure that the experiment did not go over the 60 min time budget.

5.4.2 Procedure

A pre-questionnaire was first administered to gather demographic information and determine if the participant possessed prior knowledge of chart generation software. We also collected information on whether the participant had experience working with multi-touch or pen-based user interfaces. Subsequently, the SketChart user interface was introduced, and for approximately 10 min the participant was trained on how to use the software. During this time, he or she was also introduced to all of the supported chart types and gestures. Additionally, the proctor demonstrated how questions similar to those found in Table 5.2, could be answered using the prototype application. The training involved the use of all the system functionality and a cheat-sheet with sketch examples was provided. Once the participant was comfortable with the interface, the study transitioned into the main part of the evaluation.

At this point the participant was given access to all four datasets and the set of evaluation questions. These questions were answered one at a time, and for each question, we recorded the correctness of their answer as well as how long it took to complete the task. Recognition results were also collected in order to evaluate the effectiveness of the recognizer and to determine what effect this may have had on the participant's perception of the interface. Finally, after all questions were answered, we requested feedback on the user interface using a seven-point Likert scale post questionnaire. Participants were also invited to provide additional feedback, including their overall impressions and suggestions for improvements.

[2]http://www.eia.gov/totalenergy/data/annual/showtext.cfm?t=ptb0802a—information related to energy generation. 1989–2011.

[3]http://www.census.gov/hhes/socdemo/education/data/cps/historical/tabA-1.xlsx—information related to population. 1970–2011.

Table 5.2 All tasks and questions that participants had to complete in the evaluation. These were selected to cover a wide variety of common data analysis tasks and to ensure that all chart types were utilized

#	Question
1	Use "Electricity Net Generation.csv" to draw a **Bar Graph** and indicate: (a) In which years was the "Conventional Hydroelectric Power" over 300000000? (b) Is "Solar Electricity" generation in years 2004–2005 greater than in 2006–2007
2	With the same dataset, draw a **Grouped Bar Graph** and indicate: (a) In year 2011 is "Conventional Hydroelectric Power" greater than the sum of the others? (b) In which years "Wind" electricity generation is greater than "Wood"
3	Now draw a **Stacked Bar Graph** and indicate: (a) In which years the overall electricity generation is between 300000000 and 400000000? (b) Compare overall energy generation on years 1989–1991 and years 2005–2006. Which interval generated more electricity?
4	Now draw a **100% Stacked Bar Graph** and indicate: (a) With overall generation of 1989–1990 and 2010–2011 what is the increase/decrease for Conventional Hydroelectric Power? (b) By summing up overall generation of Wind, Wood and Waste, what is the increase/decrease percentage of the three compared with Hydroelectric Power in same years as before?
5	Please load the file "Population by level of education.csv," draw a **Range Bar Graph**, and indicate: (a) Jot down which is the lowest range value for females and the year it occurs. Repeat exercise for men. (b) The year in which the population by level of education is the maximum for males and females, is it the same year?
6	With the same dataset, draw a **Box Plot** and indicate: (a) What is the approximate female population average from 1970 to 2011? (b) For tertiary education level, in which year the population is almost the same, which means that standard deviation is close to 0
7	Draw a **Paired Bar Graph** and indicate: (a) Is the data symmetric for male and female populations for each one of the education levels? (b) Show in the graph the year in which the populations almost don't change for primary level female and male
8	Please load the file "SAT scores.csv". Draw a **Histogram** and indicate: (a) Maximum frequency for math scores with 10 bins? (b) Change the number of bins to 5 and write maximum frequency
9	Please load the file "Petroleum Liquids Consumption.csv," draw a **Line Chart**, and indicate: (a) Months in which Independent Power Producers petroleum consumption decreases in respect to prior month. (b) With information from years 2012 and 2013, in which months petroleum consumption is larger in 2012 compared to 2013
10	With the same dataset draw a **Pie Chart** and indicate: (a) What is the percentage of male students in primary education in each one of the last 5 years? (b) What are the percentages of secondary education in years 1989 and 2011?

5.5 Results

In general, participants found SketChart's interface to be favorable. From the post questionnaire based on Davis's acceptance and perceived ease of use criteria [10], as shown in Fig. 5.5 (top), all aspects of the system were well received including the recognition engine, visual feedback mechanisms, interaction techniques, and graph visualizations, the latter achieving a nearly perfect score. Similarly, all of the interaction techniques scored high, see Fig. 5.5 (bottom), and participants especially enjoyed being able to select or hide specific data by interacting with the legends. These results are consistent with our findings from the post questionnaire discussions, where participants indicated that they found all gestures to be easy to use, intuitive, and enjoyable. The intuitiveness of our interface is also reflected in how participants used the software. For example, without suggestion or prompt, participants would automatically chain multiple filters together, a possibility that was never explicitly discussed during training. Another emergent behavior occurred when participants created two new data series by merging certain individual data series together using addition, and then compared their results using subtraction to determine which of the two groups were larger. This latter technique, for instance, was used by some to answer question 1b.

Recall that the tap gesture in a legend allows one to hide or show the associated data series, whether using pen or touch. As it turns out, both were utilized depending on the scenario. For instance, if a user recently used the touch interface to scroll through data, then he or she was more likely to use touch based taps, and similarly, with the stylus poised for writing, pen-based taps were more likely to be used. More importantly though, users particularly appreciated being able to interact with the legend labels period, regardless of the mode. However, as the dataset grew large, most participants found it difficult to manage the vast amount of data. One individual suggested that we should allow users to write parenthesis around the desired data in order to select the range, an idea we will consider in future work.

5.5.1 Sketch Recognition

Several chart types are recognized with 100% accuracy, though it is also clear that a small subset of chart types are very difficult to detect. The worst case occurs with box plots, where only 48% of the sketched diagrams are correctly interpreted. This is followed by 100% stacked bar graphs at 67% accuracy, and group bar graphs at 69% accuracy. Though, even with these few poor results in the mix, the system is still able to achieve an 81% overall recognition accuracy. Nevertheless, it is likely that errors in sketch recognition affect one's perception of the interface and its usability. All chart types, except box plot and 100% stacked bar graphs, received a high score for acceptance and ease of use, which contrasts slightly with the favorable (but not strong) acceptance and ease of use rating for graph drawing as an interaction technique Fig. 5.5 (bottom). It may be that the high error rate of the few charts mentioned led to a lesser impression of this interaction technique, despite that most of the individual chart types were accepted with high scores (see Fig. 5.6).

In discussions with the participants, most considered chart sketching to be an intuitive action, although some felt that the technique might be better suited as a complementary approach to toolbars or gestures. Those chart types most frequently used, for instance, may be mapped to simple gestures, and thereafter extended. In this way, visualizations can be built incrementally, and perhaps with greater speed and accuracy. Sketch errors were corrected in some cases by removing a stroke or

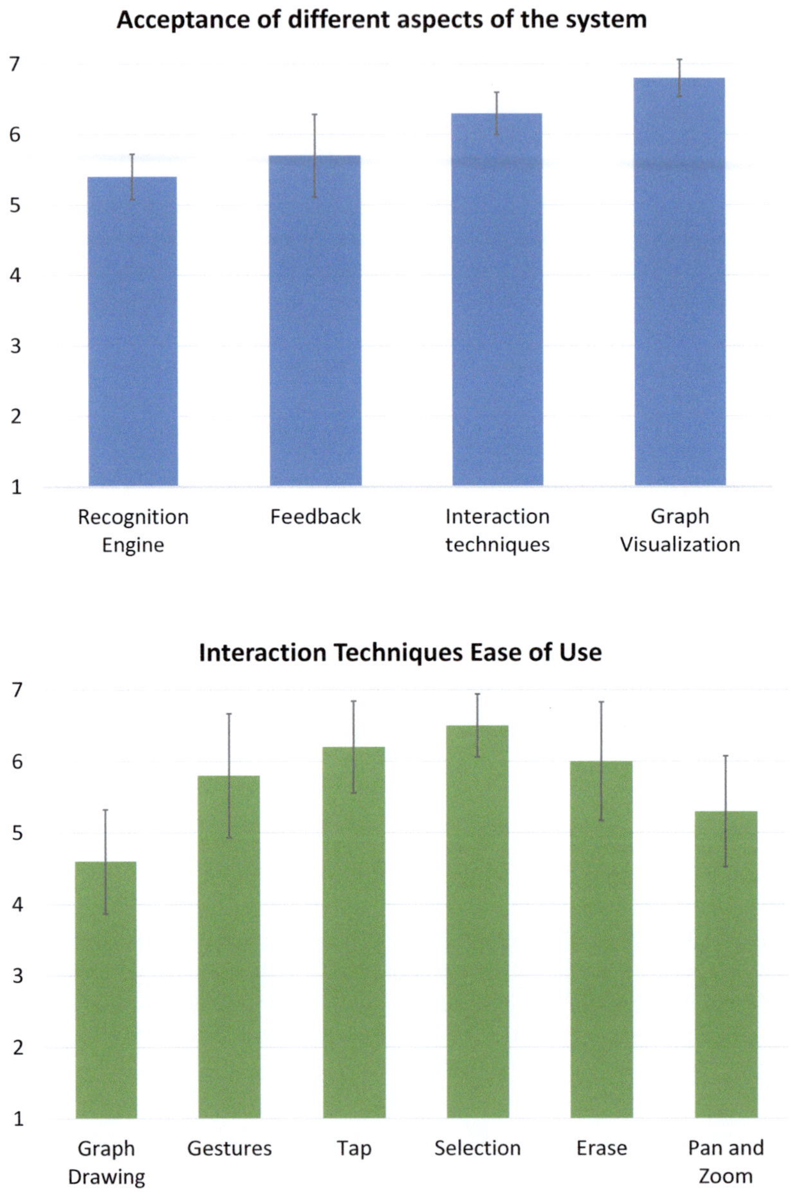

Fig. 5.5 Average Likert response (N = 10) for single questions about acceptance and ease of use of different SketChart aspects, error bars are 95% CI. Higher is better, and in all cases, participants gave favorable responses (*top*). Ease of use for various interaction techniques supported in SketChart, error bars are 95% CI and higher is better. In all cases, participants gave favorable responses, although the graph drawing response result is not strong (*bottom*)

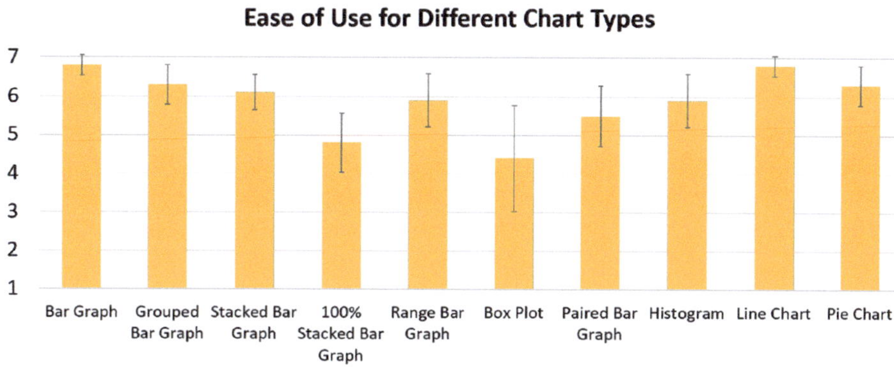

Fig. 5.6 Acceptance and ease of use for the ten chart types evaluated

set using a scribble gesture or by deleting the whole graph. Sometimes, they tend to delete the whole graph and start again since it was difficult to identify which exact stroke or set was not well recognized.

5.6 Discussion

Overall SketChart received positive feedback, with its acceptance and ease of use being highly rated. Participants claimed through discussion and a post-questionnaire evaluation that all aspects of the system are intuitive, and the proposed interactions are fluid as well as natural. The only exception to this is that chart drawing achieved only moderate acceptance. Conceptually, participants agree that chart sketching is a valid interaction technique, although it may be better suited as a complementary method to handle complex or less frequent cases. However, there are two confounding factors to consider. First, the poor recognition rate of a few chart types may have skewed participant perception when evaluating this interaction technique. Feedback from the Wizard of Oz experiment conducted by [24] suggests that this may be the case, for in their study, this form of interaction received a very positive reaction. Second, our evaluation was limited to only ten chart types, but what happens when the software supports hundreds of variants?

In this case, we believe that the expressibility of previously explored gestural interfaces [6] are limited and will require a significant learning curve as the number of supported chart types continue to increase. Gestures for rarely used charts are likely to be easily forgotten. Alternatively, chart types can be selected through a hierarchy of GUI interface elements, though the desired chart may still be difficult to locate when the software supports numerous types. With respect to data selection and manipulation, everyone agreed that SketChart benefits from its interactive legends as well as its addition, subtraction, removal, and filter gestures. Being able to tap on a single label or stroke through several labels to toggle the visibility of many data series or individual points is considered to be very helpful. The only drawback of the interactive legend however, occurs when the working dataset grows to be too large, since selecting and panning through the dataset becomes a tedious task. How to improve this aspect of the interface is one part of our future work. Element-wise addition and subtraction together with filtering were also considered to be advantageous. Overall, based on our evaluation and feedback, we believe that these small sets of gestures are quite enabling, easy to remember, as well as natural and powerful additions to data visualization software for interacting with charts.

5.7 Future Work

Even though SketChart only supports a limited number of chart types, it is already clear that recognition of hand drawn graphs needs to be significantly improved in order to work with a large array of possibilities. Presently, we plan to develop a grammar for graph visualizations so that a user can create any chart possible, and to enable users to easily switch between visualizations. Interactive techniques that allow users to format charts are also required, such as to adjust spacing, titles and legend placements, color schemes, and so on. Currently, our system only supports straight forward datasets, and methods for dealing with complex, multivariate datasets require further study. Further, by integrating handwritten mathematics into the interface, advanced data analysis may be possible. However, at this point, it remains unclear how well such a system will scale once all of these features are integrated together and what new challenges will emerge. Our evaluation explored how a student interacts with SketChart when he or she is told which chart type to use. A subsequent evaluation that examines how students explore data without this kind of guidance will also help to shed light on what additional interactions are required.

5.8 Conclusions

We presented a novel set of interactive techniques for chart generation and data manipulation, which were also implemented in our prototype application SketChart. Utilizing our proposed methods, students are able to work with data by interacting with data visualizations, thereby creating an abstraction layer that simplifies many common data analysis tasks. Interactions are performed via touch and pen-based gestures in a seamless manner, and to create new charts, users freely sketched a representation of their target visualization, which is recognized in real-time. An informal user study involving individuals from different backgrounds was conducted, and the findings were presented. We conclude that our approach provides an intuitive interface and the proposed interaction techniques are easy to remember and easy to execute. Based on the results of the study, SketChart despite some recognition errors for a subset of supported charts, show strong potential to be an initial starting point towards having a tool for students to solve descriptive statistics problems.

Acknowledgements This work is supported in part by NSF CAREER award IIS-0845921. The authors thank members of the Interactive Systems and User Experience Research Cluster of Excellence and the CPTTE community for their support and feedback throughout this project.

References

1. Alvarado, C., Davis, R.: Sketchread: a multi-domain sketch recognition engine. In: Proceedings of the 17th Annual ACM Symposium on User Interface Software and Technology, pp. 23–32. ACM (2004)
2. Anthony, L., Wobbrock, J.O.: A lightweight multistroke recognizer for user interface prototypes. In: Proceedings of Graphics Interface 2010, pp. 245–252. Canadian Information Processing Society (2010)
3. Baur, D., Lee, B., Carpendale, S.: Touchwave: kinetic multi-touch manipulation for hierarchical stacked graphs. In: Proceedings of the 2012 ACM International Conference on Interactive Tabletops and Surfaces, pp. 255–264. ACM (2012)
4. Berk, K., Carey, P.: Data Analysis with Microsoft Excel: Updated for Office 2007. Cengage Learning, Boston (2009). https://books.google.com/books?id=pRU8AAAAQBAJ
5. Brade, M., Schneider, F., Salmen, A., Groh, R.: Ontosketch: towards digital sketching as a tool for creating and extending ontologies for non-experts. In: Proceedings of the 13th International Conference on Knowledge Management and Knowledge Technologies, p. 9. ACM (2013)
6. Browne, J., Lee, B., Carpendale, S., Riche, N., Sherwood, T.: Data analysis on interactive whiteboards through sketch-based interaction. In: Proceedings of the ACM International Conference on Interactive Tabletops and Surfaces, pp. 154–157. ACM (2011)
7. Chamberlain, R., Riley, H., McManus, C., Rankin, Q., Brunswick, N.: The perceptual foundations of drawing ability. In: Thinking Through Drawing: Practice into Knowledge. Proceedings of an Interdisciplinary Symposium on Drawing, Cognition and Education, pp. 95–102. Teachers College Columbia University (2011)
8. Cheema, S., LaViola, J.: Physicsbook: a sketch-based interface for animating physics diagrams. In: Proceedings of the 2012 ACM International Conference on Intelligent User Interfaces, pp. 51–60. ACM (2012)
9. Chung, R., Mirica, P., Plimmer, B.: Inkkit: a generic design tool for the tablet pc. In: Proceedings of the 6th ACM SIGCHI New Zealand Chapter's International Conference on Computer-Human Interaction: Making CHI Natural, pp. 29–30. ACM (2005)
10. Davis, F.D.: Perceived usefulness, perceived ease of use, and user acceptance of information technology. MIS Q. **13**, 319–340 (1989)
11. Hammond, T., Davis, R.: Ladder, a sketching language for user interface developers. Comput. Graph. **29**(4), 518–532 (2005)
12. Harris, R.L.: Information Graphics: A Comprehensive Illustrated Reference. Oxford University Press, Oxford (1999)
13. LaViola, J.: Mathematical sketching: a new approach to creating and exploring dynamic illustrations. Ph.D. thesis, Department of Computer Science, Brown University (2005)
14. LaViola Jr, J.J., Zeleznik, R.C.: Mathpad 2: a system for the creation and exploration of mathematical sketches. In: ACM SIGGRAPH 2007 Courses, p. 46. ACM (2007)
15. Lee, B., Kazi, R.H., Smith, G.: Sketchstory: telling more engaging stories with data through freeform sketching. IEEE Trans. Vis. Comput. Graph. **19**(12), 2416–2425 (2013)

16. Lee, B., Smith, G., Riche, N.H., Karlson, A., Carpendale, S.: Sketchinsight: natural data exploration on interactive whiteboards leveraging pen and touch interaction. In: 2015 IEEE Pacific Visualization Symposium (PacificVis), pp. 199–206. IEEE (2015)
17. Ouyang, T.Y., Davis, R.: Chemink: a natural real-time recognition system for chemical drawings. In: Proceedings of the 16th International Conference on Intelligent User Interfaces, pp. 267–276. ACM (2011)
18. Rzeszotarski, J.M., Kittur, A.: Touchviz: (multi)touching multivariate data. In: CHI'13 Extended Abstracts on Human Factors in Computing Systems, CHI EA '13, pp. 3119–3122. ACM, New York (2013). doi:10.1145/2468356.2479625, http://doi.acm.org/10.1145/2468356.2479625
19. Rzeszotarski, J.M., Kittur, A.: Kinetica: naturalistic multi-touch data visualization. In: Proceedings of the 32nd Annual ACM Conference on Human Factors in Computing Systems, CHI'14, pp. 897–906. ACM, New York (2014). doi:10.1145/2556288.2557231, http://doi.acm.org/10.1145/2556288.2557231
20. Sadana, R., Stasko, J.: Designing and implementing an interactive scatterplot visualization for a tablet computer. In: Proceedings of the 2014 International Working Conference on Advanced Visual Interfaces, AVI'14, pp. 265–272. ACM, New York (2014). doi:10.1145/2598153.2598163, http://doi.acm.org/10.1145/2598153.2598163
21. Schmieder, P., Plimmer, B., Dobbie, G.: Sketching er diagrams. In: Proceedings of the Tenth Australasian Conference on User Interfaces, vol. 93, pp. 53–60. Australian Computer Society, Inc. (2009)
22. Sun, Z., Wang, C., Zhang, L., Zhang, L.: Free hand-drawn sketch segmentation. In: Computer Vision–ECCV 2012, pp. 626–639. Springer (2012)
23. van Dam, A.: Post-wimp user interfaces. Commun. ACM **40**(2), 63–67 (1997). doi:10.1145/253671.253708, http://doi.acm.org/10.1145/253671.253708
24. Walny, J., Lee, B., Johns, P., Riche, N.H., Carpendale, S.: Understanding pen and touch interaction for data exploration on interactive whiteboards. IEEE Trans. Vis. Comput. Graph. **18**(12), 2779–2788 (2012)
25. Xiong, Y., LaViola, J.J., Jr.: Revisiting shortstraw: improving corner finding in sketch-based interfaces. In: Proceedings of the 6th Eurographics Symposium on Sketch-Based Interfaces and Modeling, pp. 101–108. ACM (2009)
26. Zgraggen, E., Zeleznik, R., Drucker, S.: Panoramicdata: data analysis through pen and touch. IEEE Trans. Vis. Comput. Graph. **20**(12), 2112–2121 (2014). doi:10.1109/TVCG.2014.2346293

Chapter 6
Digitizing Mathematical Symbolism Through Pen and Touch Technology: A Beginning Taxonomy of Affordances for Learner Flow in Mathematics

Eric Hamilton

Abstract This essay highlights a rationale for pen and touch in K16 mathematics classrooms, arguing that digitizing mathematical cognition can promote learner flow. The personal computing and mobile technologies that have transformed society since the 1980s left a large element of mathematics learning, which relies on non-keyboard symbol systems and drawing, untouched in comparison to learning in disciplines that involve construction and manipulation of written text that can be expressed with a keyboard. Yet over the past decade pen and touch has gradually brought into the digital era that large fraction of K16 mathematics that relies on symbol and visual systems that exceed the representational capacity of the standard keyboard. It has become a portal for research in new forms of research and new forms of mathematical learning. The paper argues that learner flow can be promoted through pen and touch and identifies a three-tier initial taxonomy for classifying such applications.

6.1 Introduction

This paper highlights a rationale for pen and touch technology in mathematics classrooms. It argues that microprocessor and personal computing technologies that transformed society in the 1980s and 1990s left a large element of mathematics learning untouched, unlike changes that occurred in disciplines that involve construction and manipulation of written text. The American Standard Codes for Information Interchange (ASCII) character interface – the keyboard – became the context for mixing, manipulating and creating content. While it is not the only interface for writing, the standard keyboard is referred to in this paper as a prototypical contrast with pen and touch interfaces. In written language, cognition

E. Hamilton (✉)
Graduate School of Education and Psychology, Pepperdine University,
24255 pacific Coast Highway, Malibu, CA, 90265, USA
e-mail: eric.hamilton@pepperdine.edu

T. Hammond et al. (eds.), *Frontiers in Pen and Touch*, Human–Computer Interaction Series, DOI 10.1007/978-3-319-64239-0_6

can be embodied and mediated by converting manual keystrokes into meaning (doing so without haptic feedback specific to each character). Such interaction with keyboard interfaces are not readily possible, at least for embodying original mathematical cognition in real-time, in the symbol driven aspects of mathematics learning.

Yet in somewhat over a decade, the small but growing presence of pen and touch technology in schools has made it possible to bring into the digital era that large fraction of K16 mathematical cognition that relies on students manipulating symbol systems in ways that exceed the representational capacity of the keyboard. Pen and touch technology has become a portal for research in exciting forms of mathematical cognition concurrently embodied both by handwriting and digital technology.

This paper draws on a body of work, some of it reported in recent years in the annual Workshop on the Impact of Pen and Touch Technology in Education (WIPTTE, now renamed CPTTE, or Conference on Pen and Touch Technology in Education), to suggest the beginning of a taxonomy of pen and touch technology applications in mathematics education, particularly as those applications promote high learner engagement and classical high performance flow experiences at the outer edges of ability.

The intent is to stimulate conversation, especially within pen and touch communities, which will enable the development of a more robust classification system of pen- and digitally-embodied mathematical thought, a classification system that can be used at several levels. First and foremost, researchers and practitioners in pen and touch technology consistently report new forms of learning and understanding that the tools facilitate high engagement learning. Pen and touch technology opens a pathway into different learning experiences, and a taxonomy of applications is one means to highlight and parse those experiences. That in turn may have potential both for shaping future research directions in learning sciences and for helping educational practitioners and policy-makers understand the relative merits of investing in pen and touch technologies.

6.1.1 Limitations

While this paper is limited to mathematics learning, many elements of this overview apply to many disciplines. Additionally, obviously not all mathematics learning takes place or is mediated by handwritten expression by learners. This paper only applies to that component of mathematics learning that is mediated, on the student side, by handwriting (in contrast, for example, to various types of mathematics visualization and numerical methods software). Additionally, this paper does not reference the affordances or merits of sketch or handwriting recognition, confining itself instead to digitizing mathematical expression, with or without subsequent software recognition systems. Finally, the core interest in this paper is in pen-based systems, but combines pen and touch computing out of the practical and fortuitous reality that virtually all systems that can digitize mathematical symbols can do so with pen or touch input.

6.2 Mathematical Symbolism

Indeed, a significant component of K16 mathematics learning takes place through complex symbol systems that are represented by handwriting. Mathematical symbolism itself is the subject of a sizable literature. The role of symbolism in modern mathematics education can be traced to Hiebert's theory of the development of symbol competence [13]. Four key elements of this theory involve:

- Connecting symbols to meanings (e.g., associating the $\surd$ symbol with the meaning of an algebraic root);
- Acquiring symbol manipulations skills (e.g., expressing algebraic operations or relationships);
- Translating manipulations to express new meanings (e.g. producing mathematical solutions through symbol manipulations); and
- Developing more abstract symbol systems with new referents and meanings (e.g., the transition from symbolism used for Riemann Sums to the symbols of integral calculus).

Each of these four elements requires motor actions that fall outside of the representational functionality of the standard keyboard, which remains by wide margin the most common input interface between humans and computers. Keyboard interfaces that permit the representation of mathematical symbols (e.g., equation editing software, LaTeX, etc.) use keystroke and mouse click combinations to produce unmistakably distinct expressions, but those interfaces do not facilitate free-flowing and real-time manipulation and creation of mathematics. In contrast, symbol systems to represent mathematical thought entail complex coordination of multiple semiotic layers, including the symbols themselves, diagrams or other visual devices, and an underlying written language [19, 21]. Combining these coordinated factors with the four dimensions of Hiebert's theory of development yields the matrix in Table 6.1. This matrix, representing (competencies x coordinated semiotic systems), forms one way to classify the use of symbols, drawing, and language to represent mathematical cognition.

Table 6.1 Symbol competencies as they apply to semiotic systems in representing mathematical cognition. Each cell represents a functionality for mathematical cognition unavailable with keyboard interfaces

Coordinated semiotic systems/symbol competencies	Symbols themselves	Diagrams or visuals	Underlying written/spoken language
Connect symbols to meaning	Limited	No	No
Symbol manipulation skills	No	No	No
Manipulations translated to new meanings (using same symbol system)	No	No	No
Development of new and more abstract symbol systems	No	No	No

The proposition that keyboard systems are inadequate representational tools for expressing most real-time mathematical cognition is self-evident. The matrix, though, teases out specific aspects of that inadequacy, in that each cell in the matrix can be scrutinized with respect to whether a keyboard can readily express or represent substantive mathematical cognition as it occurs. Virtually none of the 16 cells can do so.

In real time, the representation or expression on mathematical cognition, or embodied mathematical cognition [1, 20, 22], typically takes place through handwriting [17]. Keyboards are discrete representational systems; handwriting is a continuous system. Keyboards produce a virtually limitless combination of 104 keystrokes, whereas handwriting produces a virtually limitless combination of virtually limitless gestures and symbols. The range of expression differs in both quality and quantity. Mangen and Velay [17] delve deeply into the physiology of haptics and cognition, and how what we refer to as "tangibility" is an under-theorized and under-appreciated aspect of conceptual development. Their arguments, beyond the scope of this paper, extend beyond the flexibility that hand written symbolism gives to expressing mathematical activity. They also contend that hand writing as a neuro-haptic feedback system induces a different type or quality of cognition than keyboard use.

Of course, mathematical equation editors and modern graphical interfaces represent mathematical thought well, *after* the relevant mathematical formulations have been developed. But one rarely expresses original mathematical expressions using only a keyboard. One can finalize or submit a paper using an equation editor, for example, but does not actually create the mathematics in the paper using the editor. There is a distal relationship between producing content and the subsequent process of representing that content as a sharable artifact.

This is not the case where written language is the primary representational system. There is a much more proximal relationship between producing written text and the concurrent process of representing it as a sharable artifact. One can originate, manipulate, or edit written content using the same word processing software that is used to share it. The manipulation embodiment is the same as the sharing embodiment. The representation of cognition in composing written text occurs in the same process as producing a sharable artifact of that text. The entire occupation of clerical document transcription has become progressively obsolete as composition and typing merged into a unified process. In contrast, complex problem-solving processes using mathematical symbol systems almost invariably occur before representing those processes as a sharable artifact. The embodied cognition involved in mathematical problem-solving, for example, does not occur while using a keyboard, through an equation editor. But the embodied cognition of composing narrative, in contrast, does take place while using a keyboard, through a text editor, or word processor.

Pen and touch technology digitizes the representational process and does so *concurrently* with the cognition that is represented, rather than after. Why does this make a difference? Does mathematical problem-solving cognition otherwise change when it is represented in digital rather than analog mode? In one sense, since developers of stylus technologies actively promote a "seamlessness" between handwriting on paper and handwriting on a screen (or a writing tablet or special paper, for example), the answer might appear to be "no." Yet pen and touch technology brings mathematical cognition and digital representation of mathematics into the one unified process of coordinating a richer blend of semiotic systems.

One implication for K16 education is that the affordances of the digital revolution are lost on that sizable fraction of mathematics learning that is mediated by handwriting, i.e., the "sweet spot" of embodied mathematical cognition. Just as writing has its representational interface for embodied cognition in the form of keyboards used to digitize narrative or written language, mathematics now has a representational interface for embodied cognition in the form of pen and touch computing. Bringing mathematical cognition and digital representation into one unified process can open a portal into entirely different classroom dynamics.

The following section suggests three types of important affordances that pen and touch technology provides to mathematical learning, giving examples of each from current research projects. The intent is not to be exhaustive but to stimulate discussion on development of a fuller taxonomy. One unifying theme across the taxonomy is the concept of flow [3], often treated as a "zone" of high performance at the outer reaches of one's ability. Pen and touch technology promote conditions of learner flow, a relationship first presented in the WIPTTE/CPTTE context at WIPTTE 2010 [9]. Flow states entail many conditions and effects [10, 14, 18]. Among these are the conditions of rapid feedback to actions undertaken in play or in carrying out a task, by which distractions are either minimized or suppressed in terms of attentional resources, and mental action models are followed in a smooth rhythm. Other important factors, such as an equilibrium between challenge and ability in the task or play condition, round out conditions [3]. In each of the categories below, pen and touch interfaces enable important flow-like characteristics, and promote high levels of engagement.

6.3 A Beginning Taxonomy of Affordances

The three examples below have been referenced in prior WIPTTE conferences, but outside of the context of building this taxonomy. New analyses supplement their inclusion here. They are each drawn from a single program of research, with an expectation that as the taxonomy develops it will become populated with prototypes from the fuller range of research in the field.

6.3.1 Representational Expansion

The first type of benefit or affordance of the pen and touch tools is that they permit the embodiment of mathematical cognition in real time and they do so with standard digital tools such as copying, pasting, undoing, changing colors, resizing, saving, and sharing. In the aggregate, these functions add up to sizable advantages. Pen-based computing provisions embodied mathematical cognition, and thus mathematics learning, with these powerful tools. A sizable fraction of the pedagogical advances reported in the annual WIPTTE/CPTTE meetings focus on how these affordances are leveraged to improve learning.

Research previously reported at WIPTTE and elsewhere [7, 12] enables the study of how teachers, and teacher-student collaboration teams, become highly engaged and enter flow states when producing digital media using pen and touch computing. In these studies, supported by three US agencies and by Microsoft Research, teachers and student-teacher teams in digital maker spaces producing instructional videos find the array of representational affordances of pen and touch computing conducive to high immersion states. Over 95% of more than 300 teachers and students interviewed over the past five years, in workshops held in the US, Australia, Kenya, Ghana and Namibia report experiences consistent with classical flow states, experiences that are consistent with video records of those workshops depicting complete immersion by participants in video-making. The video-making uses screen capture technologies in the form of pen-based tablet computers and writing boards to help produce a foundational track of mathematical explanation that is then burnished with editing tools to produce a final video artifact. A sandbox of a portion of the videos in the project appears at http://creationsforlearning.net.

Analysis of this consistent induction into flow states suggests it is the very representational expansiveness of pen and touch technology tools that allows individuals to self-regulate to deep and productive engagement. Affordances in pen and touch furnish media-makers with a palette of representational tools to step-up or step-down the appearance and coherence of each video frame. The rapid feedback provided by video editors is, as one participant put it, "addicting." Every mathematical idea can become more precisely formulated or connected. We observed a highly creative and multi-layered equilibration process, or balancing act, of embodying a combination of content knowledge, pedagogy, and knowledge of a student audience, embodiment that itself can be analyzed through the learner-coordinated semiotics of Table 6.1. This is only possible through the underlying mathematical pen-based representations in each video.

6.3.2 Making Mathematical Thought Visible and Assessable

A theory of personalized learning communities [11] that emphasizes high "inter-actional bandwidth" emerged from the development and testing of collaboration software that would permit a teacher to view student work remotely (e.g., [4, 6]).

Table 6.2 Primary affordance characteristics of three types of pen and touch in mathematics learning

- **Representational Expansion:** Pen and touch furnishes the learner with a broad palette of representational tools through digitizing mathematical thought in real-time
- **Making Mathematical Thought Visible and Assessable:** Embodiment of thought is sharable in real-time to teachers and other collaborators
- **Enabling Co-constructed Mathematics:** Embodiment of mathematical thought can be co-constructed with others, leading to reciprocally reinforcing socio-affective growth and mathematical progress

Later commercialized by several companies (e.g., as DyKnow and LANSchool), such collaboration software and other variations of it permit teachers to observe mathematical thought as it unfolds. This is a significant development in terms of assessing cognition (Table. 6.2).

Deep assessment of student learning requires making cognition visible through some artifacts [2, 15, 16]. These visible artifacts are proxies for cognition. A multiple-choice test produces simple artifacts that are impoverished in that they are meant to permit inferences about whether a respondent understands the material – understandings, presumably, that are far more expansive and complex than one-off answers that represent them. Because pen-based computing embodies mathematical cognition in real-time, though, tools that furnish a sight-line that makes it possible for teachers or collaborators to see that cognition in rich, real-time form. Collaborative software systems with "what you see is what I see" features provide a means to see mathematical manipulations in progress [5, 8].

A depiction of cognition that is both visual and dynamic – i.e., one that allows a teacher to observe mathematical activity in real-time through collaboration software – offers rich possibilities for a teacher to interpret and assess cognition. As discussed in prior WIPTTE conferences (e.g., [9, 12]) such affordances permit students to seek scaffolds and help in real-time problem-solving, contributing to the possibility that high engagement activities will not be inhibited by cognitive wait-states while a student is waiting for a teacher's attention.

6.3.3 Enabling Co-constructed Mathematics

Embodied cognition can produce sharable artifacts [22]. In the first example above, the artifact is a media construction such as a video or a game. In the second example, the artifact takes the form of an observable video stream of student mathematical work to permit teacher interpretation, feedback, and assessment. In this third category, the artifact is the former, a media object meant to help others learn mathematics. The affordance originates if the artifact making is collaborative, and, in particular, co-constructed by student-teacher teams. Pen and touch computing, with the affordance of representing complex symbol systems across multiple semiotic layers, allows an individual to "unpack" mathematical thought in a cognitively powerful process of self-explanation. This, in turn, implies an obvious potential for individuals to co-construct artifacts. We have found in our research that co-creation of pen-based mathematical artifacts may have important pro-social aspects that reciprocally reinforce academic growth.

Our research context is one in which teachers and students collaborate to produce instructional videos, with the prospect that students and teachers helping each other embody mathematical cognition to help others learn, brings important value to each. We refer to this as "participatory teaching" by which students assume a limited subset of instructional responsibilities from their teachers.

Co-creating a pen-based media artifact has been a primary means (along with co-planning lessons and curating online video resources) to structure this collaborative activity. Table 6.3 provides examples of comments by students when working with teachers. Among the socio-emotional factors emerging are a deeper awareness of the perspectives of teachers, appreciation of "how the brain works" in peers, the immense satisfaction in helping others learn, and the need to be creative to foster new understandings by peers. Student reporting of flow experience (i.e., by self-reporting factors such as autotelic motivation and time dilation) remains at over 90% when students co-create media, but the range of experiences that they report expands to encompass potent socio-affective dimensions.

Table 6.3 Sample utterances by high schools students participating in summer 2015 participatory teaching depicting important socio-affective growth through affordance of pen-based collaborative media-making

- I felt excited because I've always loved to help people, and I do love math, so in order to do two things that I love... it was just incredible
- I have learned that you need to be really specific and detailed with what you are trying to teach to other students, (who) are eventually going to learn this, because if you don't explain well they won't understand and be confused as I was
- I notice that I can see how other students are thinking about these math concepts when I give them feedback... I'm making a video to teach others about a concept I didn't get last year. I'm seeing where I got lost now that I have to explain this to others
- This is pretty cool. I can see how everyone's brains work by the way they make their videos. It's good that there are so many of us (in our club) making these videos so that the students who watch them will be able to find one that matches how their brains work
- As a student, what surprised me the most from this experience was that I got to see the point of view from my teacher on how they teach us, the students
- I learned the value of team work and what it's like to be in the teacher's shoes
- That's what I try and do with my videos most of the time... Instead of just teaching a person, I try to show them in a way that if someone has never seen it before, they can solve it

6.4 Conclusion

The field of pen and touch computing has opened a means for digital embodiment of real-time mathematical cognition. The implications are expansive in consideration of the profound need to transform mathematics learning and to render mathematics into an accessible domain in which all learners can develop competence and exercise creativity and imagination. In our research, we seek means for learners to experience high engagement and flow levels of immersion in mathematics. We have found that the representational palette of pen and touch computing tools provides a portal to that type of immersion. The design of experience has fallen into three categories. *Producing artifacts with pen and touch*, with the aim of helping others acquire mathematical understandings, is intrinsically captivating to the subjects with whom we have worked. *Exposing embodied mathematical cognition*, i.e., making it visible, is a means to permit high levels of nuanced and precise feedback by teachers, furnishing them an unrivaled sight-line that extend beyond one-off judgments of whether an idea is understood. *Co-creating mathematical artifacts* has now been shown to reciprocally reinforce valuable, pro-social growth and mathematics learning. Each of these three categories has analogs across the growing field of pen and touch technology. The intent of this paper is to begin a conversation leading to development of a substantial research agenda around how and why pen and touch technologies can activate learning-sensitive mechanisms.

Acknowledgements The author acknowledges the generous support of the US National Science Foundation, the US State Department, the US Department of Education, Microsoft Research, and Pepperdine University in the research reported here, which reflects the views only of the author.

References

1. Anderson, M.L.: Embodied cognition: a field guide. Artif. Intel. **149**(1), 91–130 (2003)
2. Collins, A., Brown, J.S., Holum, A.: Cognitive apprenticeship: making thinking visible. Am. Educ. **15**(3), 6–11 (1991)
3. Csikszentmihalyi, M.: Flow and the Psychology of Discovery and Invention. Harper Collins, New York (1996)
4. Hamilton, E.R.: Computer assisted instructional delivery system and method. US Patent 5,176,520 (1993)
5. Hamilton, E.: Remote sharing of electronic paper (slatemate®). In: International Conference on Technology in Education (ICTE), London (1994)
6. Hamilton, E.R.: Computer assisted instructional delivery system and method. United Kingdom Patent 479408 (1999)
7. Hamilton, E.: Advancing a complex systems approach to personalized learning communities: bandwidth, sightlines, and teacher generativity. J. Interact. Learn. Res. **26**(1), 89–104 (2015)
8. Hamilton, E., Kapenda, L., Mbasu, Z., Miranda, H., Ngololo, E., Ogwel, J.C., Stern, D.: A model and research agenda for teacher and student collaboration using pen-based tablets in digital media making in sub-Saharan Africa. In: Revolutionizing Education with Digital Ink, pp. 223–230. Springer International Publishing (2016)

9. Hamilton, E., Harding, N., Berque, D., Reed, R.: Tablet computing, creativity and teachers as applied microgenetic analysts: a paradigm shift in math teacher professional development. In: Impact of Pen-Based Technology on Education, pp. 47–56 (2010)
10. Hamilton, E., Hurford, A.: Combining collaborative workspaces with tablet computing: research in learner engagement and conditions of flow. In: 2007 37th Annual Frontiers In Education Conference-Global Engineering: Knowledge Without Borders, Opportunities Without Passports, pp. T3C–3. IEEE (2007)
11. Hamilton, E., Jago, M.: Toward a theory of personalized learning communities. In: Jacobson, M.J., Reimann, P. (eds.) Designs for Learning Environments of the Future, pp. 263–281. Springer, Boston (2010)
12. Hamilton, E., Kapenda, L., Mbasu, Z., Miranda, H., Ngololo, E., Ogwel, J.C., Stern, D.: A model and research agenda for teacher and student collaboration using pen-based tablets in digital media making in sub-Saharan Africa. In: Revolutionizing Education with Digital Ink, pp. 223–230. Springer International Publishing (2016)
13. Hiebert, J.: A theory of developing competence with written mathematical symbols. Educ. Stud. Math. **19**(3), 333–355 (1988)
14. LeFevre, J.: Flow and the quality of experience during work and leisure. In: Csikszentmihalyi, M., Csikszentmihalyi, I.S. (eds.) Optimal Experience: Psychological Studies of Flow in Consciousness, pp. 307–318. Cambridge University Press, New York (1988)
15. Lesh, R., Hoover, M. , Hole, B., Kelly, A., Post, T.: Principles for developing thought-revealing activities for students and teachers. In: Kelly, A.E., Lesh, R.A. (eds.) Handbook of Research Design in Mathematics and Science Education. Routledge, Abingdon (2000). Accessed 10 Nov 2017, Routledge Handbooks Online
16. Ma, J., Cole, R., Pellom, B., Ward, W., Wise, B.: Accurate visible speech synthesis based on concatenating variable length motion capture data. IEEE Trans. Vis. Comput. Graph. **12**(2), 266–276 (2006)
17. Mangen, A., Velay, J.L.: The haptics of writing: cross-disciplinary explorations of the impact of writing technologies on the cognitive-sensorimotor processes involved in writing. In: Learning to Write Effectively: Current Trends in European Research, pp. 405–407. Brill, Bradford (2012)
18. Nakamura, J., Csikszentmihalyi, M.: The concept of flow. In: Csikszentmihalyi, M. (ed.) Flow and the Foundations of Positive Psychology, pp. 239–263. Springer, Dordrecht (2014)
19. O'Halloran, K.L.: Interdependence, interaction and metaphor in multisemiotic texts. Soc. Semiot. **9**(3), 317–354 (1999)
20. O'Malley, C.: Design considerations for tangible and embodied learning. In: Proceedings of the 31st European Conference on Cognitive Ergonomics, p. 2. ACM (2013)
21. Radford, L.: The ethics of being and knowing: towards a cultural theory of learning. In: Semiotics in Mathematics Education: Epistemology, History, Classroom, and Culture, pp. 215–234. Sense, Rotterdam/Taipei (2008)
22. Shapiro, L.: Embodied Cognition. Routledge, New York (2010)

Chapter 7
Sketching Graphs in a Calculus MOOC: Preliminary Results

Jennifer French, Martin A. Segado, and Phillip Z. Ai

Abstract We developed an online sketch-based input and grading tool, which can be integrated into existing MOOC platforms; this tool allows students to construct graphs using mouse, touch, or pen input and provides immediate targeted feedback. This paper presents some preliminary results from use of this sketching tool in *18.01.1x Calculus 1A: Differentiation*, an online calculus course offered by MIT on the edX platform.

7.1 Introduction

The ability to assess student graphs of mathematical functions is an important part of introductory calculus curricula. Moreover, the use of qualitative hand drawn sketches is a standard expert problem solving technique in a variety of fields: science, technology, engineering, and math. Pedagogically, tasks such as graph sketching—which require students to construct responses from scratch—offer increased cognitive engagement over passive recognition tasks [3], and have the potential to measure skills that cannot be adequately measured by multiple-choice questions [8].

Mathematical sketching problems are typically completed on paper and graded manually. However, the feedback students receive is delayed due to the logistics of human grading, and may not be detailed. Automatic grading of free-form student sketches can provide immediate, targeted feedback at early stages in the learning process to provide the most beneficial learning gains [6, 11]. Additionally, there has been an explosion in the popularity of Massively Open Online Courses

J. French (✉)
Department of Mathematics, MIT, Cambridge, MA, 02139, USA
e-mail: jfrench@mit.edu

M.A. Segado
Department of Mechanical Engineering, MIT, Cambridge, MA, 02139, USA
e-mail: msegado@mit.edu

P.Z. Ai
MIT, Cambridge, MA, 02139, USA

T. Hammond et al. (eds.), *Frontiers in Pen and Touch*, Human–Computer Interaction Series, DOI 10.1007/978-3-319-64239-0_7

(MOOCs) amongst both learners and universities. Assessing graphical fluency in such environments is challenging: students submit graphs from the wide range of devices used to access MOOC courseware (e.g. mobile phones, tablets, shared desktops, laptops) and grading must be accomplished at scale.

Based on the perceived learning gains awarded to both on-campus and MOOC students, we created a flexible sketch input and grading tool to support the functionality needed for introductory calculus assessment and deployed this tool in a calculus MOOC [10]. At a high level, the tool comprises a front-end HTML5 web application, which may be embedded in MOOC content (Fig. 7.1), and a back-end grading library, which provides instructors with pre-built functions that use common calculus language to check features of submitted graphs over specified points or intervals (value, slope, domain where increasing or decreasing, concavity over intervals, placement of critical points, locations of asymptotes, and the like). This tool has the following novel characteristics:

- *MOOC compatibility.* The tool integrates with the edX platform (though it may also be used independently) and is designed to support all major browsers without requiring additional plugins. While it may be used with touch or pen devices, it also uses smoothing in order to remain usable with ordinary mouse input to support the diversity of platforms used by learners. To the best of the authors' knowledge, this is the first sketching tool to be used in a MOOC.
- *Extensibility.* The tool has a modular architecture and will be open-sourced by mid-2016. New front-end plugins may be added to support additional drawing functionality, and different back-end grading libraries may be developed to support domains other than calculus. All components share a common data format to maximize interoperability and the potential for widespread analysis. To the best of the authors' knowledge, all related tools are proprietary.
- *Semantic multi-part data entry.* Drawing plugins in the toolbar are given domain-specific labels based on what they are intended to represent (e.g., "Function f(x)". or "Asymptote"), and learners use these plugins to enter multiple types of data on the same graph. The choice of drawing plugins and their associated labels (as well as axis scaling and other variables) are configured by course content creators on a per-problem basis. While semantic labeling is supported at least tangentially in one other application (ETS's proprietary Graph Editor; see below) it was a central feature in both the design of our tool and its use in calculus problems.

More information on the development of the sketch input tool may be found in [10].

7.2 Related Work

A number of researchers have explored the area of automatic mathematical graph grading. One project with an approach similar to ours is *BeSocratic* [1, 2]—specifically, its *SocraticGraphs* module. This standalone tool also allows students

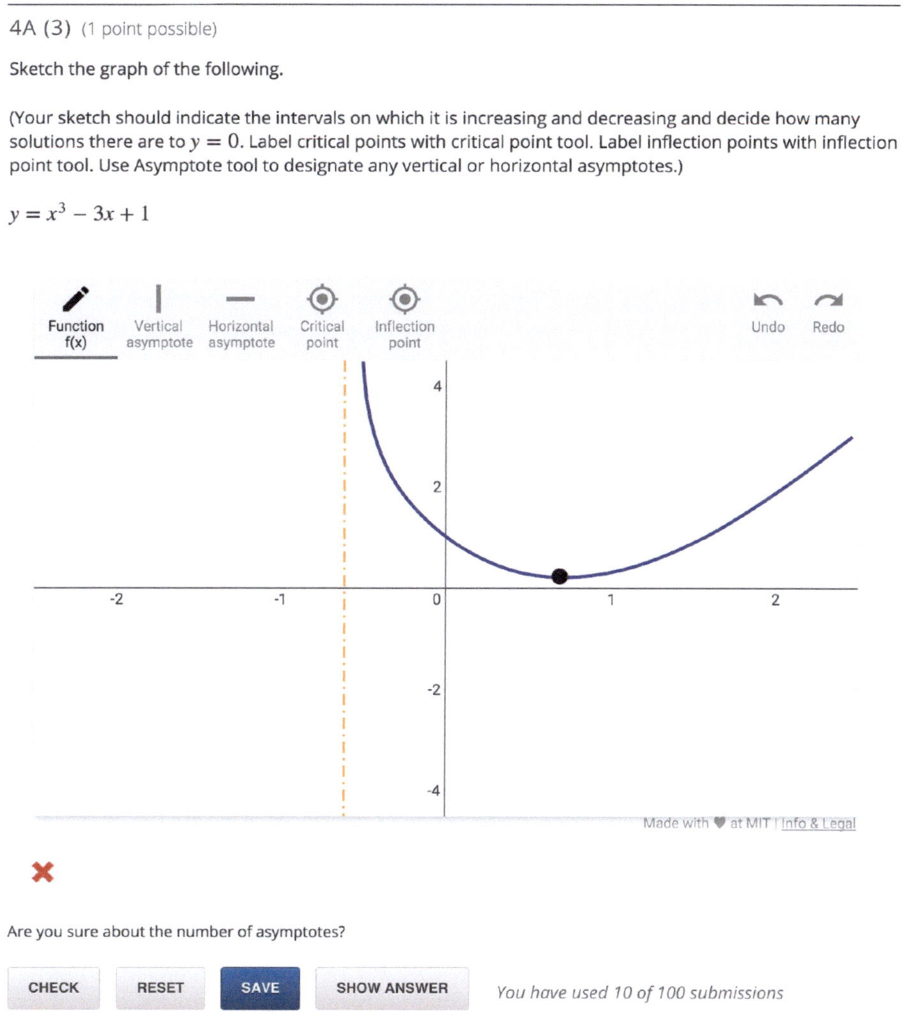

Fig. 7.1 A graphing problem in MIT's 18.01.1x calculus MOOC using the sketch input tool (*rectangular*, *center*). The tool is embedded into an edX problem block and uses the standard edX platform mechanisms to handle submissions, state persistence, and feedback display. Note the semantically-labeled drawing plugins in the toolbar; the types, order, and labeling of these may be configured on a per-problem basis by course content authors

to draw curved-line graphs using either mouse or touch input; the tool checks these sketches against a set of instructor-configured grading rules (specified using a graphical authoring interface) and displays feedback to the student. *BeSocratic* has been deployed in chemistry, molecular biology, and computer science courses at several universities.

Another similar project is the web-based function plot response tool developed at Michigan State University [7] as a way of teaching kinematics graphs. The tool is designed to be embedded directly into online course platforms (LON-CAPA in this case) and is also capable of providing immediate targeted feedback. In contrast to both *SocraticGraphs* and our own sketch input tool, this system does not permit direct unconstrained sketching, but instead allows users to construct plots by dragging the control points of cubic Hermite splines. A longitudinal study is underway to assess the formative benefits of this tool and, hopefully, provide more data on the relative merits of graph construction and graph interpretation problems.

A pair of tools which focus more on summative assessment are the *m-rater* scoring engine and accompanying Graph Editor developed by ETS [4, 8]. As with the function plot response tool above, this editor does not support direct sketching, but instead allows users to graph points, straight lines, and smooth curves (splines) by dragging points onto a set of axes. The *m-rater* scores straight lines and quadratic curves analytically, and scores more general curves by checking them at several locations. The ETS Graph Editor shares our support for multi-part data entry and also appears to support semantic labeling.[1]

A completely different approach to grading sketched graphs was evaluated by Lukoff as part of his doctoral dissertation [9]; instead of evaluating the graph according to a fixed rubric, Lukoff trained support-vector-machine classifiers on a set of manually-labeled graph sketches (which were produced and labeled online using Amazon Mechanical Turk). The technique showed promise, but does not yet appear to be in use in any educational environments.

7.3 Methodology

A set of 13 graphical input problems[2] were created and included in the 18.01.1x calculus MOOC [5]. The problems are summarized in Table 7.1. In the hopes of avoiding frustration with the new tool, most problems allowed effectively unlimited tries with no penalty; only the two final exam problems had an imposed limit of five attempts each.

The sketch input tool itself was built and refined in parallel with course content to ensure that its functionality aligned with course goals and assessments. The initial grader tolerances and behavior were based on data from an earlier prototype of the tool, but updated based on student reports of graphs that did not grade correctly (e.g., due to unforeseen drawing styles). Such graphs were examined manually to ensure their correctness and, when possible, the grading library was updated to support them.

[1]See the "Graph Editor" figure on page 3 of ETS's automatic scoring technologies brochure: https://www.ets.org/Media/Home/pdf/AutomatedScoring.pdf

[2]A tutorial problem and several problems created with an earlier prototype of the sketching tool are not include in these results.

Table 7.1 Summary of problems using the sketch input tool in 18.01.1x. For clarity, inessential details have been omitted from most problem statements

ID	Problem statement
app1-14-2	Sketch a function f on the interval $-4 < x < 6$ that has the extrema and inflection points determined by the previous problem[a]
app2-7-1	Sketch the rational function $f(x)$ defined below. $f(x) = \frac{2x^2+1}{x^2-1}$, $f'(x) = \frac{-6x}{(x^2-1)^2}$, $f''(x) = \frac{6(3x^2+1)}{(x^2-1)^3}$
app2-17-1	Sketch e^x/x on the interval $-\infty < x < \infty$
hw4A-4-1	Sketch the graph of the following: $y = x^3 - 3x + 1$
hw4A-5-1	Sketch the graph of the following: $y = x^4 - 4x + 1$
hw4A-6-1	Sketch the graph of the following: $y = x^2/(x-1)$
hw4A-7-1	Sketch the graph of the following: $y = e^{-x^2}$
hw4A-8-1	Sketch the graph of the derivative of the function displayed below.[b] Include zero crossings and label critical points
hw4B-2-1	Sketch the graph of $f(x) = \frac{1}{x^2-1}$
hw4B-2-2	Sketch the graph of f' and f'' given $f(x) = \frac{1}{x^2-1}$
hw4B-4-4	In the case where A is critically damped[c] and $a = 1$, sketch the graph of A. (The function sketch can be qualitative. Label any asymptotes, critical points, and inflection points qualitatively)
final-4-1	Draw the graph of the function (qualitatively accurate). $y = \sqrt{x+1}/(x-b)$, $1 < b < 2$ (Your sketch should indicate the intervals on which it is increasing and decreasing and decide how many solutions there are to $y = 0$. In this problem, you do not need to test the second derivative to identify the concavity. There are no inflection points. Label critical point and asymptotes using the appropriate drawing tools)
final-8-2	Sketch the graph of $f(x) = 1 + \frac{a}{x} + \frac{a}{x^2}$ (same function as above[d]) on $-\infty < x < \infty$ showing the horizontal and vertical asymptotes. While the sketch need not be to scale, please make sure the x–coordinate of all critical point(s) and inflection point(s) are accurate

[a]The problem preceding this one asked students to find critical points and inflection points of a function given the signs of its first and second derivatives

[b]The original problem statement included a graph of a rational function with labeled asymptotes, one critical point, and one inflection point

[c]An earlier problem detailed a mathematical model for a heavy swinging door with angle A

[d]An earlier problem asked students to find the intervals on which the given function was increasing, decreasing, concave up or concave down

7.4 Preliminary Results

A total of 8187 non-blank final responses were submitted by 871 unique users for the sketch problems described in Table 7.1. Of these, 7311 (89.3%) were classified as correct by our grader code. A breakdown by problem is provided in Table 7.2. Overall, the data are encouraging; most students who submitted responses were able to arrive at the correct answer, many in their first attempt, with 2–3 attempts

Table 7.2 Summary of non-blank submissions for graphing problems described in Table 7.1

ID	No. final responses	Percent correct	No. attempts (by percentile)					
			Correct			Incorrect		
			10th	50th	90th	10th	50th	90th
app1-14-2	700	89.3	1	2	5	1	2	6
app2-7-1	669	87.9	1	2	7	1	3	9
app2-17-1	603	87.9	1	2	5	1	2	8
hw4A-4-1	674	94.1	1	2	5	1	3.5	9.2
hw4A-5-1	646	95.2	1	2	4	1	2	6
hw4A-6-1	623	95.2	1	2	4	1	2	6.1
hw4A-7-1	616	94.0	1	3	6	1	3	10
hw4A-8-1	575	87.0	2	3	9	1	3	13.4
hw4B-2-1	654	95.3	1	2	4	1	2	8
hw4B-2-2	617	88.3	1	3	7	1	2	7.9
hw4B-4-4	506	92.9	2	3	7	1	3	9
final-4-1	648	73.5	1	2	4	1	5	5
final-8-2	656	81.2	1	1	4	1	5	5

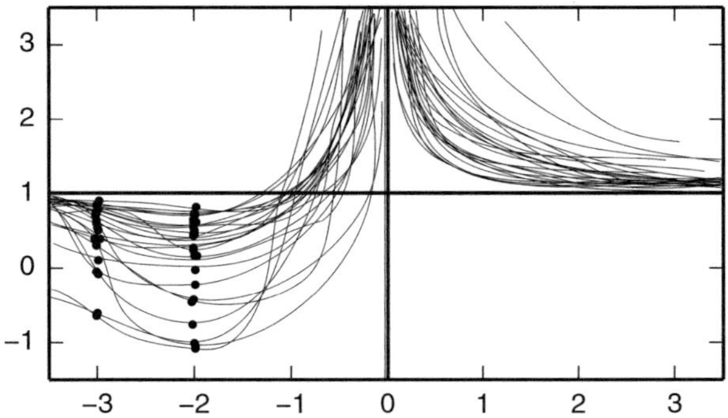

Fig. 7.2 Subset of 25 correct submissions for problem *final-8-2*. The answers vary considerably within the parameters permitted by the problem statement but all appear qualitatively correct

being the norm. The median number of attempts taken for incorrect submissions was similar, except in the two final exam problems, where most students demonstrated enough persistence to use up all five permitted attempts.

The graphs of individual submissions may also be examined for insight. Figure 7.2 shows a subset of 25 such submissions, drawn randomly from the pool of correct responses to the *final-8-2* problem. One notable feature is the degree of variability in these submissions: all indeed appear correct given the problem statement, but details such as the vertical position of critical points and the way the curves approach the vertical asymptote vary considerably between graphs.

Submissions that were graded as incorrect proved more difficult to examine in aggregate. Since each submission consists of multiple points and line segments, it becomes difficult to identify which of these belong together when a large number are overlaid, even when submissions are first grouped by the feedback message that was displayed to students. Nevertheless, some of these grouped visualizations

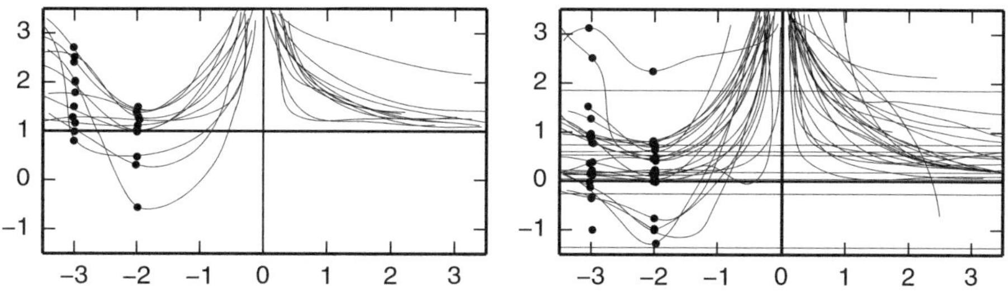

Fig. 7.3 Subset of incorrect submissions for problem *final-8-2*. Submissions *on the left* were rejected with feedback "What is the relationship of your function to the horizontal asymptote?" while those *on the right* were rejected with feedback "Check the location of your horizontal asymptote"

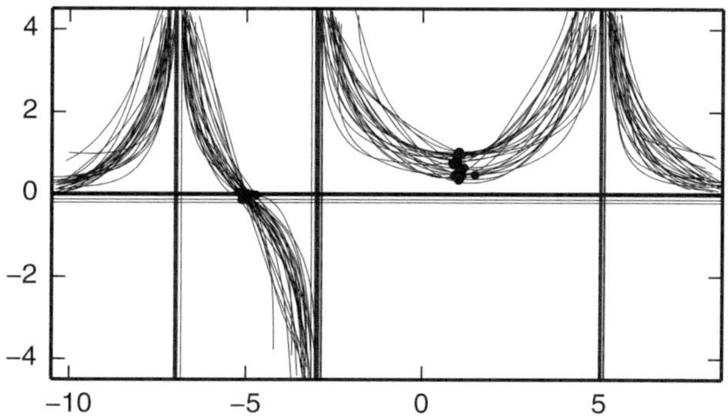

Fig. 7.4 Subset of 25 correct submissions for problem *hw4A-8-1*, one of the more complex graphs students were asked to sketch in 18.01.1x

did convey interesting trends; the submissions shown in Fig. 7.3 clearly illustrates some common misconceptions, with the subplot on the left identifying students who didn't realize that $f(x) \to 1^-$ as $x \to -\infty$ and many of those on the right identifying $y = 0$ instead of $y = 1$ as the position of the horizontal asymptote.

Another takeaway from our data is that MOOC students are capable of entering complex plots quite accurately with this tool (e.g., Fig. 7.4), despite our expectation that at least some of these were drawn using mouse input instead of the more natural touch or pen modalities.[3]

7.5 Future Work

A variety of opportunities exist for future work. These may be loosely divided into four categories:

Data Analysis There is considerable room for analysis of existing results. Manual grading of individual responses could identify grading errors and lead to improve-

[3]While information on the devices used by students is indeed logged by edX platform, it is not exported by edX's instructor-facing interface. We hope to obtain this information in the future.

ments in the grader library default tolerances. Clustering approaches for identifying common misconceptions in solutions may also be valuable; in particular, it would be interesting to identify features most often missing from nearly-correct graphs and find ways to improve our feedback mechanism to guarantee student success. It would also be interesting to use the built-in tolerances in our grading library to determine confidence intervals for correctly- and incorrectly-graded student sketches. These borderline cases could then be flagged for review, which would lead to further improvements of the grading library algorithms.

We are also collecting new data from a second MOOC course (*18.01.2x Calculus 1B: Integration*), as well as two on-campus courses currently running at MIT: (*18.01 Calculus I* and *2.001 Mechanics and Materials I*). The tool is being used in new ways in these courses. In the new MOOC, students sketch areas and regions between curves in order to set up integration problems. This allows us to provide scaffolding and feedback on integration problems that would be impossible without the use of graphing. All of the MOOC calculus sketching problems are being used on-campus, which gives us the opportunity to interview students directly about their experience using the tool. In the mechanics course, the tool is being used to grade axial force diagrams: students are shown an image of a beam with forces along its length and asked to draw the axial force resultant directly on top of the image (as well as label any discontinuities).

Grading Improvements Certain limitations of the grading library were identified through the use of this tool in 18.01.1x. One particular challenge is checking the curvature of functions in a way that is consistent with a human grader. We are implementing more robust grading algorithms and using the student data to determine the appropriate tolerances. Another limitation lies in the typical structure of our grading scripts: to grade a complex graph, it is often easiest to check features of the free-form curve relative to student labeled landmarks such as asymptotes, critical points, and inflection points. If a student fails to provide these landmarks, the grader stops immediately and gives students feedback on the missing items; however, providing feedback in this order may be in conflict with the optimal learning process.

We hope to simplify the instructor interface with the grading library. Currently, an instructor writes simple Python scripts specifying the conditions to check; for problems in a wide range of disciplines, however, this process could be automated (e.g., by having grading conditions for the time response of an electrical circuit or mechanical system auto-generated from the problem specifications). This also opens up the exciting possibility of auto-generating entire randomized *problems*: for example, the number and locations of forces on a mechanical beam could be randomized, thus providing students with infinite practice opportunities for a particular type of mechanics problem.

It may also be interesting to explore peer grading or instructor-assisted grading approaches for graphical problems, though this is not on our immediate roadmap.

Learner Interface Improvements The addition of common editing features (e.g., dragging, deleting, and clipboard functions) would improve overall usability. We hope to add a spline tool similar to that in [1] and [8], making the tool more usable for learners lacking a pen or touch device. Such a spline tool could also feature a keyboard-and-screenreader interface to enable its use by visually- and mobility-impaired students.

Collaboration Finally, once the tool is stable and accessible, we will be releasing it under an open-source license so that others can both use and improve it. Open access is key, and important so that we can begin to forge collaborations both in and out of MIT.

7.6 Conclusions

A free-form sketch-input and automatic grading tool designed for use with pen, touch, and mouse input was deployed in a calculus MOOC delivered online through edX. In addition to focusing on MOOC compatibility, the key features of the tool include a modular architecture designed for open-source collaboration and the gradeability of multiple customizable, semantically-labelled plugins in a single student sketch based on standard calculus terminology.

An analysis of 8187 responses collected with the graphical input tool showed that the median total attempts for students who do not get a correct answer is 2 or 3, suggesting that if we have not given a student useful feedback quickly, they may not persist to get a correct response. Those who get correct responses tend to do so on average in 2 or 3 attempts, with the 90th percentile using between 4 and 9 attempts depending on the difficulty of the problem, despite essentially unlimited attempts. Qualitatively, our grader generally performed as expected; however, analysis of the incorrect graphs showed that the feedback was not always correlated with all of the misconceptions in incorrect graphs, and a finer level of feedback based on student errors may be helpful to improve learning gains.

Preliminary results are encouraging. While considerable work remains to be done, we expect that the use of such graphical tools in both MOOCs and classroom environments enhances learning gains in student ability to sketch functions and provides a rich source of data for future study.

References

1. Bryfczynski, S., Pargas, R., Cooper, M., Kylmkowsky, M.: BeSocratic: graphically assessing student knowledge. In: Proceedings of the IADIS International Conference Mobile Learning, pp. 3–10. IADIS (2012)
2. Bryfczynski, S., Pargas, R.P., Cooper, M.M., Michael Klymkowsky, J.H., Grove, N.P.: Classroom uses for BeSocratic. In: Adler, A., Hammond, T., Payton, M., Valentine, S. (eds.) The Impact of Pen and Touch Technology on Education, pp. 127–136. Springer, Cham (2015)

3. Chi, M.T.H., Wylie, R.: The ICAP framework: linking cognitive engagement to active learning outcomes. Educ. Psychol. **49**(4), 219–243 (2014)
4. Fife, J.H.: Automated scoring of mathematics tasks in the Common Core era: enhancements to m-rater in support of CBAL^{TM} mathematics and the Common Core assessments. ETS Res. Rep. Ser. **2013**(2), i–35 (2013)
5. French, J., Wang, S., Jerison, D., Staffilani, G.: Calculus 1A: differentiation (2015). Online course. https://www.edx.org/course/calculus-1a-differentiation-mitx-18-01-1x. Accessed 30 Jan 2016
6. Hattie, J., Timperley, H.: The power of feedback. Rev. Educ. Res. **77**, 81–112 (2007)
7. Laverty, J., Kortemeyer, G.: Function plot response: a scalable system for teaching kinematics graphs. Am. J. Phys. **80**(8), 724–733 (2012)
8. Livingston, S.A.: Constructed-response test questions: why we use them; how we score them. Princeton: R&D Connections, no. 11. Educational Testing Service (2009)
9. Lukoff, B.: The design and validation of an automatically-scored constructed-response item type of measuring graphical representation skill. Ph.D. thesis, Stanford University (2010)
10. Segado, M., French, J.: Interactive sketch input tool for MOOCs (2015). Paper presented at the learning with MOOCs II workshop, Columbia University, 2–3 Oct 2015. http://web.mit.edu/~jfrench/www/pdf/sketchinput_paper.pdf. Accessed 29 Jan 2016
11. Shute, V.J.: Focus on formative feedback. Rev. Educ. Res. **78**, 153–189 (2008)

Part IV
Personalized Feedback and Sketching in Science and Engineering

Chapter 8
Tablet PC Goal-Directed Practice Coupled with Real-Time Targeted Feedback Enhances Food Chemistry Learning

Enrique Palou, Judith V. Gutierrez Cuba, Nelly Ramirez-Corona, and Aurelio Lopez-Malo

Abstract The *How People Learn* framework was utilized to redesign the courses Food Chemistry and Advanced Food Chemistry in order to improve undergraduate and graduate food chemistry teaching and learning by creating learning environments that promote an interactive classroom while integrating formative assessments into classroom practices by means of Tablet PCs and associated software. By means of *InkSurvey* and *Classroom Presenter* we were able to gauge student learning during goal-directed practice in real-time, provide instantaneous targeted feedback, and make immediate pedagogical adjustments as needed. Course redesign increased student participation and formative assessments while instructors utilized the information gained through real-time assessments to tailor instruction to meet student needs. In this chapter, quantitative and qualitative data will be shared, as well as insights and conclusions from our experiences that indicate the effectiveness of goal-directed practice coupled with real-time targeted feedback to enhance food chemistry learning.

8.1 Introduction

In the National Research Council's seminal summaries of current knowledge with regards to *How People Learn* [7, 8] there are convincing and recurrent calls to embed formative assessments during teaching and learning processes. Based on a comprehensive theoretical foundation, a noteworthy body of research indicates

E. Palou (✉) • J.V. Gutierrez Cuba • N. Ramirez-Corona • A. Lopez-Malo
Department of Chemical and Food Engineering, Center for Science, Engineering, and Technology Education, Universidad de las Americas Puebla, Cholula, Puebla, 72810, Mexico
e-mail: enrique.palou@udlap.mx; judith.gutierrezca@udlap.mx; nelly.ramirez@udlap.mx; aurelio.lopezm@udlap.mx

T. Hammond et al. (eds.), *Frontiers in Pen and Touch*, Human–Computer Interaction Series, DOI 10.1007/978-3-319-64239-0_8

that frequent formative assessments can actively engage learners, effectively inform instruction, and increase student metacognition. However, educators' attempts to gather formative assessment often prove not only burdensome, but include excruciating delays until the instructor is able to respond to (mis)conceptions revealed in these formative assessments [24]. Furthermore, Ambrose et al. [2] emphasized that goal-directed practice coupled with targeted feedback are critical to learning. This is one of the research-based principles of learning established by the Eberly Center for Teaching Excellence [14], which reports that "learning and performance are best fostered when students engage in practice that focuses on a specific goal or criterion, targets an appropriate level of challenge, and is of sufficient quantity and frequency to meet the performance criteria. Practice must be coupled with feedback that explicitly communicates about some aspect(s) of students' performance relative to specific target criteria, provides information to help students progress in meeting those criteria, and is given at a time and frequency that allows it to be useful." There are broad theoretical foundations for embedding goal-directed practice coupled with real-time targeted feedback in classroom instruction. However, educators need an efficient, robust method for implementing its application in undergraduate and graduate classrooms as well as evidence supporting its effectiveness. Our goal was to enhance food chemistry teaching and learning by creating high-quality learning environments that promote an interactive classroom while integrating formative assessments into undergraduate and graduate classroom practices by means of Tablet PCs and associated software in order to gauge student learning during goal-directed practice in real-time, provide instantaneous targeted feedback, and make immediate pedagogical adjustments as needed.

8.2 Previous Work

Several faculty members and graduate students at Universidad de las Américas Puebla[1] (UDLAP) have been examining mobile technology since 2007 to improve engineering teaching and learning. We have learned a lot about the potential of pen and touch technologies and associated software to create learning environments that are knowledge-, learner-, community-, and assessment-centered as highlighted by the *How People Learn* (HPL) framework [7, 8]. Results for several undergraduate and graduate courses are available elsewhere [11, 16–19, 21, 27–29]. The studied courses were improved taking into account technological advances and recent research on human learning.

[1] www.udlap.mx

8.2.1 Using Information About How People Learn

A key publication that informed our course redesign is *How People Learn* [7]. An organizing structure used in this volume is the HPL framework. It highlights a set of four overlapping lenses that can be used to analyze the degree to which learning environments are: "*knowledge centered, learner centered, community centered*, and *assessment centered*, in the sense of providing multiple opportunities to make students' thinking visible so they can receive feedback and be given chances to revise" [8].

8.2.2 Goal-Directed Practice Coupled with Targeted Feedback

Ambrose et al. [2] define "*practice* as any activity in which students engage their knowledge or skills (for example, creating an argument, solving a problem, or writing a paper) and *feedback* as information given to students about their performance that guides future behavior"; while emphasizing that the full potential of practice and feedback could not be accomplished unless the two are effectively combined. Brent and Felder [9] proposed that "students should learn problem-solving strategies and improve skills by initially attempting small tasks that require the strategies and skills, getting feedback on their attempts, trying again with better results, and gradually moving to increasingly complex problems. Their improvement would be accelerated if they fully understand the instructor's learning goals and the feedback they get is clearly related to the targeted skills"; then suggested several pedagogical strategies (which we utilized in the studied courses) in order to promote goal-directed practice coupled with targeted feedback.

According to Ambrose et al. [2] implications of the body of research on practice and feedback are that to achieve the most effective learning, "students need sufficient practice that is focused on a specific goal or set of goals and is at an appropriate level of challenge" and highlight the benefits of using a given amount of practice time more efficiently by "focusing students' efforts on what they need to learn (rather than what they already know or may be more comfortable doing) and setting their goals for performance at a reasonable and productive level of challenge"; while stating three key implications on what makes feedback more effective, "the feedback must: (1) focus students on the key knowledge and skills you want them to learn, (2) be provided at a time and frequency when students will be most likely to use it, and (3) be linked to additional practice opportunities for students." Then present several pedagogical strategies (which we applied in the studied courses) that can help provide students with goal-directed practice and targeted feedback. It is clear that if practice and feedback are carefully designed with the above mentioned features in mind, we would make the teaching and learning processes not only more effective but also more efficient.

8.2.3 Formative Assessment

The purpose of *Assessment of Learning* is usually summative and is typically performed at the end of a task, unit of work, etc.; it is designed to provide evidence of achievement to parents, other educators, the students themselves, and sometimes to outside groups such as employers or other educational institutions. The emphasis shifts from summative to formative assessment in *Assessment for Learning*, which happens during the learning process, often more than once, rather than at the end. In *Assessment for Learning*, instructors use assessment as an investigate-able tool to find out as much as they can about what their students know and can do, and what confusions, preconceptions, or gaps they might have. Through *Assessment as Learning* students are able to learn about themselves as learners and become aware of how they learn, become metacognitive, by reflecting on their work on a regular basis, usually through self- and peer-assessment and decide (often with the help of the instructor, particularly in the early stages) what their next learning will be [13].

Black and Wiliam [5] identified a number of features of effective formative assessment. Possibly, the most important was that, to be effective, formative assessment had to be integrated into classroom practice, requiring an important reorganization of classroom actions; they also noticed that for assessment to function formatively, feedback had to be utilized by students, and thus the differential treatments that are incorporated in response to the feedback are at the heart of effective teaching and learning processes [32]. Black and Wiliam [5] stated that "practice in a classroom is formative to the extent that evidence about student achievement is elicited, interpreted, and used by teachers, learners, or their peers, to make decisions about the next steps in instruction that are likely to be better, or better founded, than the decisions they would have taken in the absence of the evidence that was elicited." Then, "two features appear to be particularly important in designing assessments that will support learning: (1) the evidence generated is instructionally tractable, i.e., is more than information about the presence of a gap between current and desired performance. The evidence must also provide information about what kinds of instructional activities are likely to result in improving performance; (2) the learner engages in actions to improve learning; this may be undertaking the remedial activities provided by the teacher, asking a peer for specific help, or reflecting on different ways to move his/her own learning forward" [32]. The involvement of learners, and their peers, is explicitly incorporated in the proposal that formative assessment could be conceived of as involving three main processes (identifying where learners are in their learning, where they are going, and how to get there) exercised by three categories of actors (instructor, learner, and peers). The resulting matrix of nine cells, could be organized as five key strategies of formative assessment, as suggested by Black and Wiliam [6], Leahy et al. [26], and Wiliam [32] as shown in Table 8.1. Wiliam [32] reviewed recent attempts to theorize formative assessment as well as *Assessment for Learning* and *Assessment as Learning* in a number of ways, specifically in terms of classroom strategies and practical techniques that we utilized in studied classrooms.

Table 8.1 Formative assessment "key strategies" [6, 26, 32]

	Where the learner is going	Where the learner is now	How to get there
Teacher	Clarifying learning intentions and sharing criteria for success	Engineering effective classroom discussions, activities, and tasks that elicit evidence of learning	Providing feedback that moves learners forward
Peer	Understanding and sharing learning intentions and criteria for success	Activating students as learning resources for one another	
Learner	Understanding learning intentions and criteria for success	Activating learners as the owners of their own learning	

8.2.4 Tablet PC Associated Software

Tablet PCs can be easily introduced in diverse courses, particularly when coupled with selected software could promote an interactive classroom, integrating formative assessments into classroom practices in order to gauge student learning in real-time, provide immediate feedback, and make instantaneous pedagogical adjustments [18, 19, 24, 25]; associated software used in this work were *InkSurvey*[2] and *Classroom Presenter*.[3]

8.2.4.1 InkSurvey

InkSurvey is a free web-based software designed to facilitate the collection of real-time formative assessment [24]. Using this tool, the instructor can embed formative assessments during instruction by posing an open-format question. Students equipped with touch- or pen-enabled mobile devices (Tablet PCs, iPads, iPhones, Android devices including *most* smartphones) are then actively engaged in their learning as they utilize the keyboard or digital ink to draw, sketch, or graph their responses. The instructor receives these responses instantaneously providing insights into student thinking and what they do and do not know. Subsequent instruction can then repair and refine student understanding in a very timely manner. Kowalski et al. [24] illustrated the wide applicability of *InkSurvey* by means of a series of vignettes featuring instructors of diverse subjects, with students

[2]ticc.mines.edu

[3]classroompresenter.cs.washington.edu

using various touch- and/or pen-enabled mobile devices in multiple educational environments at different countries. In a companion chapter [24, 25], instructors shared selected data, insights, and conclusions from their experiences regarding the effectiveness of this tool.

8.2.4.2 Classroom Presenter

A diversity of Tablet PC compatible software is being utilized to facilitate communication within the classroom, such as *Classroom Presenter* [3]. We identified classroom assessment techniques (CATs) appropriate [4] to each question or activity of the studied courses and then adapted them into a Tablet PC/*Classroom Presenter* environment. The instructor also made use of CATs that are already features within *Classroom Presenter*, like the polling features [19, 27]. Instructor utilized CATs to appraise student learning in real-time and perform immediate pedagogical adjustments as necessary. *Classroom Presenter* broadcasts the instructors' screen content to the entire class by means of wireless networking. In this mode, students are capable to receive application output and instructor's annotations as well as add their own notes to every presentation [3, 27].

8.2.5 Capturing Differences of Learning Environments

The VaNTH observation system (VOS) includes a series of instruments developed to capture qualitative and quantitative *classroom observation* data from teaching and learning experiences, which incorporates *several* elements of the HPL framework and uses four recurring methods of collecting classroom data: recording student-teacher interactions, recording student academic engagement, recording narrative notes of classroom events, and rating specific indicators of effective teaching [12, 16]. VOS provides information regarding pedagogical strategies, interactions occurring within a class, and evidence with regards to levels of student engagement. Main components of the VOS are: (i) the Classroom Interaction Observation (CIO), sampled real-time, which records student and faculty interactions; (ii) a time-sampled Student Engagement Observation (SEO), which notes whether students are (un)engaged in academic tasks; (iii) qualitative Narrative Notes (NN) on the lesson content, lesson context, mitigating circumstances, and additional information about the classroom; and (iv) Global Ratings (GR), which provide summative information about major aspects of the pedagogical strategies underlying the class session [20, 28].

While there remains much more work to be done to integrate research on goal-directed practice coupled with targeted feedback with fundamental research regarding instructional design, feedback, self-regulated learning, and motivation; nowadays there is a robust body of theoretical and empirical work that suggests that integrating real-time targeted feedback with goal-directed practice may well have unprecedented power to increase student engagement and improve learning. Main research question of this particular study was: "does goal-directed practice coupled with real-time targeted feedback by means of Tablet PCs and tested associated software enhances food chemistry learning?"

8.3 Methodology

8.3.1 Context

UDLAP is a Mexican private institution of higher learning committed to first-class teaching, public service, research, and learning in a wide range of academic disciplines including engineering, business and economics, physical and social sciences, humanities, and arts. Since 1959, the Commission on Colleges of the Southern Association of Colleges and Schools (SACSCOC) has accredited UDLAP in the US [28]. UDLAP's Chemical and Food Engineering Department offers several graduate programs (accredited as of high quality by the Mexican National Council of Science and Technology), as well as three undergraduate programs: the Food Engineering undergraduate program is approved by the Institute of Food Technologists (IFT) in the US and accredited by the *Consejo de Acreditación de la Enseñanza de la Ingeniería* (CACEI) that is the peer-accrediting agency of the US ABET in Mexico; the Chemical Engineering undergraduate program is also accredited by CACEI, while the Environmental Engineering is an undergraduate program that started in fall 2012 [1].

8.3.2 Participants and Setting

The studied courses, Food Chemistry and Advanced Food Chemistry, are a junior level, 3 credit required course for the food engineering program and a first semester 3 credit required course for the food science MSc and PhD programs at UDLAP, respectively. Studied courses' major goal is to help students think about the way a food chemist does. Therefore, students are involved in answering two key questions: (i) how the composition, structure, and properties (especially in terms of quality and safety) of foods are affected by chemical changes the food experiences? and (ii) how the understanding of key chemical and biochemical reactions can be applied to many situations encountered during formulation, processing, and storage of food? Detailed description of courses is available elsewhere [19]. Studied courses are taught alternately during the year. In the spring semester the undergraduate course (Food Chemistry) is offered while the graduate course (Advanced Food Chemistry) is taught in the fall semester. Therefore, this study was carried out during two periods per year for 7 years. Most of the data were collected from spring 2008 (when

Table 8.2 Studied courses' participants by cohort

	Class	Total students	Gender (%)	
			Male	Female
Undergraduate (n = 125)	2008	25	16.0	84.0
	2009	23	34.8	65.2
	2010	10	10.0	90.0
	2011	15	33.3	66.7
	2012	19	21.1	78.9
	2013	14	26.7	73.3
	2014	15	25.0	75.0
	2015	14	20.0	80.0
Graduate (n = 69)	2008	5	20.0	80.0
	2009	7	0	100
	2010	7	42.8	57.2
	2011	15	26.7	73.3
	2012	16	43.8	56.2
	2013	11	36.4	63.6
	2014	3	0	100
	2015	5	20.0	80.0

courses' redesign took place) to fall 2015. The same instructor has been in charge of the studied courses for the last 17 years; therefore, redesign of the studied courses was based on his expertise related to food chemistry as well as to engineering education, particularly with the HPL framework and its implementation in both undergraduate and graduate courses. The studied populations (that included the entire class of the food engineering undergraduate program or graduate programs, respectively) are presented in Table 8.2. UDLAP's Institutional Review Board approved the research and every participant signed an informed consent form.

8.3.3 Course Redesign and Tablet PCs

Prior to spring 2008, the studied courses were redesigned taking into account research on human learning and cognitive processes that underlie expert performances in order to create learning environments that were knowledge-, learner-, community-, and assessment-centered as highlighted by the HPL framework [7, 8] while starting in spring 2009 technological advances [24, 25, 28] were introduced since thanks to two Hewlett-Packard (HP) grants, UDLAP received 84 HP Tablet PCs to utilize them in selected courses (including the studied ones). In particular, we were interested in using Tablet PC and associated software to encourage active and cooperative learning, enhance classroom interactive engagement while probe student understanding through frequent formative assessments as described in detail elsewhere [16, 19, 27, 28]. Tablet-PCs have been utilized in every session of studied

Fig. 8.1 Student goal-directed practice (*left*) coupled with real-time targeted feedback for the whole group (*center*) or a particular team (*right*) in the studied undergraduate classroom

courses since spring 2009 by each student enrolled. The instructor utilized one of the Tablet PCs (wirelessly connected to the classroom projector, so he was able to move with its tablet) and the students utilized each one a Tablet PC.

8.3.4 Goal-Directed Practice Coupled with Real-Time Targeted Feedback

In brief, the sequence of the activities, as can be seen in Fig. 8.1, realized during goal-directed practice coupled with real-time targeted feedback for this study was that the instructor by means of *InkSurvey* or *Classroom Presenter* posted a question (from a previous semester quiz or exam) or an activity focused on a specific course goal and/or learning outcome carefully selected so as to target an appropriate level of challenge, and being of sufficient quantity and frequency in order to meet the performance criteria for that particular goal or outcome.

Questions were asked mainly before every class began in order to reveal students' prior knowledge and at the end of every class to elicit how much they learned. Furthermore, challenge-based activities (at least two every class) were designed as problem-solving learning environments following Jonassen [22] and included a variety of approaches to instruction such as case-based instruction, problem-based learning, learning by design, inquiry learning, and anchored instruction. There are important differences among these approaches, but important commonalities as well [15, 16, 27]. Once students received the questions or activities on their Tablet PCs, they wrote their answers (individually or as teams) and then send them back to the instructor Tablet PC. Undergraduate and graduate students utilized their Tablet PCs to respond to the challenging questions or activities (posed by the instructor) with their own words/sentences/paragraphs entered manually via the keyboard, or with digital ink (pen/touch) that allow them handwriting, as well as input of sketches, equations, graphs, diagrams, chemical structures, etc., while the instructor received an instantaneous compilation of these student responses either on its *InkSurvey* or *Classroom Presenter* platform [19, 27, 28].

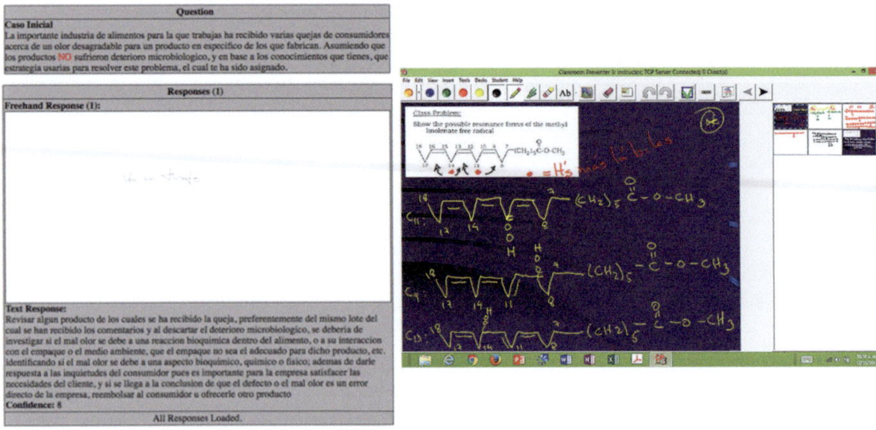

Fig. 8.2 Screenshots (in Spanish) of selected student answers using *InkSurvey* (*left*) or *Classroom Presenter* (*right*)

Since practice must be coupled with feedback that explicitly communicates about some aspect(s) of students' performance relative to specific target criteria, it provides information to help students progress in meeting those criteria, and is given at a time and frequency that allows it to be useful [2, 14]; thus, the instructor displayed selected responses (very good or poor ones as well as in the case of misconceptions) to the rest of the class on the classroom presentation screen to make students' thinking visible (for themselves, their peers, and the instructor) and give them chances to revise (and resubmit their work), as well as to provide opportunities for "what if" thinking, given variations on the challenging question or activity and for new problems that also involved the lesson's concepts (Fig. 8.2). Furthermore, in response to students' or teams' answers (and sometimes questions), the instructor and/or the students (individually or as teams) provided feedback, comments, and/or change teaching and learning accordingly [16, 19, 27–29]. Attempts to help students reflect on their own processes as learners (to be metacognitive) were also emphasized during real-time targeted feedback [27–29]. Course pace was not affected since redesigned courses already included many practice-feedback activities; moreover, Tablet PCs and associated software allowed inclusion of even more of these activities.

The main goal of tested formative assessment exercises with *InkSurvey* and *Classroom Presenter* was to (by means of these goal-directed practices coupled with real-time targeted feedback) enable students to acquire attitudes, skills, and knowledge enhancing their learning of food chemistry. Further, these attitudes, skills, and knowledge would be useful for their formal summative assessments, because finally, the questions asked in these exercises were similar to those that appeared in studied courses' quizzes and exams during the semester, which had a direct effect on their final grades [19, 27, 28].

8.3.5 Data Collection and Analysis

For this study, we analyzed the scores of summative quizzes applied during the semester that had an impact on students' grades and compared them with the scores (assessed only for this study and not part of the final grade of the course) of the formative assessment exercises that utilized *InkSurvey* or *Classroom Presenter* during the classes [19]. The VaNTH Observation System (VOS) was used to systematically assess HPL framework implementation in studied classrooms as previously described in detail by Gazca et al. [16]. Observations were made over a year, the first course observed during spring semester was Food Chemistry; the graduate course (Advanced Food Chemistry) was observed during the fall semester. The two observers achieved a 70% inter-rater reliability in using the VOS. Studied courses were observed during an entire semester in two weekly sessions, each 90 min long. This allowed us to obtain the following records: using CIO, NN, and GR a total of 138 observations for each of the three instruments due to a scheme of six observations per class. In the case of SEO we had 3174 observations [17]. Furthermore, beginning in the spring 2009 we applied two online surveys at the studied courses. The first survey (applied at the beginning of the semester) had the purpose of knowing students' expectations regarding the use of Tablet PCs. The second survey (applied at the end of the semester) was designed to understand the academic experience that the students had with the use of Tablet PCs and associated software [19].

In order to examine how students perceived the use of Tablet PCs and associated software for goal-directed practice coupled with real-time targeted feedback, we conducted semi-structured interviews (lasting on average 20 min each) with selected undergraduate (6 men and 15 women) and graduate (four men and eight women) students that had completed the studied courses. Students were selected from the pool of potential participants using maximum variation, choosing them to adequately represent studied courses' demographics [18]. Instructor reflections and insights are also included.

8.4 Results and Discussion

8.4.1 Quantitative Data

In Table 8.3 the means of grades (out of 10) of undergraduate and graduate students in the two studied courses' three summative quizzes for 2008 (redesigned courses but no Tablet PCs and associated software were utilized) as well as 2009–2015 classes (after Tablet PCs' implementation) can be observed. With the use of Tablet PCs and associated software (*InkSurvey* and *Classroom Presenter*) during several formative assessments (Tablet PC goal-directed practice coupled with real-time targeted feedback) prior to the summative quiz, students improved their learning,

Table 8.3 Mean grades (out of 10) of undergraduate or graduate students in summative quizzes for 2008 (before studied courses' redesign and Tablet PC implementation) as well as 2009–2015 classes (after their redesign and implementation of Tablet PC goal-directed practice coupled with real-time targeted feedback). Same teacher taught every one of the classes

	Year	Quiz-1	Quiz-2	Quiz-3
Undergraduate	2008	7.8	7.4	8.9
	2009	8.0	8.0	9.8
	2010	9.8	7.9	9.3
	2011	5.8	8.5	8.7
	2012	7.0	8.9	7.9
	2013	9.2	9.5	9.5
	2014	9.3	9.6	9.5
	2015	9.4	9.6	9.5
Graduate	2008	7.7	7.5	9.0
	2009	8.0	8.0	9.1
	2010	9.0	8.0	9.6
	2011	9.4	9.9	9.1
	2012	8.1	9.2	9.3
	2013	8.7	9.3	9.4
	2014	8.9	9.4	9.5
	2015	9.1	9.5	9.6

which is more than evident when comparing 2009–2015 grades in these summative assessments (quizzes) with respect to students' grades after course redesign but before Tablet PC implementation (2008).

Another important finding is that as the instructor of studied courses became more knowledgeable and confident with the use of goal-directed practice coupled with real-time targeted feedback pedagogical strategies, food chemistry student learning (undergraduate and graduate) was further enhanced as summative quizzes' grades for the latter years of implementation are greater than for initial years of implementation. Similar results have been reported for diverse subjects (physics, mathematics, chemical engineering, and biology), with students using diverse pen-enabled mobile devices (Tablet PCs, iPads, and Android 4.0 tablets/smartphones), indicating the effectiveness of *InkSurvey* and similar pedagogical strategies (focused on real-time formative assessments) in diverse educational environments (K-12, community college, publicly-funded engineering university, private university, and graduate school) [24, 25].

Furthermore, Fig. 8.3 exhibits the grades of graduate or undergraduate students in formative assessments (assessed only for this study and not part of the final grade of the course) using *InkSurvey* and those obtained at the corresponding summative quiz (exam result in the figure), which had a direct effect on students' final grade. In both cases, undergraduate and graduate courses, the formative assessment exercises (goal-directed practice coupled with real-time targeted feedback performed with the Tablet PCs and *InkSurvey*) had a positive and significant ($p < 0.05$) impact on the grades of the summative quizzes.

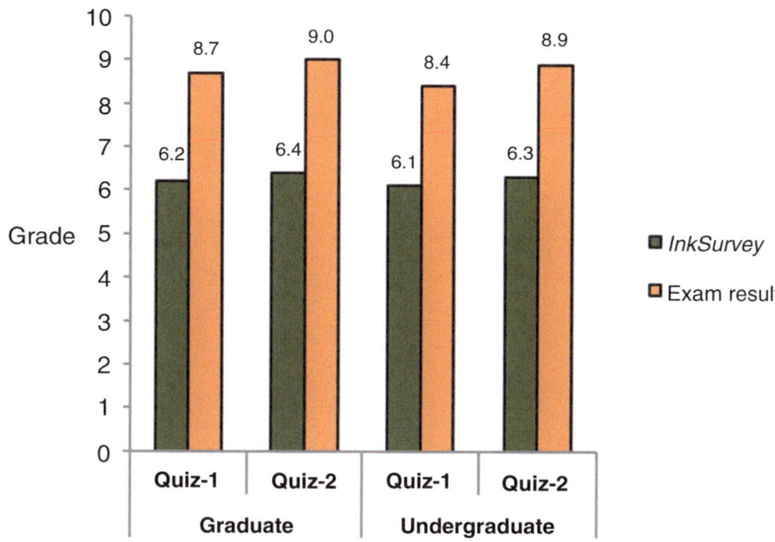

Fig. 8.3 Mean grades (out of 10) of graduate or undergraduate students in formative (by means of *InkSurvey*) or summative (exam result) assessment of the two first quizzes of studied courses studied courses after their redesign and implementation of Tablet PC goal-directed practice coupled with real-time targeted feedback (2009–2015)

Figure 8.4 exhibits the mean grades (out of 10) of studied courses' three summative quizzes taken by undergraduate or graduate students in the quizzes applied 4 years before course redesign and Tablet PC implementation (2005–2008) that were compared with results (after) obtained during the next 7 years (2009–2015) when the courses were redesigned and Tablet PC goal-directed practice coupled with real-time targeted feedback was frequently utilized. As can be seen, a conclusive and significant ($p < 0.05$) impact on the grades of the summative quizzes was confirmed.

8.4.2 *Qualitative Data*

VOS was used to systematically assess HPL framework implementation in studied courses' classrooms as well as in selected "HPL framework taught" and "traditionally taught" courses to assess the redesign of studied courses. Further details can be observed in Gazca et al. [16, 17] and Palou et al. [27, 28]. Observations demonstrated that the studied courses are aligned with the HPL framework and therefore course redesigns were successful in that regard.

By means of VOS' CIO, SEO, and GR components [12, 20] it was documented that Food Chemistry and Advanced Food Chemistry redesign significantly ($p < 0.05$) increased student participation [16, 19, 27, 28]. VOS' CIO, NN, and GR com-

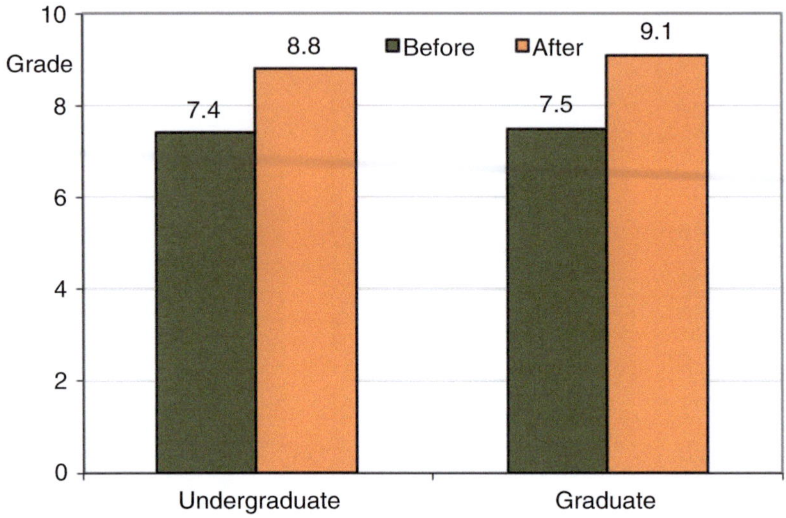

Fig. 8.4 Mean grades (out of 10) of three summative quizzes taken by undergraduate or graduate students before (2005–2008) and after (2009–2015) implementation of Tablet PC goal-directed practice coupled with real-time targeted feedback

ponents [12, 20] demonstrated that formative assessments and feedback were more common in the redesigned studied courses by means of Tablet PC goal-directed practice coupled with real-time targeted feedback [16, 19, 27, 28]. Instructors in the redesigned studied courses (as well as in the "HPL framework taught") utilized the information gained through real-time formative assessment to tailor instruction and meet student needs [12, 16, 19, 20, 27, 28]. VOS captured important differences between redesigned (and "HPL framework taught") and "traditionally taught" classroom experiences; which can be utilized to measure levels of "HPLness" of a lesson [10, 18].

Additionally, VOS noticeably captured differences among instructors' teaching styles and generated detailed feedback that they may draw on to self-assess [12, 16, 17, 19, 20, 27, 28]. Student final grades in redesigned studied courses (as well as in the "HPL framework taught") were higher than those found in "traditionally taught" courses. Further, fewer students failed (less than 5%) the Food Chemistry and Advanced Food Chemistry courses and the percentage of students who stayed in these courses (as in the "HPL framework taught" ones) until the end (close to 95%) was higher than for "traditionally taught" courses [16, 17, 28].

With regards to the online surveys, detailed descriptions are available in Gutiérrez Cuba et al. [19]. Less than 40% of students in each of the studied courses have had previous contact with a Tablet PC; further, many of the students did not use or were not allowed to use their laptops or mobile devices during class in many other courses. After using the Tablet PC and associated software in the redesigned courses, students were asked if they considered that a Tablet PC had advantages or disadvantages with respect to a regular laptop; they highlighted the following features: the possibility to take notes directly on the slides (*Classroom Presenter*), greater interaction between students and instructor (*InkSurvey* and *Classroom Presenter*), quickly make “sketches, drawings, and handwriting” (Tablet PC), enhancing their learning processes. At the end of the courses every surveyed student felt that the use of Tablet PCs and associated software in the redesigned courses improved their learning experiences and they would like to utilize a Tablet PC in other courses. Every semester, students made suggestions to expand the benefits of Tablet PC goal-directed practice coupled with real-time targeted feedback that were taken into account to enhance the following semester course, these include: more interaction among students (and teams), instructor could ask more questions for which we have to draw, illustrate, or sketch, include more activities for which we could state our points of view and/or opinions.

On the other hand, semi-structured interviews with students that had completed the studied courses indicated a number of themes that consistently appeared within the interview sessions and were addressed by undergraduate and graduate students from diverse perspectives. Five overall themes emerged: student participation in class by means of Tablet PCs, impact on learning, potential of Tablet PCs and associated software, formative assessments by means of goal-directed practice coupled with real-time targeted feedback, as well as advantages and disadvantages of using the Tablet PCs in Food Chemistry and Advanced Food Chemistry classrooms. Gutiérrez Cuba et al. [18] reported upon these themes for the graduate studied course and explicitly tied the HPL framework with the results of graduate student interviews by means of the structure developed by Carney [10] who identified goals and sub-goals of each lens of the HPL framework as well as classroom practices and pedagogical strategies associated with every lens and how the redesigned learning environments operate at the intersection of the four lenses. With regards to the Food Chemistry course, our findings of the semi-structured interviews demonstrated that undergraduate students think (as was the case for graduate students [18]) that Tablet PC goal-directed practice coupled with real-time targeted feedback: uplifted their motivation to participate in class as well as their scores in summative assessments (graded work-products); made the classroom climate more active and them constantly thinking, thus their learning increased; promoted a great deal of real-time feedback by the instructor and their peers making their thinking visible; allowed many chances to revise their thinking and thus their work-products. Among the disadvantages, undergraduate students think that the instructor should be advised the many chances they have of checking their e-mail and social networks while using the Tablet PCs.

Instructor reflections made clear that: (i) Because of the anonymity afforded by *InkSurvey* and *Classroom Presenter*, students felt very comfortable sharing their thinking. This enabled him to frequently assess student understanding during instruction, problem solving, and peer interactions to identify in real-time students' most common difficulties, provide immediate and targeted feedback, redirect classroom actions, and/or refine instruction as previously reported by several authors in diverse educational environments and utilizing selected software [18, 19, 23–25, 27, 28, 31]; (ii) Tablet PC goal-directed practice coupled with real-time targeted feedback generated many possibilities for self- and peer-assessment, enabling students to anonymously analyze their own and classmates thinking; (iii) Another positive result in studied courses was a visible increase in student motivation to participate in class discussions and problem-solving activities mediated through *InkSurvey* and *Classroom Presenter*; (iv) Further, the redesigned Food Chemistry and Advanced Food Chemistry courses enhanced student understanding of the engineering design approach to problem solving as well as undergraduate and graduate students' skills to solve practical problems and complete real world food chemistry projects as assessed in several other student work products (not part of this manuscript) such as problem-based learning projects, homework, assignments, exams, and journals [18, 19]; (v) Students' initial conceptions provided the foundation on which more formal understanding of the subject matter was built. Frequent formative assessment helped make students' thinking visible to themselves, their peers, and the instructor. Facilitated by Tablet PCs and associated software, feedback (in both courses) that guided modification/refinement in student and instructor thinking increased.

Additionally, some other important impacts were evident, particularly on the instructor itself: (i) He was able to identify the most common difficulties and misconceptions that undergraduate and graduate students have in food chemistry courses; (ii) Became proficient in providing real-time and targeted feedback to students; (iii) Purposeful helped students reflect on their own processes as learners; and (iv) Understood how through Tablet PC goal-directed practice coupled with real-time targeted feedback, student thinking can be revealed. Therefore the student learning experience in the studied classrooms was enhanced causing improvements in both instruction and student academic success as previously reported in various educational settings [16–19, 23–25, 27, 28, 30] where readers can find further practical advice for others considering using a similar set up with tablets or other mobile devices (android or iOS based), how approaches changed over the course of application, and lessons learned.

8.5 Conclusion

Quantitative and qualitative data, summarized in this paper, supports the effectiveness of goal-directed practice coupled with real-time targeted feedback to enhance food chemistry learning. This includes evidence of learning gains and positive student attitudes, as well as instructor insights from undergraduate and graduate educational environments while using pen and touch enabled mobile devices (Tablet PCs). It is clear that InkSurvey and Classroom Presenter are efficient and robust software for implementing application of goal-directed practice coupled with real-time targeted feedback in both undergraduate and graduate classrooms by means of Tablet PCs in order to gauge student learning during goal-directed practice in real-time, provide instantaneous targeted feedback, and make immediate pedagogical adjustments as needed.

Acknowledgements We acknowledge financial support from HEWLETT-PACKARD through the HP Technology for Teaching Higher Education Grant Initiative for Latin America for the project "High-Quality Learning Environments for Engineering Design: Using Tablet PCs and Guidelines from Research on *How People Learn*" as well as within the HP Catalyst Grant Initiative for the project "Critical Support Systems to Enhance the Development of Twenty-First Century Expertise in Engineering Students: Using Tablet PCs and Associated Technologies, the *Framework for 21st Century Learning*, and Guidelines from Research on *How People Learn*". Author Gutiérrez Cuba gratefully acknowledges financial support for her PhD studies from Universidad de las Américas Puebla and the National Council for Science and Technology of Mexico (CONACYT) as well as a postdoctoral fellowship from CONACYT.

References

1. Altamirano, E., Lopez-Malo, A., Palou, E.: Direct and indirect assessment of Universidad de las Américas Puebla's food engineering program outcomes. In: Proceedings of the American Society for Engineering Education Annual Conference and Exposition (2013)
2. Ambrose, S.A., Bridges, M.W., DiPietro, M., Lovett, M.C., Norman, M.K.: How Learning Works: Seven Research-Based Principles for Smart Teaching. Wiley, New York (2010)
3. Anderson, R., McDowell, L., Simon, B.: Use of classroom presenter in engineering courses. In: Proceedings 35th Annual Conference on Frontiers in Education, FIE'05, pp. T2G–13. IEEE (2005)
4. Angelo, T., Cross, K.: A Handbook of Classroom Assessment Techniques for College Teachers. Jossey-Bass Publishers, San Francisco (1993)
5. Black, P.J., Wiliam, D.: Inside the black box: raising standards through classroom assessment. *Phi Delta Kappan* **80**(2), 139–148 (1998)
6. Black, P., Wiliam, D.: Developing the theory of formative assessment. Educ. Assess. Eval. Account. (Formerly: J. Pers. Eval. Educ.) **21**(1), 5–31 (2009)
7. Bransford, J.D., Brown, A.L., Cocking, R.R., et al.: How People Learn. National Academy Press, Washington DC (2000)
8. Bransford, J., Vye, N., Bateman, H.: Creating high-quality learning environments: guidelines from research on how people learn. In: The Knowledge Economy and Postsecondary Education: Report of Workshop, pp. 159–198 (2002)
9. Brent, R., Felder, R.: How learning works. Chem. Eng. Educ. **45**(4), 257–258 (2011)
10. Carney, K.: Toward a Definition of HPL-Ness. Center for the Study of Learning, Instruction, and Teacher Development. University of Illinois, Chicago (2015)
11. Chávez-Torrejón, G., Husted, S., Ramirez-Corona, N., Lopez-Malo, A., Palou, E.: Fostering the development of critical thinking in an introduction to chemical process engineering design course. In: Proceedings of the American Society for Engineering Education Annual Conference and Exposition (2014)
12. Cox, M.F., Cordray, D.S.: Assessing pedagogy in bioengineering classrooms: quantifying elements of the "how people learn" model using the VaNTH observation system (VOS). J. Eng. Educ. **97**(4), 413 (2008)
13. Earl, L.M., Katz, M.S.: Rethinking Classroom Assessment with Purpose in Mind: Assessment for Learning, Assessment as Learning, Assessment of Learning. Manitoba Education, Citizenship & Youth, Winnipeg (2006)

14. Eberly Center for Teaching Excellence: Theory and Research-Based Principles of Learning. Carnegie Mellon University, Pittsburgh (2015)
15. Felder, R.M., Woods, D.R., Stice, J.E., Rugarcia, A.: The future of engineering education II. Teaching methods that work. Chem. Eng. Educ. **34**(1), 26–39 (2000)
16. Gazca, L., Palou, E., Lopez-Malo, A., Garibay, J.M.: Capturing differences of engineering design learning environments by means of VaNTH observation system. In: Proceedings of the American Society for Engineering Education Annual Conference and Exposition (2009)
17. Gazca, L., López-Malo, A., Palou, E.: Analysis of the implementation of the how people learn framework through direct classroom observation in selected food engineering courses. In: Proceedings of the American Society for Engineering Education Annual Conference and Exposition (2011)
18. Gutiérrez Cuba, J., López-Malo, Palou, E.: Graduate student perspectives on using Tablet PCs and associated technologies. In: Proceedings of the American Society for Engineering Education Annual Conference and Exposition (2012)
19. Gutiérrez Cuba, J., López-Malo, A., Palou, E.: Using tablet PCs and associated technologies to reveal undergraduate and graduate student thinking. In: Proceedings of the American Society for Engineering Education Annual Conference and Exposition (2011)
20. Harris, A.H., Cox, M.F.: Developing an observation system to capture instructional differences in engineering classrooms. J. Eng. Educ. **92**(4), 329–336 (2003)
21. Husted, S., Gutiérrez Cuba, J., Ramirez-Corona, N., López-Malo, A., Palou, E.: Multidimensional assessment of creativity in an introduction to engineering design course. In: Proceedings of the American Society for Engineering Education Annual Conference and Exposition (2014)
22. Jonassen, D.H.: Learning to Solve Problems: A Handbook for Designing Problem-Solving Learning Environments. Routledge, New York (2010)
23. Kowalski, F., Kowalski, S., Hoover, E.: Using inksurvey: a free web-based tool for open-ended questioning to promote active learning and real-time formative assessment of tablet PC-equipped engineering students. In: Proceedings of the American Society for Engineering Education Annual Conference and Exposition (2007)
24. Kowalski, F.V., Kowalski, S.E., Colling, T.J., Cuba, J.G., Gardner, T.Q., Greivel, G., Palou, E., Ruskell, T.: Using inksurvey with pen-enabled mobile devices for real-time formative assessment: I applications in diverse educational environments. In: Adler, A., Hammond, T., Payton, M., Valentine, S. (eds.) The Impact of Pen and Touch Technology on Education, pp. 297–305. Springer, Cham (2015)
25. Kowalski, F.V., Kowalski, S.E., Colling, T.J., Cuba, J.G., Gardner, T.Q., Greivel, G., Palou, E., Ruskell, T.: Using inksurvey with pen-enabled mobile devices for real-time formative assessment II. Indications of effectiveness in diverse educational environments. In: Adler, A., Hammond, T., Payton, M., Valentine, S. (eds.) The Impact of Pen and Touch Technology on Education, pp. 307–314. Springer, Cham (2015)
26. Leahy, S., Lyon, C., Thompson, M., Wiliam, D.: Classroom assessment: minute by minute, day by day. Asses. Promot. Learn. **63**(3), 19–24 (2005)
27. Palou, E., Gazca, L., García, J.A.D., Lobato, J.A.R., Ojeda, L.G.G., Arnal, J.F.T., Munguía, M.T.J., López-Malo, A., Garibay, J.M.: High-quality learning environments for engineering design: using tablet PCs and guidelines from research on how people learn. Int. J. Food Stud. **1**(1), 1–16 (2012)
28. Palou, E., Husted, S., Chávez-Torrejón, G., Apud, Z.R., Gazca, L., Cuba, J.V.G., Ramírez-Corona, N., López-Malo, A.: Critical support systems to enhance the development and assessment of 21st century expertise in engineering students. In: Ge, X., Ifenthaler, D., Michael Spector, J. (eds.) Emerging Technologies for STEAM Education, pp. 217–243. Springer, Cham (2015)
29. Ramirez-Corona, N., Zaira, R., Lopez-Malo, A., Palou, E.: Assessing metacognitive awareness during problem solving in a kinetics and homogeneous reactor design course. In: Proceedings of the American Society for Engineering Education Annual Conference and Exposition, pp. 1–14 (2013)

30. Simon, B., Anderson, R., Hoyer, C., Su, J.: Preliminary experiences with a tablet PC based system to support active learning in computer science courses. ACM SIGCSE Bull. **36**(3), 213–217 (2004)
31. Tront, J.G., Eligeti, V., Prey, J.: Classroom presentations using tablet PCs and writeon. In: 36th Annual Frontiers in Education Conference, pp. 1–5. IEEE (2006)
32. Wiliam, D.: What is assessment for learning? Stud. Educ. Eval. **37**(1), 3–14 (2011)

Chapter 9
Score Improvement Distribution When Using Sketch Recognition Software (Mechanix) as a Tutor: Assessment of a High School Classroom Pilot

Randy Brooks, Jung In Koh, Seth Polsley, and Tracy Hammond

Abstract Effectively communicating engineering concepts in high school classrooms is an important goal of many curricula in secondary education since the material can help prepare students for future study in Science, Technology, Engineering, and Math (STEM) fields. However, the material can be challenging. This work describes the deployment of Mechanix to a high school environment with the intent of improving the effectiveness of student practice regarding free body diagrams, vector analysis, and truss problem solving. Mechanix is a globally-available, Internet-connected digital tutor that provides immediate, constructive feedback to the learner while also providing student-level metrics to the instructor. It was made available to students in a STEM-infused classroom as part of Project Lead the Way (PLTW). The focus of this study is to evaluate the progress realized by differing academic levels of students. We used pre- and post-testing to assess progress, which showed an average increase of 1.65 points on a 12 point scale ($p < 0.005$). A greater increase was found in the 'A-level' high school students while historical, college-level studies suggest significant progress may be realized at all levels versus current tutoring techniques as the students continue to utilize Mechanix.

R. Brooks (✉)
Lovejoy Independent School District, Lucas, TX, 75002, USA
e-mail: randy_brooks@lovejoyisd.net

J.I. Koh • S. Polsley • T. Hammond
Sketch Recognition Lab, Department of Computer Science & Engineering, Texas A&M University, College Station, TX, 77843, USA
e-mail: jungin@tamu.edu; spolsley@tamu.edu; hammond@tamu.edu

T. Hammond et al. (eds.), *Frontiers in Pen and Touch*, Human–Computer Interaction Series, DOI 10.1007/978-3-319-64239-0_9

9.1 Introduction

Secondary education is increasingly adopting STEM-infused curricula to encourage students to pursue advanced degrees in STEM fields. This material is also intended to help prepare students for future study, so educators may wish to include as much advanced engineering material as possible. However, they face the same difficulties always posed by teaching such advanced knowledge. In particular, a challenge for any course is providing learners with timely and effective feedback on out-of-classroom practice. A recent instructional enhancement promulgating through math and science classrooms attempting to address this is described as the "flipped classroom." Bergmann and Sams [4] coined this term to describe a course delivery redesign whereby students learn from video lectures outside of class as 'homework' and then spend the time in class working through practice and extension activities with instructors providing immediate feedback.

While flipped classrooms are proving effective, another movement is underway. In Disrupting Class [6], the authors project that by 2019, 50% of all high school courses will be delivered online. Building on that vision, a 2015 presentation at the Texas Computer Educator Association (TCEA) Convention[1] described a path for transitioning from flipped classrooms to online classrooms with recreating the rigorous and engaging classroom environment at the center.

Though flipping the classroom does provide more learner interaction with peers and the instructor regarding conceptual questions and practice, the impending movement to online courses requires the discovery of a way to replicate this feedback loop in the digital space. As options are investigated, the digital world venue requires revisiting instructional design for a method that is most effective in this new environment. Tripp and Bichelmeyer [12] provide a basis for employing Intelligent Tutoring Systems (ITS) as a key component of the Instructional Design Strategy (IDS) Rapid Prototyping. Their analysis denotes this design method supporting the production of the TAMU SRL Mechanix software as that of an early adapter in the quality online tool movement by building tools for a new space. They describe the heart of the model being the parallel processes of design and research, or construction and utilization which best address the greater complexity of a human factors-intensive field such as the process of instruction.

Bishop and Verleger [5] studied this design after a few years of expanding deployment and found video lectures to be performing better than in class lectures, while interactive videos were setting a new, higher standard. They found that adoption of ITS in the classroom was consistently as effective as human instructors. VanLehn [17] provides supporting data regarding digital ITS based on a 4-year study covering the benefits to be gained by moving paper-based homework to a technology-based system. He found considerable success (11% improvement versus control group) in a physics course which he attributed to the use of an

---

[1]Presented by Randy Brooks in TCEA 2015 http://www.slideshare.net/RandyBrooks3/tcea-the-next-giant-leapfrom-flipped-to-online.

ITS. VanLehn's study reasoned that because homework practice creates significant student learning, the benefits of a quality ITS are substantial.

Engaging, rigorous, and interactive tutorials are the crux of quality online courses. Though improvement is realized through well-constructed digital lectures versus their in-class counterparts, studies centered around quality practice supports are providing opportunities for significant learning. Mechanix is an online tutorial tool providing immediate, constructive feedback to learners as they practice while also providing individual learner metrics to the instructor [11]. Mechanix provides the model of the tool that Tripp and Bichelmeyer describe both in construction and in learner use, while differentiating itself with a guiding design focus on the human element. The foundation for Mechanix is based in LADDER (Language for Describing Drawing, Display, and Editing in Recognition) [9], a generic sketching programming language, with a goal to build a sketch recognition system that allows sketchers to draw naturally without having to learn a new set of stylized symbols. Sketching the free body and vector diagrams is a key layout element, but each student may develop a custom path to create the final design. The goal from the instructor perspective is that the student produces a sketch matching norm that supports student learning of the target concepts. Previous studies have evaluated the deployment of the Mechanix in first year college engineering courses with promising results. The purpose of this study is the evaluation of a Mechanix deployment in a high school engineering preparation course with a focus on the improvement differences realized among students grouped by previous year mathematics course grades.

9.2 Prior Work

While they have recently started to acquire a great deal of attention, Intelligent Tutoring Systems have been in development for some time. From a certain perspective, most educational software can be viewed as an ITS, but the increasing popularity of online supplements and course materials has made ITS a more available option in today's classroom. For instance, Pearson Education offers a curriculum for teaching STEM to primary and secondary level students, called Project STEM, but this curriculum is specifically designed for K-8th grade age groups and may not provide the flexibility a high school teacher would want in a STEM-infused classroom.[2] Similarly, Pearson Education also offers Mastering Physics at the post-secondary level alongside McGraw Hill's Connect software, but this material may be too advanced and rigid for adoption in high school curriculum. The U.S. Military Academy West Point has designed a tool called West Point Bridge Designer, an interactive system for building and simulating bridge designs. This tool is used by the Engineering Encounters Bridge Design Contest (EEBDC) to help

[2]http://www.pearsonschool.com.

encourage students in middle school and high school to engage in an applied STEM exercise.[3] Unfortunately, this software, while being very powerful, is specifically limited to bridge building and entails a fairly substantial learning curve.

To develop a STEM interest in younger students, we see a need for ITS capable of illustrating concepts in ways similar to some of the existing tools but providing them in a natural, easy-to-use interface. For this reason, we look to sketch-based interaction as a powerful yet simple means of furthering ITS adoption in the classroom, especially for high school students. There are a number of educational tools built around sketching as the primary interaction modality. A couple of tools similar to Mechanix, the primary tool used in this work, are WinTruss and Sketch Worksheets. Mechanix has already been thoroughly tested against several existing tools, including WinTruss, and has been found to provide a more scaffolded development process in addition to greater flexibility on the part of the instructor [3]. Sketch Worksheets is a more general system that allows the instructor to develop almost any sketch-based problems they wish [18]. While this retains maximum control over the curriculum for the instructor, it again introduces more features at some cost of an increased learning curve and reduces some of the domain-specific understanding that is included in Mechanix. For instance, Mechanix understands that an arrow is always a force and provides students with live feedback in the form of color-coding to help direct them to the solution; this type of understanding makes Mechanix very easy to work with when dealing with truss and free body diagram problems [11].

Another system which bears mention is Newton's Pen, developed by Lee et al. [10]. Newton's Pen is specifically designed for learning how to solve free body diagram problems through sketch-based interaction. This is very similar to the goal of Mechanix, but they differ in several regards. Firstly, Newton's Pen focuses almost entirely on free body diagrams, considering statics problems from their most abstract representations. While this is a useful approach, Mechanix considers the design of general closed body shapes and trusses as more easily-visualized problems with clear applications [1, 2, 7, 13]. Also, while Newton's Pen uses audio feedback to direct students, Mechanix relies on a combination of color-coding and message-based feedback to direct students live and on-demand. The digitizer used by Newton's Pen is also very specific, while Mechanix is intended to be used on any computer that can run Java.

Mechanix was selected for use in this work because it includes the desired flexibility on the instructor side for curriculum design without being too general or difficult to understand. Additionally, on the student side, it uses the domain-specific knowledge paired with sketch recognition to make a natural and intuitive interface. It is an ITS designed to help teach free body diagram and truss problem solving skills, primarily for post-secondary engineering students. Mechanix has a well-established history in the classroom and continuous development in terms of design, feature set, and algorithmic capability [1, 11, 14–16]. Many of these prior works on Mechanix

[3]http://www.bridgecontest.org.

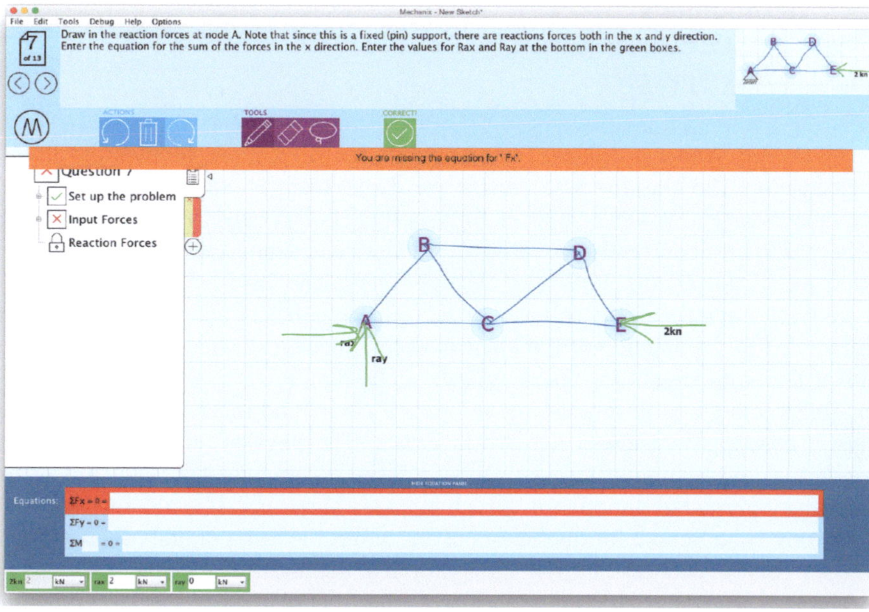

Fig. 9.1 Screenshot of feedback from Mechanix regarding a student's truss submission

describe how the tool compares student sketches to instructor models and begins a feedback loop with the student guiding them to attain objectives through recurring student modification of their own design. Figure 9.1 shows how Mechanix provides feedback to students regarding their truss analysis submissions.

The deployment of Mechanix in a portion of the engineering classes at LeTourneau University provides some basis for study comparison and a source for prediction [8]. From a scoring perspective, the 73 LeTourneau students involved in the study performed equally well whether in the control (textbook software) or experimental (Mechanix) group. Yet the key items from the LeTourneau study are found in the learner evaluation follow-up analysis where student observations suggest that Mechanix's main benefits were early feedback, promoting visualization, and teaching a good problem solving process. Mechanix leads the students, as would an in-person instructor, to adhere to the process in order to be successful.

A furthered analysis is presented in the AAAI article by Valentine et al. [14] describing their vision as advancing artificial intelligence such that the automated instruction emulates the expert performance achieved by human instructors. In this study they describe many unique features deliberately constructed to interpret typical human action. For example, rather than requiring use of an eraser selection, the user may simply 'mark out' an entry and Mechanix understands that this is user input for erase. Pursuit of this level of human action inferential detail is clearly a core design principle of Mechanix.

9.3 Methodology

9.3.1 *Class Specifics*

Lovejoy High School in Lucas, Texas is an all Pre-AP College Preparatory Public High School. The classroom venue for the deployment of Mechanix is a Project Lead the Way (PLTW) Principles of Engineering (PoE) course. PLTW is a STEM-focused non-profit currently providing curriculum and support to instructors in over 8,000 American schools. PoE is an elective for students investigating their interest in studying engineering in college. In fact, many colleges are now offering first year credits for completion of PoE. The curriculum covers a wide array of activities with a focus on ingraining the Engineering Design Process in the student skill set, constructing and programming automated compound machines, and applying physics concepts to address real world issues. The physical classroom provides significant construction space as well as a desktop computer for each student for accessing a myriad of digital tools which includes Mechanix. It is a normal practice at Lovejoy High School for instructors of any subject to bring digital tools, whether time-tested or in beta-test, into their classroom as support for students.

The PoE student population is a mix of high school class levels with previous Math and Science knowledge ranging from concurrent Geometry and Biology to previous completion of AP Calculus and AP Physics courses. The class target is sophomores, yet it is open to advanced freshmen and interested juniors and seniors. Consequently, there is significant differentiation applied to all lessons. The current grade-level distribution is: Freshman—2, Sophomores—19, Juniors—6, Seniors—9. Of the 36 students in the class, 31 students chose to participate in the study. Participating student data was subsequently grouped according to their performance level in their previous semester mathematics course. Nineteen students earned an A, 8 students earned a B, and 4 students earned a C. Proficiency in mathematics is one of the determinants for suggesting that students consider this engineering course.

9.3.2 *Deployment*

The platform for vector analysis practice was student choice. Traditional paper-based practice forms were provided as well as Mechanix logins for tool access. The same practice problems existed in each option. Mechanix was positioned as simply another digital tool that the instructors of Lovejoy High School have discovered and were offering for student support. There was no grade taken in regard to Mechanix use. As students interested in engineering have a propensity to be curious about new offerings, all students used Mechanix for their practice, including those that chose not to participate in the study.

The Mechanix tutorial was constructed with progressively reduced scaffolding across eight practice problems. The first practice action was designed to familiarize the students with Mechanix by providing solutions along with direction about how to sketch and document the solutions. In an effort to get the student knowledge levels more in-line before having the students work in Mechanix, the Mechanix deployment began with a day of lecture and practice regarding free body diagrams and vector analysis.

As expected, the combination of a summer break and limited or no learner prior exposure to vector analysis of the majority of the students generated low scores on the pre-assessments (quizzes) used for this study. Armed with this data, the instructor then developed interventions targeting the identified weaknesses of the specific student population involved. New practice problems in Mechanix would be one of the interventions implemented going forward.

The Mechanix tool proved a stable platform throughout the study. Students encountered issues related to the Java version on the high school computers driving some unplanned program closure, yet a solution was quickly identified and applied. Figure 9.2 is a screenshot of the instructor's template for Practice Problem #4. This is provided as a sample of the visual element of Mechanix. Note that students must sketch the free body diagram depicted and then determine the x and y components of each vector in order to determine a force level required to keep the design static.

Table 9.1 depicts the timeline for vector analysis instruction to include the Mechanix tool as an available support in the LHS PLTW PoE classes. The Mechanix impact evaluation involves comparing assessment (quiz) scores before using Mechanix for a week of practice to the scores immediately following that week of the use of Mechanix.

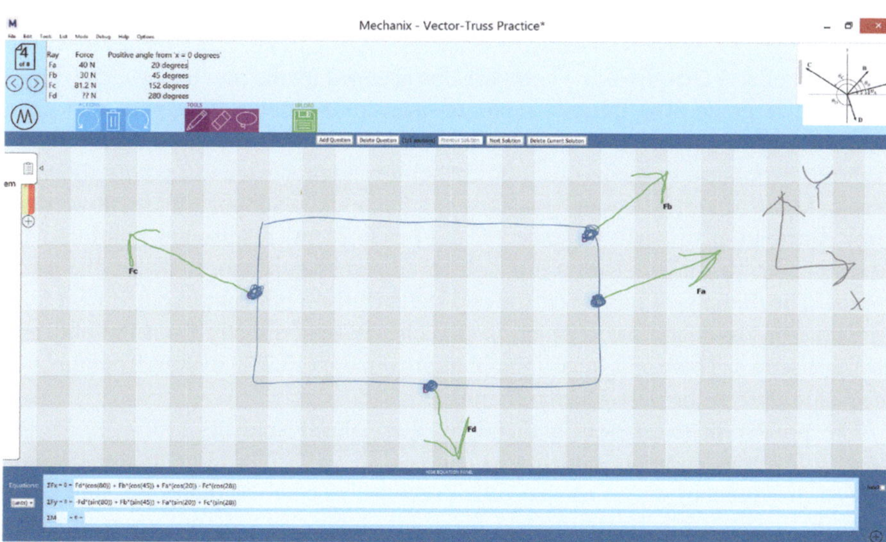

Fig. 9.2 An image of the Mechanix software showing one of the assigned vector analysis problems

Table 9.1 Timeline of the Mechanix deployment plan

Date	Activity
Monday, August 24, 2015	School begins
Wednesday, September 02, 2015	Forty-five minute lecture regarding construction of free body diagrams and analyzing vectors
	Four paper-based practice problems were assigned to be completed before the pre-Mechanix quiz on the following Wednesday
	PowerPoint presentations and the key to the four practice problems were posted digitally for student reference
Wednesday, September 09, 2015	Students complete four question paper-based pre-Mechanix vector analysis quiz during class
Thursday, September 10, 2015	Short Mechanix demonstration for the class
	Students were assigned eight vector analysis practice problems with student choice to use paper-based, Mechanix (an ID/PW was provided to each student), or a combination of the two
	No score was given regarding practice completion
	Students are given some class time during the following week to work on the practice problems
Friday, September 18, 2015	Students complete four question paper-based post-Mechanix vector analysis quiz during class

9.4 Results and Discussion

Table 9.2 reflects the quiz score comparisons grouped by the previous year's overall mathematics grade. The quizzes each consisted of 4 similar challenges and were scored out of 12 points with 4 points related to accurately completing the free body diagrams, 6 points allocated to accuracy of set-up and process, and 2 points allocated to calculation accuracy. Though, as expected due to the wide incoming knowledge-level distribution, a high variance (11.7+) is present, results are solidified by group-paired, two-tailed T-tests that compared across-quiz scores within the same groups. The findings of this T-test showed statistically significant improvement with a p value of less than 0.005, which is also clearly seen directly from the pre- and post-test scores. There was measureable improvement (13.7%) overall from the pre-Mechanix quiz to the post-Mechanix quiz.

The greater progress was realized by the A-level students (17.3%). As the engineering course targets students which have shown acuity in math courses, this A-level group, many of which had little previous exposure to the vector language, spent the week in Mechanix making the connections that they are accustomed to making in their math courses. Consequently, many significantly increased their understanding of vector analysis. Though training and scaffolding was built into the Mechanix practice problems, more direction and instruction is required to better support the B-level and C-level student populations. Though quite intuitive software,

Table 9.2 Summary of student mean score improvement data

Analysis category	Number of students	PreMechanix, quiz score	PostMechanix, quiz score	quiz score improvement	Percent improvement on 12 total point quiz (%)	(p < t) of two-tailed, paired two sample t-test
Full study	31	5.47	7.11	1.65	13.7	0.005
'A' students	19	4.93	7.01	2.08	17.3	0.020
'B' students	8	7.84	8.84	1.00	8.3	0.100
'C' students	4	3.25	4.13	0.88	7.3	0.470

there is a learning curve associated with adapting Mechanix for tutoring support which needs to be addressed by the instructor when building a course timeline.

It is worth noting that the B-level students ultimately outperformed the A-level students in this study, a result which is perhaps surprising. This is due to the distribution of students in the STEM-infused class considered in this work. Many of the B-level students also included higher-level students, e.g. juniors and seniors, while A-level students were predominantly freshmen and sophomores. The juniors and seniors would have slightly more background in pre-calculus and physics than their younger peers, which provided enough prior knowledge that they could score higher in their initial assessments. Regardless of the specific backgrounds of the students, we hope that the results clearly show the gains made by all students in terms of their pre- and post-assessment after using Mechanix.

Increased adherence to the problem-solving process was noted in Green's reporting of student observations [8] that Mechanix, by deliberate design, emphasizes the benefits of learning to follow a problem-solving process. Mechanix is expected to flourish in an environment where pen and touch technology is widespread, yet this population, with the exception of two outliers, was content to work exclusively with their standard computer mouse. The vector practice in Mechanix was followed by participation in a Georgia Tech MOOC on Statics and an online bridge building contest with a focus on vector application in truss design. Following these exercises, the students will utilize Mechanix for truss analysis practice where, similar to the previously identified findings [14], they are expected to show significant progress upon subsequent uses of Mechanix when compared with the impact at initial deployment.

An important point of discussion is that Mechanix was adopted by all students in the classroom for this study. With the absence of a control group of students, we were less interested in observing the impacts of Mechanix versus traditional solutions in the classroom, a point which has already been analyzed and discussed thoroughly through previous works, and more concerned with the viability of adopting Mechanix, an ITS designed for college-level students, in the high school classroom. Students showed significant improvement, and based on these findings, Mechanix is sufficiently intuitive and open to be a useful tool to more populations than just the original target group. Furthermore, even though in this study, the students used traditional mouse-based inputs, ITS tools like Mechanix offer additional benefits through their natural sketch-based interaction since a growing number of students own or have available touch and stylus enabled interfaces. We look forward to potential further adoption of Mechanix in a high school environment based on the results of this study.

9.5 Conclusion

With the increasing digitization of the classroom and the push for more STEM-inclusive high school curricula, teaching tools like ITS are becoming powerful options for instructors and students alike. Mechanix is an ideal tutoring and support tool for project-based learning courses such as the high school PLTW PoE course evaluated in this study because it can help students learn the process of problem-solving, especially in the STEM domain of truss analysis, by leading them to the solution through its personalized feedback. Once students embrace Mechanix, it allows for impactful, student-centered learning at any time with immediate constructive feedback. Mechanix drives the student to complete key analysis steps, which may often be skipped when using non-sketch-based practice methods, to help reinforce concepts during the learning stage. From the instructor perspective, Mechanix provides a great deal of flexibility in designing solutions for the students to pursue, as well as access to a growing array of student metrics, so that problems may be tailored to a specific curriculum. Mechanix has been widely-studied in post-secondary engineering classrooms, and with the growing number of applications to new domains outside college classrooms, Mechanix shows a lot of potential for empowering high school students to better master STEM concepts.

Acknowledgements The authors would like to thank the Lovejoy Independent School District, Texas A&M University, and the Sketch Recognition Lab for their unceasing support and assistance throughout this study. This work was supported in part by an NSF RET Supplement on project NSF EEC 1129525.

References

1. Atilola, O., Field, M., McTigue, E., Hammond, T., Linsey, J.: Mechanix: a sketch recognition truss tutoring system. In: ASME 2011 International Design Engineering Technical Conferences and Computers and Information in Engineering Conference, pp. 645–654. American Society of Mechanical Engineers (2011)
2. Atilola, O., Vides, F., Mctigue, E.M., Linsey, J.S., Hammond, T.A.: Automatic identification of student misconceptions and errors for truss analysis. In: American Society for Engineering Education. American Society for Engineering Education (2012)

3. Atilola, O., McTigue, E.M., Hammond, T., Linsey, J.: Mechanix: evaluating the effectiveness of a sketch recognition truss tutoring program against other truss programs. In: 120th American Society for Engineering Education Annual Conference & Exposition (ASEE) (2013)
4. Bergmann, J., Sams, A.: Flip Your Classroom: Reach Every Student in Every Class Every Day. International Society for Technology in Education, Eugene/Arlington (2012)
5. Bishop, J.L., Verleger, M.A.: The flipped classroom: a survey of the research. In: ASEE National Conference Proceedings, Atlanta (2013)
6. Christensen, C.M., Horn, M.B., Johnson, C.W.: Disrupting Class: How Disruptive Innovation Will Change the Way the World Learns, vol. 98. McGraw-Hill, New York (2008)
7. Field, M., Valentine, S., Linsey, J., Hammond, T.: Sketch recognition algorithms for comparing complex and unpredictable shapes. In: IJCAI, vol. 11, pp. 2436–2441 (2011)
8. Green, M.G., Caldwell, B.W., Helms, M., Linsey, J.S., Hammond, T.A.: Using natural sketch recognition software to provide instant feedback on statics homework (truss free body diagrams): assessment of a classroom pilot. In: 2015 ASEE Annual Conference and Exposition. ASEE Conferences, Seattle (2015). 10.18260/p.25007, https://peer.asee.org/25007
9. Hammond, T.A.: Ladder: a perceptually-based language to simplify sketch recognition user interface development. Ph.D. thesis, Massachusetts Institute of Technology (2007)
10. Lee, W., de Silva, R., Peterson, E.J., Calfee, R.C., Stahovich, T.F.: Newton's pen: a pen-based tutoring system for statics. Comput. Graph. **32**(5), 511–524 (2008)
11. Nelligan, T., Polsley, S., Ray, J., Helms, M., Linsey, J., Hammond, T.: Mechanix: a sketch-based educational interface. In: Proceedings of the 20th International Conference on Intelligent User Interfaces Companion, pp. 53–56. ACM (2015)
12. Tripp, S.D., Bichelmeyer, B.: Rapid prototyping: an alternative instructional design strategy. Educ. Technol. Res. Dev. **38**(1), 31–44 (1990)
13. Valentine, S., Field, M., Smith, A., Hammond, T.: A shape comparison technique for use in sketch-based tutoring systems. In: Proceedings of the 2011 Intelligent User Interfaces Workshop on Sketch Recognition, IUI 2011, Palo Alto, vol. 11, no. 5 (2011)
14. Valentine, S., Vides, F., Lucchese, G., Turner, D., Kim, H.H., Li, W., Linsey, J., Hammond, T.: Mechanix: a sketch-based tutoring and grading system for free-body diagrams. AI Mag. **34**(1), 55 (2012)
15. Valentine, S., Vides, F., Lucchese, G., Turner, D., Kim, H.H., Li, W., Linsey, J., Hammond, T.: Mechanix: a sketch-based tutoring system for statics courses. In: IAAI (2012)
16. Valentine, S., Lara-Garduno, R., Linsey, J., Hammond, T.: Mechanix: a sketch based tutoring system that automatically corrects hand-sketched statics homework. In: Hammond, T., Valentine, S., Adler, A., Payton, M. (eds.) The Impact of Pen and Touch Technology on Education. Springer, New York (2015)
17. VanLehn, K.: The relative effectiveness of human tutoring, intelligent tutoring systems, and other tutoring systems. Educ. Psychol. **46**(4), 197–221 (2011)
18. Yin, P., Forbus, K.D., Usher, J.M., Sageman, B., Jee, B.D.: Sketch worksheets: a sketch-based educational software system. In: IAAI (2010)

Chapter 10
DCSR: A Digital Circuit Sketch Recognition System for Education

Shuo Ma, Yongbin Sun, Pengchen Lyu, Seth Polsley, and Tracy Hammond

Abstract Digital logic is an important part of any engineering curriculum in today's digital era, and it is often taught visually through circuit diagrams. However, for students just learning logic, this process can be non-interactive, with students typically drawing and solving diagrams that will only be evaluated by a human grader later. This paper presents DCSR (Digital Circuit Sketch Recognition), a system that recognizes hand-drawn digital logic circuits through a web interface and calculates the truth value of its output based on students' input. It allows users to draw freely and gives immediate feedback; DCSR aims to provide an interactive, sketch-based approach for educators to assist students in learning digital logic. It was evaluated by 15 electrical engineering students.

10.1 Introduction

Digital circuits are fundamental knowledge for students majoring in electronic engineering and play an important role in other engineering fields. Sometimes, students may find it difficult to evaluate their work if they draw digital circuits on paper. As electronic and paperless study becomes increasingly popular in today's technology-driven classrooms, there have been some digital circuit studying applications developed. These applications can provide immediate feedback for both student and instructor, which is the most significant advantage of computer-aided study. However, traditional circuit simulation tools can be too complex, requiring additional effort for the student to master. One alternative to traditional circuit simulation is using a natural interaction method like sketching to build simulations.

Sketch-based technologies have the potential to revolutionize education, and support students' learning processes. There are already sketch-based tools in some domains targeted at enhancing student learning, like Mechanix for truss diagrams [14, 26, 30, 34], Chemical Diagrams [28], Mathpad [15], Analogue Circuit

S. Ma (✉) • Y. Sun • P. Lyu • S. Polsley • T. Hammond
Sketch Recognition Lab, Department of Computer Science & Engineering, Texas A&M University, College Station, TX, 77843, USA
e-mail: mashuo_chn@tamu.edu; salmon89@tamu.edu; lvpengchen@tamu.edu; spolsley@tamu.edu; hammond@tamu.edu

T. Hammond et al. (eds.), *Frontiers in Pen and Touch*, Human–Computer Interaction Series, DOI 10.1007/978-3-319-64239-0_10

Recognition System [11, 32], iCanDraw for drawing [10], etc. Our focus in this work is to build a system which helps students learn digital circuits more interactively through sketching. We present Digital Circuit Sketch Recognition (DCSR), which combines the flexibility of paper with a computer's evaluation capabilities through a web-based interface that is easy for students to access and use.

10.2 Previous Work

Digital circuit recognition and simulation are relatively mature domains. One of the most relevant prior works is SketchySPICE [20]. Based on the SATIN toolkit for sketch recognition tasks, SketchySPICE was an effort to bring some of the functionality of the CAD-like tool SPICE to a sketch-based interface. However, SketchySPICE is only a recognition system for AND, OR, and NOT gates with no simulation aspect. Feng introduced multiple techniques for rapidly recognizing many different circuit elements, including capacitors, inductors, sources, and more, but these too are focused on algorithm rather than application [12, 13].

In general, many of the related works explore the underlying algorithms in the sketch recognition space, supporting hiearchical processing [23], multi-stroke recognition [9, 19], multi-domain understanding [2, 16], and localization recognizers inspired by computer vision [27]. These and related algorithms have been included in sketch-based applications, like DEMIN [24], LADDER [18], Tahuti [17], and others [22]. While some of these tools include circuit recognition, even limited simulation, they are largely focused on different domains than circuits, such as wire frame design, UML diagrams, and cross-domain descriptions of sketches.

An earlier work for recognizing logic circuits and providing simulation was introduced by Liwicki [25]. This approach used neural networks to achieve recognition of gates, connections, and input values (0 or 1). Unfortunately, users were again only permitted to use AND, OR, and NOT gates. DCSR is constructed around a very simple approach but designed to support a wider variety of logic circuits with support for more gates.

There are two existing types of digital circuit educational systems closely related to DCSR which we wish to address specifically. The first lets users choose circuit elements from a menu and connect them to create a digital circuit. LogicPad [21] is one such system. It is a pen-based application for boolean algebra visualization that lets users manipulate boolean function representations through handwritten symbol and gesture recognition coupled with a drag-and-drop interface. In essence, LogicPad focuses more on boolean algebra than circuit recognition. Although this system will not misunderstand users' "drawing" like other approaches might, it still relies on drag-and-drop interaction. While drag-and-drop is a useful interaction technique in some applications, it does not support students' learning processes as directly as sketching, which has been demonstrated across several studies on Mechanix [7, 8, 29, 33, 34].

The other digital circuit recognition approach uses drawing. One representative system is LogiSketch [1, 3]. LogiSketch is one of few complete sketch recognition systems that allows the students to draw freely, without drawing style constraints [4]. It made some contributions to the field of pen-based educational tools, being the first pen-based simulation tool for the domain of digital circuits. It implements delayed recognition of unconstrained sketches in this complex domain and incorporates behind-the-scenes user-targeted learning that improves recognition as a student uses the system.

These systems have solved some parts of the problems, but DCSR seeks to be a more complete sketch-based system by supporting freehand drawing and also boolean algebra computations. Furthermore, since our system is on-line, users do not have to spend time downloading and installing.

10.3 Implementation

10.3.1 Interface

As recommended in most sketching software, DCSR's main page contains a large canvas. Users are able to draw freely on this canvas, which has been styled to mimic a blackboard with white-on-black ink. Logic gates are automatically recognized and beautified once drawn. The input and output (I/O) connections are specified simply by drawing. The large canvas is intended to encourage the student to explore learning through sketching and help promote an easy, interactive environment.

Additional features of DCSR are accessible along the bottom of the canvas. These include options like erasing, specifying input values, providing the student's proposed output, and running the simulation of the drawn circuit. Erasing and placing inputs are relatively straightforward features; the simulation option relies heavily on the recognition stage before it is able to run.

10.3.2 Logic Gate Recognition

The core of DCSR's logic gate recognition uses the $P algorithm [35]. $P is the most recent version of the $ family recognizers, which are a set of algorithms based on matching new sketches against a library of template sketches. Older version include the $N [5] and $1 algorithms [37], with $1 being the earliest version and dealing only with uni-stroke shapes. The $N algorithm can handle multi-stroke templates, but it generates all possible permutations of a given multi-stroke shape to match the templates, which is not computationally efficient. We considered the $N-Protractor algorithm, but it is a revised version of $P that does not make sufficient speed enhancements [6]. Ultimately, we selected $P because it records strokes as point clouds to achieve much more efficient comparison.

While we wanted to use a $ family recognizer to keep the system small and portable, we did encounter some difficulty recognizing shapes that are visually similar, for instance, "OR" and "AND" gates. Because of these visual similarities, we extended the algorithm with a decision tree [39]. By comparing all seven logic gates in our system, we divided them into two types according to their visual attributes. "AND", "NAND", and "NOT" belong to type one, each containing at least one straight line. "OR", "NOR", "XOR", "XNOR", and "I/O" – we regard I/O as a logic gate in DCSR for recognition purposes – belong to type two, gates which do not contain straight lines. While a user is drawing, we perform a line test on each stroke using Rubine's features of stroke length and distance between endpoints [31]. If the ratio between the distance between endpoints and the stroke length is less than 0.95, we assume the stroke is not a line. This threshold was drawn from Wolin et al.'s work on the ShortStraw algorithm [38]. The $P algorithm runs on the different types of logic gates based on the decision tree's line detection results (Table 10.1).

According to our survey, students prefer to get feedback every time they finish sketching a logic gate. For that reason, DCSR recognizes the circuit as the user draws. Recognition occurs on strokes after 1.5 s of non-action [36] or manually when a student right clicks. This approach also helps us avoid the accumulated error problem encountered by Logisketch's multi-stage recognition [3].

In order to provide users with immediate feedback as well as neat workspaces, we beautify users' drawings after recognition. Users' sketches are substituted with a vectorized representation of the recognition element, which is scaled according to the size of the input sketch. After beautification, users can see standard logic gates displayed with input and output pins, see Fig. 10.1.

10.3.3 Wire Recognition

Because every component of the circuit diagram is added by sketching, we allow users to connect logic gates just by drawing wires. To find wire connections, DCSR checks the start points and end points of each stroke. If those points are within some threshold distance of pins on a recognized gate, the stroke will be recognized as a wire. The threshold was determined empirically from our tests, and we selected a distance of one tenth the height of the beautified gate.

10.3.4 Truth Value Calculation

Besides circuit recognition and beautification, our system can also check whether users give the correct output truth value. Due to the similarity between the structures of a digital circuit and a tree, we calculate the truth value on each wire recursively; see Fig. 10.2.

Table 10.1 Supported gates with sample user inputs and resulting beautified output. Colors have been inverted for printing purposes

Gate	User sketch	Beautified result
AND		
NAND		
OR		
NOR	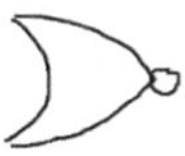	
NOT	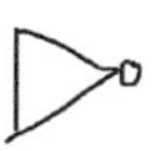	
XOR		
XNOR	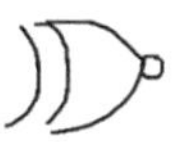	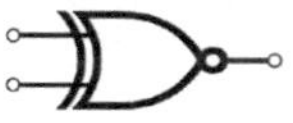

```
function calc(node)
{
  if node is a leaf then
    return node.value;
  else
    return node.logicCalc(calc(node.left), calc(node.right));
}
function logicCalc()
{
  for each wire that connects an output symbol
    calc(wire);
}
```

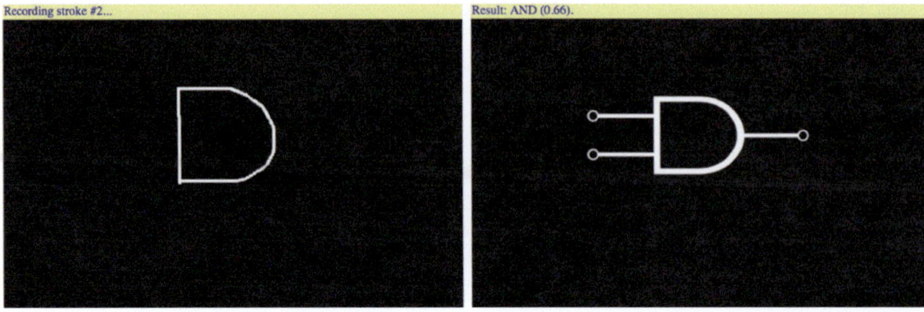

Fig. 10.1 An example of single logic gate recognition, user's drawing (*left*) and result of recognition as well as beautification (*right*)

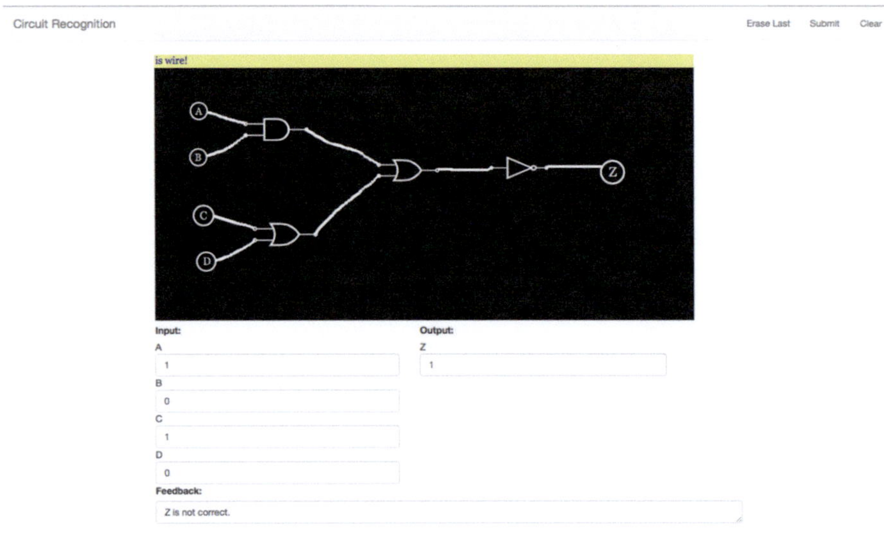

Fig. 10.2 An example of how the system checks user's input and output

10.4 Results and Evaluation

We invited fifteen students from electrical engineering to test our system. In order to find the accuracy of a single logic gate, we asked the participants to draw each gate five times. See Table 10.2 for the resulting recognition accuracy for each gate.

We found an overall recognition accuracy of 93.19% in these tests. The accuracy of "XNOR" is the lowest at 84%, while "NOR" is the highest with a perfect score. Recall that line-checking was added to avoid confusion between type one and type two gates. The decision tree helps overall accuracy greatly since no misidentifying occurs among "NOT", "AND", and "OR". However, there is still some incorrect recognition within types, between "XNOR" and "XOR".

Table 10.2 Test result of single logic gates

Gates	Correctly recognized	Mistakenly recognized	Accuracy
AND	70	5	93.99%
NAND	65	10	86.86%
OR	73	2	97.33%
NOR	75	0	100%
NOT	70	5	93.99%
XOR	68	7	90.67%
XNOR	63	12	84%
I/O	75	0	100%
WIRE	70	5	93.99%
TOTAL	629	46	93.19%

Discussion with test participants revealed several limitations in the current system. One is that while drawing gates, pins cannot be drawn as they are added later during beautification. Similarly, logic gates "AND", "NAND", and "NOT" are currently drawn in multiple strokes for line-checking purposes. Users wished to have the ability to draw gates in one stroke, and they also wished for more freedom over drawing pins. One final point, though less important, is that users must follow the direction of signal flow when drawing wires in order for the boolean algebra operations to be computed correctly.

10.5 Conclusion

We have presented DCSR, an on-line, pen-based digital circuit sketch recognition system that supports boolean function verification. DCSR provides users with immediate feedback of their drawing, which allows students to correct their work when the result of recognition is not what they expected. We also revised the $P algorithm to fit the domain better by adding a decision tree layer. The result of our experiments found that DCSR performs adequately overall and does not suffer from accumulating errors in large circuits. However, a few users commented on wanting more drawing freedom over a couple constraints. In the future, we plan to extend DCSR with more features and removed drawing constraints so that it can be an easy, online, beautifying digital circuit system.

Acknowledgements We would like to thank all the volunteers who tested our system and provided feedback. We would also like to thank the Computer Science and Engineering department at Texas A&M University and the members of the Sketch Recognition Lab.

References

1. Alvarado, C.: Sketch recognition for digital circuit design in the classroom. In: 2007 Invited Workshop on Pen-Centric Computing Research. Citeseer (2007)
2. Alvarado, C., Davis, R.: Dynamically constructed bayes nets for multi-domain sketch understanding. In: ACM SIGGRAPH 2006 Courses, p. 32. ACM (2006)
3. Alvarado, C., Kearney, A., Keizur, A., Loncaric, C., Parker, M., Peck, J., Sobel, K., Tay, F.: Logisketch: a free-sketch digital circuit design and simulation system. In: Proceedings of the Workshop on the Impact of Pen and Touch Technologies in Education, WIPTTE'13 (2013)
4. Alvarado, C., Lazzareschi, M.: Properties of real-world digital logic diagrams. In: First International Workshop on Pen-Based Learning Technologies (PLT 2007), pp. 1–6. IEEE (2007)
5. Anthony, L., Wobbrock, J.O.: A lightweight multistroke recognizer for user interface prototypes. In: Proceedings of Graphics Interface 2010, pp. 245–252. Canadian Information Processing Society (2010)
6. Anthony, L., Wobbrock, J.O.: $ n-protractor: a fast and accurate multistroke recognizer. In: Proceedings of Graphics Interface 2012, pp. 117–120. Canadian Information Processing Society (2012)
7. Atilola, O., McTigue, E.M., Hammond, T., Linsey, J.: Mechanix: evaluating the effectiveness of a sketch recognition truss tutorin program against other truss programs. In: 120th American Society for Engineering Education Annual Conference & Exposition (ASEE), ASEE, Atlanta, 23–26 June 2013, 15 pp.
8. Atilola, O., Valentine, S., Kim, H.H., Turner, D., McTigue, E., Hammond, T., Linsey, J.: Mechanix: a natural sketch interface tool for teaching truss analysis and free-body diagrams. Artif. Intell. Eng. Des. Anal. Manuf. **28**(2), 169–192 (2014)
9. Calhoun, C., Stahovich, T.F., Kurtoglu, T., Kara, L.B.: Recognizing multi-stroke symbols. In: AAAI Spring Symposium on Sketch Understanding, pp. 15–23 (2002)
10. Dixon, D., Prasad, M., Hammond, T.: icandraw: using sketch recognition and corrective feedback to assist a user in drawing human faces. In: Proceedings of the SIGCHI Conference on Human Factors in Computing Systems, pp. 897–906. ACM (2010)
11. Edwards, B., Chandran, V.: Machine recognition of hand-drawn circuit diagrams. In: Proceedings 2000 IEEE International Conference on Acoustics, Speech, and Signal Processing (ICASSP 2000), vol. 6, pp. 3618–3621. IEEE (2000)
12. Feng, G., Sun, Z., Viard-Gaudin, C.: Hand-drawn electric circuit diagram understanding using 2D dynamic programming. In: IAPR, pp. 493–498 (2008)
13. Feng, G., Viard-Gaudin, C., Sun, Z.: On-line hand-drawn electric circuit diagram recognition using 2D dynamic programming. Pattern Recogn. **42**(12), 3215–3223 (2009)
14. Field, M., Valentine, S., Linsey, J., Hammond, T.: Sketch recognition algorithms for comparing complex and unpredictable shapes. In: IJCAI, pp. 2436–2441 (2011)
15. Fitzgerald, J., Geiselbrechtinger, F., Kechadi, T., et al.: Mathpad: a fuzzy logic-based recognition system for handwritten mathematics. In: Ninth International Conference on Document Analysis and Recognition (ICDAR 2007), vol. 2, pp. 694–698. IEEE (2007)
16. Hammond, T.: Ladder: a perceptually-based language to simplify sketch recognition user interface development. Ph.D. Doctoral Dissertation, Massachusetts Institute of Technology (MIT), Cambridge (2007). Advisor: Randall Davis, 495 pp.
17. Hammond, T., Davis, R.: Tahuti: a geometrical sketch recognition system for UML class diagrams. In: Technical Report SS-02-08: Papers from the 2002 Association for the Advancement of Artificial Intelligence (AAAI) Spring Symposium on Sketch Understanding. AAAI, Menlo Park, 8 p., (2002)
18. Hammond, T., Davis, R.: Ladder, a sketching language for user interface developers. Comput. Graph. **29**(4), 518–532 (2005)

19. Hammond, T., Paulson, B.: Recognizing sketched multistroke primitives. ACM Trans. Interact. Intell. Syst. **1**(1), 4:1–4:34 (2011). doi:10.1145/2030365.2030369. http://doi.acm.org/10.1145/2030365.2030369
20. Hong, J.I., Landay, J.A.: SATIN: A Toolkit for Informal Ink-Based Applications. ACM, New York (2007)
21. Kang, B., LaViola, J.: Logicpad: a pen-based application for visualization and verification of boolean algebra. In: Proceedings of the 2012 ACM International Conference on Intelligent User Interfaces, pp. 265–268. ACM (2012)
22. Kara, L.B., Gennari, L., Stahovich, T.F.: A sketch-based interface for the design and analysis of simple vibratory mechanical systems. In: ASME 2004 International Design Engineering Technical Conferences and Computers and Information in Engineering Conference, pp. 1047–1057. American Society of Mechanical Engineers (2004)
23. Kara, L.B., Stahovich, T.F.: Hierarchical parsing and recognition of hand-sketched diagrams. In: ACM SIGGRAPH 2007 Courses, p. 17. ACM (2007)
24. Lin, J., Newman, M.W., Hong, J.I., Landay, J.A.: Denim: finding a tighter fit between tools and practice for web site design. In: Proceedings of the SIGCHI Conference on Human Factors in Computing Systems, pp. 510–517. ACM (2000)
25. Liwicki, M., Knipping, L.: Recognizing and simulating sketched logic circuits. In: International Conference on Knowledge-Based and Intelligent Information and Engineering Systems, pp. 588–594. Springer (2005)
26. Nelligan, T., Polsley, S., Ray, J., Helms, M., Linsey, J., Hammond, T.: Mechanix: a sketch-based educational interface. In: Proceedings of the 2015 ACM International Conference on Intelligent User Interface (2015)
27. Oltmans, M.: Envisioning sketch recognition: a local feature based approach to recognizing informal sketches. Ph.D. thesis, Massachusetts Institute of Technology (2007)
28. Ouyang, T.Y., Davis, R.: Recognition of hand drawn chemical diagrams. In: Proceedings of the National Conference on Artificial Intelligence, vol. 22, p. 846. Menlo Park, CA; Cambridge, MA; London; AAAI Press; MIT Press; 1999 (2007)
29. Polsley, S., Ray, J., Nelligan, T., Helms, M., Linsey, J., Hammond, T.: Leveraging trends in student interaction to enhance the effectiveness of sketch-based educational software. In: Hammond, A.A.T., Valentine, S., Payton, M. (eds.) The Impact of Pen and Touch Technology on Education. Part II, Chapter 6. Springer, New York (2015)
30. Polsley, S., Ray, J., Nelligan, T., Helms, M., Linsey, J., Hammond, T.: Leveraging trends in student interaction to enhance the effectiveness of sketch-based educational software. In: Revolutionizing Education with Digital Ink, pp. 103–114. Springer (2016)
31. Rubine, D.: Specifying Gestures by Example, vol. 25. ACM (1991)
32. Sezgin, T.M., Davis, R.: Sketch recognition in interspersed drawings using time-based graphical models. Comput. Graph. **32**(5), 500–510 (2008)
33. Valentine, S., Vides, F., Lucchese, G., Turner, D., Kim, H.h., Li, W., Linsey, J., Hammond, T.: Mechanix: a sketch-based tutoring and grading system for free-body diagrams. AI Mag. **34**(1), 55 (2012)
34. Valentine, S., Vides, F., Lucchese, G., Turner, D., Kim, H.h., Li, W., Linsey, J., Hammond, T.: Mechanix: a sketch-based tutoring system for statics courses. In: Proceedings of the Twenty-Fourth Innovative Applications of Artificial Intelligence Conference (IAAI), pp. 2253–2260. AAAI, Toronto (2012)
35. Vatavu, R.D., Anthony, L., Wobbrock, J.O.: Gestures as point clouds: a $ p recognizer for user interface prototypes. In: Proceedings of the 14th ACM International Conference on Multimodal Interaction, pp. 273–280. ACM (2012)
36. Wais, P., Wolin, A., Alvarado, C.: Designing a sketch recognition front-end: user perception of interface elements. In: Proceedings of the 4th Eurographics Workshop on Sketch-Based Interfaces and Modeling, pp. 99–106. ACM (2007)
37. Wobbrock, J.O., Wilson, A.D., Li, Y.: Gestures without libraries, toolkits or training: a $1 recognizer for user interface prototypes. In: Proceedings of the 20th Annual ACM Symposium on User Interface Software and Technology, pp. 159–168. ACM (2007)

38. Wolin, A., Eoff, B., Hammond, T.: Shortstraw: a simple and effective corner finder for polylines. In: SBM, pp. 33–40 (2008)
39. Zesheng, S., Jing, Y., Chunhong, J., Yonggui, W.: Symbol recognition in electronic diagrams using decision tree. In: Proceedings of the IEEE International Conference on Industrial Technology, 1994, pp. 719–723. IEEE (1994)

Chapter 11
An Intelligent Sketching Interface for Education Using Geographic Information Systems

Aqib Niaz Bhat, Girish Kasiviswanathan, Christy Maria Mathew, Seth Polsley, Erik Prout, Daniel W. Goldberg, and Tracy Hammond

Abstract Students learning geography aim to be familiar with a variety of geographic features and be able to identify them on maps. Drawing the geographic entities on a map would, ideally, be a better measure of the recall of the characteristics of such entities and comprehension of the various concepts in geography. However, for teachers trying to evaluate the drawings of a large number of students, this can pose a challenge. In this work, we present a sketch recognition system designed for aiding learning in geography, a field mostly unexplored by the expansive body of work in sketch recognition and education. Our web application allows users to draw rivers on a map and uses a similarity measure to evaluate students' work. Our main idea is to combine shape and location information of a sketch and check this against the shape information from our data set of geographic features. We evaluated the developed system with 10 users across multiple tests, and the findings reinforce our hope of helping students gain geographic knowledge in an intuitive and effective way through sketching.

A.N. Bhat (✉) • G. Kasiviswanathan • C.M. Mathew • S. Polsley • T. Hammond
Sketch Recognition Lab, Department of Computer Science & Engineering,
Texas A&M University, College Station, TX, 77843, USA
e-mail: aqib_bhat@tamu.edu; girishk14@tamu.edu; cmaria_91@tamu.edu; spolsley@tamu.edu; hammond@tamu.edu

E. Prout
Department of Geography, Texas A&M University, College Station, TX, 77843, USA
e-mail: prout@geog.tamu.edu

D.W. Goldberg
Department of Geography, Department of Computer Science & Engineering,
Texas A&M University, College Station, TX, 77843, USA
e-mail: daniel.goldberg@tamu.edu

T. Hammond et al. (eds.), *Frontiers in Pen and Touch*, Human–Computer Interaction Series, DOI 10.1007/978-3-319-64239-0_11

11.1 Introduction

Pen-based input devices are becoming increasingly popular in the field of education owing to their capability of teaching and evaluating students by giving them a platform to express their ideas by drawing. Various metrics and tools have been developed to understand psychological interaction patterns [3] and evaluate their school readiness. KimCHI [25] evaluated the effectiveness of gesture and sketch based learning systems to assess and improve the cognitive skills of children. The coming of age of these devices has been a boon for teachers, who are able to improve the curriculum, exploiting the potential of distance education by setting automatically graded assignments. In cases where the student-to-teacher ratio is large, intelligent interfaces and visual aids play a significant role in the learning process of students.

For education systems across the world, a major challenge is to improve the geographic knowledge of students. In most schools, students are evaluated based on multiple choice questions on a labelled map (Fig. 11.1), which tends to drastically confine the conceptual understanding to a mere mental image of the features.

On the other hand, freehand sketching encourages learners to apply their knowledge and intuition by dynamically constructing visual perspective of the geographic features, resulting in a highly involved process of *learning by doing*. More importantly, the interactive process helps students to appreciate subtle and peripheral aspects of a feature, such as the countries that have common borders or the drainage patterns of certain rivers (e.g., the Euphrates river flows from the Syrian border right through Iraq). The goal of our system is to recognize the user's intent by extracting a set of shape descriptors from the user's sketch, and validate it against some pre-defined knowledge about the geographical feature using a reasonable notion of 'correctness', along with providing interactive feedback.

Geographic Information Systems such as Google Maps, ArcGIS and OpenLayers have become extremely powerful in recent times, and are well augmented by the efforts of the open source community to create a cartological equivalent of a crowd sourced semantic web model, such as DBPedia [6], where massive efforts are underway to make structured information on map features available to the entire online community. Our system makes use of the Natural Earth data set [29], stored using Google Fusion Tables [15] which allows visualization of the data in shapefile (ESRI vector format) or KML (Keyhole Markup Language) format. With this highly reliable data set of geographical features and mapping tools at our disposal, we do not require any explicit training examples or solution templates, unless some teacher wishes to include an extremely localized feature which is not present in our extensive data set.

Our initial system aims at giving students a platform to practice drawing rivers on a world map. The system computes a similarity metric between the student's

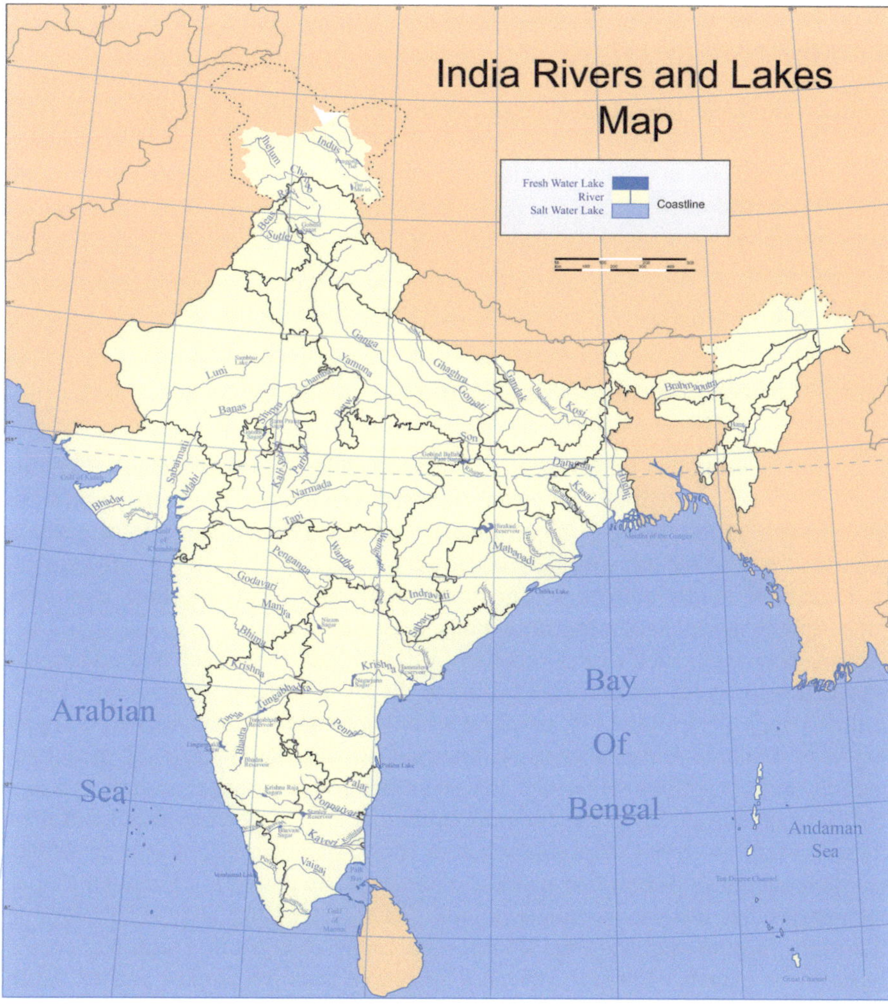

Fig. 11.1 The conventional method of teaching maps, where students are asked to label already drawn features

sketch and the template[1] of that physical feature sourced from cartographic sources, and accepts the sketch if it exceeds an empirical threshold. The intent here is to check the recall of students with respect to the geographic knowledge they gain from their geography education rather than making them learn how to draw rivers or gain

[1]Rivers are not of any standard or typical shape but the actual river shapes can serve as a guide for students learning to draw them accurately. Hence, we refer to the river shapes sourced from standard maps as templates. We give the student the option of seeing the actual river shape on the map in the learning mode. This feature can be disabled in the evaluation mode.

expert cartographic skills. Our main contribution towards this goal is the integration of a sketch recognition system with a mapping interface backed by authoritative geographic data to make useful interpretations which should also provide the ability to scale automatic grading for large classes. In the following sections, we will explore some related work before detailing the implementation of the presented system. We include an evaluation with 10 users and close with a discussion of results and final remarks.

11.2 Prior Work

11.2.1 Sketch Recognition Methods

Sketch recognition can be broadly summarized under 3 classes, namely Gesture-based, Geometry-based, and Appearance-based. Gesture-based algorithms [8, 32, 38, 39] such as the $ family algorithms are powerful discriminators for gestures independent of affine transformations. However, one can perform the same gesture on a map in multiple strokes, directions, sizes and locations, thus making it impossible to model geographical features as gestures. Geometry-based algorithms combine low-level, such as PaleoSketch [30], and high-level, such as Ladder [17], recognizers to detect and combine primitive shapes in strokes. Alvarado and Davis, in their paper SketchRead [1], describe a domain independent recognition engine that uses dynamic Bayesian nets to interpret strokes based on a set of defined patterns and constraints. However, this approach requires a user to define a high level language to describe the symbols in the domain. This method is helpful for domains such as circuit diagrams where a set of standard shapes can be defined based on relationships between strokes. While these are powerful for regular shapes, our domain has little scope for regularity. Curvature graphs [34, 41] provide an interesting way of analysing the k-largest curvature changes in a template against those in the sketch. However, the need to merge multiple strokes and explicitly define stroke segments makes this approach too complex for our context since the sketch has to be represented as a point cloud rather than as temporally ordered strokes. Vision-based algorithms [7, 23, 27, 33] focus on the visual appearance of the image rather than its geometric features. These were found to be useful in our context.

11.2.2 Sketching in Education

The use of touch interfaces in education has also been vastly studied. Mathpad [26] is useful for recognizing mathematical equations. Mechanix [4, 5, 12, 24, 36, 37] is an automated system that has been adopted by physics teachers for setting interactive assignments in solid dynamics and mechanics. Feng et al. [11] have built a system that can recognize electrical circuit symbols. CogSketch [13] with its Sketch Worksheets [40] model allows teachers to create domain specific worksheets for students to complete, and analyses the assignments using spatial sparsity, and does not automatically evaluate the solutions. The design of CogSketch is generalizable but equally hinders the use of available datasets because they have to be recoded as per CogSketch's encoding strategy. We have focused on geography education and not aimed at evaluating the cognition using parameters such as pauses, etc. CogSketch revolves around the understanding spatial phenomena but in case of maps, other aspects such as direction, shape, etc. can also be critical. We have to take inspiration from CogSketch on how students could demonstrate their understanding of causal and relationship information. The application of sketch recognition algorithms to the geography domain however, is novel and has not been attempted before. While many GIS applications as well as third party tools provide a rich drawing toolkit to doodle on maps, none of them have a recognition engine that can understand a user's gestures on a map beyond the conventional ones for zooming and panning.

11.2.3 Sketching in GIS

Godwin et al. [14] describes a visualization tool used by the Atlanta police department for making queries by drawing paths and boxes on a map, which does not use any recognition component beyond superimposing pre-defined data points against a drawn path. Drawing on paper maps has also been studied [35], where a digital pen is used to get coordinates of symbols drawn on a map and then transmit them to a database. Geographical mining [2] has also been subject to a huge body of work, which is a possible extension of our approach. Military course-of-action sketch recognition systems [16, 18, 21] to allow commanders to draw their plans or UAV instructions [9, 10] on maps to be automatically recognized by making use of a primitive shape recognizer [19, 31] combined with a high level statistic-based recognizer [22] to return the top interpretations of symbols and characters drawn on a map. However, this approach operates on a canvas over a regular map, and does not explore the possibility of integrating the symbol information with GIS data.

11.3 Methodology

The current implementation of our system focuses on rivers, helping students learn to identify and draw them based on visual appearance. The visual appearance of a river is defined by the following two factors:

1. **Shape**: This primarily refers to the visual appearance of the feature. More specifically, it refers to the relative position of each point to every other point in an image. Such shape descriptors can be used to gauge the visual similarity between two objects.
2. **Location**: The location information is available through the latitude-longitude information of a sketch and can directly be compared.

We employ a recognition method that exploits these two distinctive features, by combining two techniques, namely shape context [7] and Hausdorff distance [23, 33]. Prior to this, we examine the option of flagging incorrect sketches early, by preprocessing the candidate sketch using metrics including bounding box, and path length. The final similarity result is a weighted average of the results returned by these classifiers.

When building a sketch recognition system in this domain, it is also important to consider some of the common constraints, especially in regard to accommodating the trade-off between user's drawing freedom and ease of recognition. This system, in particular, requires that the students get complete freedom to reproduce the mental map in mind. The challenges in building such a robust recognizer are therefore manifold.

- Unlike regular sketches which are drawn in an iterative improvement manner [20], we don't expect rivers to be drawn using touch-up strokes. However, there would be continuation strokes as river can be long and have converging or diverging branches. The recognition therefore has to be stroke-independent.
- The system needs to be receptive even towards poor and messy sketches, as long as the essential aspects of a feature are sufficiently captured.
- The sampling rate is heavily dependent on the input device used, and is influenced by the zoom level of the map while sketching.
- Sketches need to be stored as strokes in the geographic coordinate system. However, we do not need to use geodesic distance as we consider 2D maps.
- Geographical features are extremely irregular, and the resulting sketches will vary greatly with drawing style. In fact, there is little uniformity even within the same class of physical features.

The remainder of this section details the three main stages of our system and its algorithms, as shaped according to our goals and recognized constraints.

11.3.1 User Interaction

The system consists of a UI which allows the user to select a river, draw the river, and get the similarity between the actual river information (shape, size, and location) and the drawn river. Comparison with the original river is made possible with a data set of rivers. The details of the UI and data set are provided in the following section.

11.3.1.1 Data Set

Natural Earth [29] is an open source community funded by some of the leading cartographical research institutions, and hosts a vast pool of map data that is free to use.[2] For this application, we use the `Lake Centerlines and Rivers` data set that contains the information for over a 1000 rivers world wide. This data is generally in shapefile format which can be directly used on the Google maps professional version but not the free one. It is, therefore, first converted using Quantum GIS to an array of strokes, where each stroke is represented by an array of points. A lot of the data is also hosted on Google Fusion Tables [15], which is used for gathering, visualizing and sharing data. This data is in SVG (Scalable Vector Graphics) and KML format, thus making conversion easier. The final strokes for all the rivers are then converted to an internal representation of strokes.

11.3.1.2 UI Design

The user interface is built using AJAX, PHP, and the Google maps API, with styling controls to define map appearance, behavior and its event model. First, the user has to select a geographical feature from the list of known features, following which the template is loaded from the database. Once the user starts sketching, the stroke points are generated as the pen is moved, depending on the sampling rate of the device. If the pen is lifted, a new stroke is recorded and appended to the list of strokes for the session. The user also has an option to clear the sketch and restart his drawing. For practicing by tracing, an option called "Show Solution" is also provided in the "learning" mode of the system, to serve as a memory aid for the user before he or she attempts to sketch in the "exam" mode. Finally, on submitting of an attempt by a user for evaluation, their sketch is compared with the template, and the accuracy of the drawing is determined.

11.3.2 Preprocessing

Preprocessing can help reduce the computation to a large extent by identifying grossly incorrect attempts as early as possible. Although currently our system doesn't require high computation power, preprocessing would certainly be beneficial as our system evolves to incorporate more sophisticated calculations to improve accuracy. However, we would also explore in future, broader studies whether

[2]The complete data set of all physical features including reefs, coasts, geographic lines and glaciers can be obtained at http://www.naturalearthdata.com/downloads/10m-physical-vectors/

Fig. 11.2 The subfigures **(a,b)** show possible stroke length errors. Bounding box errors are shown in the figures **(c)** and **(d)** (The template is shown in *red,* and the user's drawing in *blue*).

students would still prefer to see their detailed scores in such cases. The following cases can arise when the student draws a river. Figure 11.2 shows a few examples of these cases.

- The total stroke length of the candidate stroke varies drastically from that of the template stroke
- The candidate is drawn nowhere close to the template, i.e. there is a little or no overlap between their bounding boxes.

In the first case, the candidate is removed from consideration based on the difference in stroke length. The total stroke length or path length is computed by summing the Euclidean distances between every pair of consecutive stroke points.

$$d = \|TotalStrokeLength_{candidate} - TotalStrokeLength_{template}\|$$

The sketch will be flagged as being incorrect if $d \geq 0.5 \times TotalStrokeLength_{template}$. This ensures that the template stroke is not smaller than half of the stroke length of the template and not larger than 1.5 times of the stroke length of the template.

In the second case, the candidate sketch is marked as being wrong if the percentage of the points of the candidate stroke in the bounding box of the template is less than 10%, i.e., if the sketch is too far from the template.

We did not use the root mean square error (RMSE) as an error or performance metric as in commonly done in geospatial analysis because the number of points in a user's attempt could potentially differ a lot from that of the template and hence, the one-to-one correspondence between the two groups of points would be hard to judge. In the future, we would also like to explore RMSE as a similarity measure if we approximate the templates as polylines and check how a user's attempt fits with the same.

11.3.3 Recognition

Vision based approaches are good for recognizing shapes that are irregular, and they primarily try to find the intra-point and inter-point relationships, given two images.

11.3.3.1 Shape Similarity Using Shape Context

The Shape Context algorithm proposed by Belongnie, Malik, and Puzicha [7] and first used in sketching by Oltmans [28] gives a measure of similarity between shapes. It tries to find which point in one shape corresponds to which point in another one with the use of a descriptor named 'shape context' which is calculated for each point. The shape context for a point in a shape gives an indication of how the remaining points in the shape are distributed relative to it. An aligning transform is estimated using the agreement between the shape contexts. The matching errors between the points on the two shapes, namely, the template and the attempt of the user, are summed and combined with the magnitude of the aligning transform to give the the dissimilarity between those two shapes.

For each point on the shape, a coarse histogram of the relative coordinates of the remaining points is computed by taking the vector distance of the point with respect to all other points. The histogram is defined to be the shape context of p(i). Bins that are uniform in log-polar space are used (similar to the Bullseye features diagram), resulting in shape context being affected by the positions of nearby sample points rather than away ones. Chi square test statistic is used to get a measure of the cost of matching two points. After getting the set of costs between all pairs of points on the two shapes, bipartite graph matching is done to minimize the total cost of matching, i.e. we need to find the permutation of the points of the second shape such that the matching has minimum cost.

The sketched input is in the form of an array of points with stroke identifiers. The Shape Context method is used for object recognition and hence stroke information is not needed. We use a helper method to convert our input data into SVG format to pass onto the method which gives a similarity measure based on the Shape Context method.

11.3.3.2 Location Similarity Using Hausdorff Distance

The location similarity is an important aspect for sketches drawn on maps. If the shapes are similar but the location is incorrect, it will result in a low similarity ratio. The Hausdorff distance measure is used to find the similarity in terms of distance. In this paper, we have used a modified version of Hausdorff distance. For calculating the Hausdorff distance we calculate two distance vectors D_A and D_B such that

$$D_A = \min_{b \in P_B} |a - b|, a \in P_A$$

where P_A is a vector of the points in A and P_B is a vector of points in B. Similar to D_A, D_B is calculated. The Hausdroff distance is the maximum value from the average of D_A and D_B. The modification from the original Hausdorff method is that it uses the average of the minimum values to find h(A,B), rather than taking the maximum. The reason to use the modified version of Hausdorff is to avoid large h(A,B) or h(B,A) value caused due to a small number of outliers. By taking the average we make sure that the distance from all the points, i.e. all the values in D_A and D_B are contributing to h(A,B) and h(B,A) respectively.

$$h(A, B) = \frac{\sum(D_A)}{N_A}$$

where N_A is the number of points in A, defines the distance h(A,B). We calculate h(B,A) similarly using D_B. The Hausdorff distance H(A,B) is the maximum value from h(A,B) and h(B,A). The maximum is taken in order to take into consideration the difference in stroke length of the candidate and the template.

We restricted the Hausdorff distance to a maximum of 10 while finding the similarity. If the Hausdorff distance value is more than 10, we consider that it to have zero similarity with respect to location. The similarity is measured as:

$$similarity_{Hausdorff} = 1 - H(A, B)/10$$

11.3.3.3 Stroke Length Ratio

The stroke length also is an important feature while we consider similarity between two sketches. To take this into consideration, we take into account the ratio of the difference in stroke lengths to the total stroke length of the template. The following gives the similarity measure based on stroke length:

$$similarity_{strokeLength} = 1 - \frac{d}{TotalStrokeLength_{template}}$$

11.3.3.4 Combining Classifiers

The final similarity is computed by a weighted voting based classifier. The percentage accuracy is taken as the summation of similarity from Hausdorff distance, similarity from shape context, and the similarity from stroke length. We denote the final similarity by:

$$\gamma = similarity_{ShapeContext} * 0.5 + similarity_{Hausdorff} * 0.4 + similarity_{strokeLength} * 0.1,$$

where $similarity_{ShapeContext}$ is the similarity result from Shape Context algorithm, $similarity_{Hausdorff}$ is the similarity measure from Hausdorff distance, and $similarity_{strokeLength}$ is the measure obtained using stroke lengths.

These similarity measures are important because they give a method to measure the recall of geographic objects, as would be common on human and physical geography exams, and also allow for the automated assessment of student's recall answers which would enable the use of these types of interactive answers on assessments in large classes.

11.4 Experiment and Results

We conducted a user study with 10 users collecting 12 samples from each user for 6 rivers. This included 2 drawing per river per user. The users were allowed to see the river on the map and then sketch the river. This was to measure the similarity in good cases. After this sketch, the users were asked to sketch without seeing the river. This was to get the different ways the users actually draw the river without seeing the actual river. This also helped us to get a similarity measure for different styles of drawing. Table 11.1 gives the similarity measure of users drawing, seeing the actual river, and Table 11.2 gives the similarity measure of users drawing without seeing the actual river.

Table 11.1 Accuracy when users were allowed draw while looking at the solution

User	Nile	Danube	Amazonas	Yangtze	Euphrates	Ganges
1	0.8553	0.9059	0.9153	0.8158	0.8767	0.8468
2	0.8146	0.8398	0.9823	0.8738	0.9536	0.8730
3	0.8627	0.8555	0.9664	0.9119	0.8673	0.8621
4	0.9004	0.7823	0.9812	0.9103	0.9647	0.9083
5	0.8958	0.9139	0.9768	0.9089	0.8884	0.8698
6	0.9005	0.8336	0.9629	0.9649	0.8878	0.8390
7	0.9136	0.8606	0.9215	0.9010	0.9270	0.8711
8	0.9137	0.8886	0.9215	0.9995	0.9423	0.9107
9	0.9078	0.8589	0.9741	0.9820	0.9087	0.9534
10	0.9255	0.8819	0.9667	0.9415	0.9397	0.9531

Table 11.2 Accuracy when users were not shown the solution

User	Nile	Danube	Amazonas	Yangtze	Euphrates	Ganges
1	0.7050	0.6341	0.7773	0.8756	0.6088	0.7737
2	0.6387	0.5761	Bounding Box Error	0.6047	0.5785	0.6957
3	0.7389	0.7644	0.8544	0.5929	0.8674	0.7452
4	0.7463	0.7748	0.8705	0.8437	0.8418	0.7572
5	0.7243	0.7117	0.9035	0.8511	0.8713	0.8932
6	0.8480	0.7799	0.8576	0.7602	0.9440	Stroke Length Error
7	0.7127	0.7459	0.8361	0.7995	0.8732	0.7579
8	0.7836	0.7583	0.8901	0.7423	0.9458	0.8457
9	0.8155	0.7294	0.8996	0.8830	0.9055	0.8457
10	0.6916	Bounding Box error	Stroke Length Error	0.9531	Stroke Length Error	0.7609

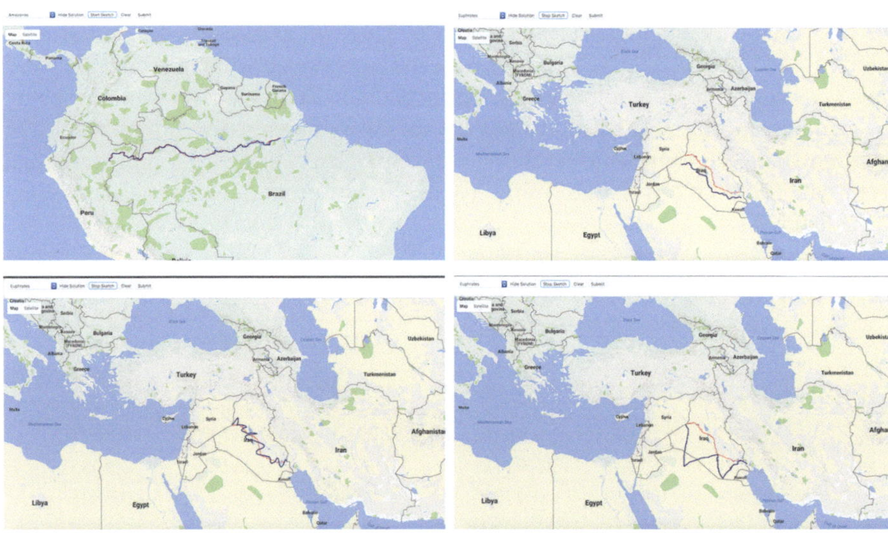

Fig. 11.3 The *top left* image shows the Amazon river drawn almost perfectly, yielding an accuracy of over 90%. The *top right* shows the Euphrates river drawn in the right way, but at the wrong location. The *bottom left* shows the Euphrates river drawn wrongly, but at approximately the right location. The *bottom right* shows the Euphrates river drawn poorly and completely off position

11.5 Discussion

As we analyzed the user study we came up with 4 scenarios based on the location and the shape. Figure 11.3 shows the 4 scenarios

- Similar Shape and Similar Location: Our system was able to produce a similarity measure greater than 85% in most of the cases for this condition.
- Similar Shape but Dissimilar Location: Our system gave a lower similarity as the distance measure is higher. The similarity will depend on how far the template and the candidate sketches are.
- Similar Location but Dissimilar Shapes: Our system gave a similarity lower than the best case. The similarity in this case will depend mostly on how dissimilar the two sketches are.
- Dissimilar Location and Dissimilar Shape: Our system gave a lower similarity compared to the best case. The similarity in this case depends on how far is the sketch and how dissimilar is the sketch compared to the template sketch.

In addition to the above results, we noticed that preprocessing using bounding box and stroke length helped obtain some true negative results when users were asked to draw rivers without seeing the actual river.

The users who took part in our study expressed their satisfaction with the system, and believed it has the potential to be a very useful tool in mapping education. They also agreed that the system helped them relearn some of the basics of geography.

We interviewed a faculty member at Texas A&M University, College Station, who has taught introductory geography courses. In his experience, students are evaluated about their knowledge of the spatial relationships and other properties of geographic features by questions which ask them to label geographic entities on maps already having them in place. Asking students to draw rivers in exams may be too time-consuming and difficult, but it could certainly help the students learn and remember geographic features if implemented and utilized as a learning aid. As an expert in geography education, the Professor explained that he would assign different weights for the various aspects of an attempt by a student to judge the understanding of the crucial details for a river. For grading the attempts by students he considered the location of the drawn river to be the most significant. Knowing the origin or source location and the end where a river drains into a delta is also very important. Hence, length and shape are not enough by themselves. At river deltas, distributaries are formed. Students should be able to approximate such distributions for important deltas such as those of the Nile, Danube, Ganges/Brahmaputra, etc.

Students would also need references to help associate rivers with their locations. e.g. political boundaries, elevation, vegetation, etc. As an example, the Yangtze river in China would be hard to draw in case just the international borders are shown. Immediate feedback is clearly needed as an assessment of a learner's comprehension of the material (where geographic objects are) and their ability to recognize, interpret, and utilize spatial relationships. In case of an attempt being judged as wrong, the Professor expressed that the detailed breakdown of scores on various metrics should also be available if a student wishes to dig deeper and that the details in terms of percentages are indicative of how right an attempt was but explanatory notes would be more useful, e.g. "too far North/South/East/West".

One feature that is taught to students about rivers which we could incorporate in future versions is that of direction. For example, while the Mississippi flows South, the Nile flows North. This makes the students understand that the topography of a region influences the direction that a river takes and that the North or South directions shown on maps are simply arbitrary choices. Students could also be asked to draw or mark island groups, individual islands like Greenland, regions like southeast Asia, Baja California peninsula, major cities, broad wind patterns or ocean currents e.g. the Gulf stream, El Niño and La Niña currents, etc. Besides sketching, students could add meta-data in their answers, for example, the starting and end points of a river. Similarly warm or cold currents could be indicated with different colors and the direction of rivers, currents, etc. could be indicated with arrows.

There is a lot of latent potential in sketch-based map interfaces that is yet to be tapped. From our experiment, we determined several possible paths of advancement we hope to pursue. First would be to add support for a more diverse range of physical features, such as contours, shaded regions of deserts and seas, coastlines, and state boundaries. While our classifiers can be extended to some of these, many new domains may require specific heuristics to comprehend the entire range of possible geometric features. This leads to discussion of more generalized classifiers based on optimal feature sets that better characterize the template of any physical feature. For instance, curvature and feature-area based algorithms may prove to be useful if we compromise on stroke independence.

There are broader applications as well considering that besides being used to gauge a user's recall of geographic features, sketching on maps can also give a glimpse at a person's mental map of the absolute and relative positions of various places and objects. Beyond maps and education, the combination of stroke information and GIS data can be used in various other areas such as disaster management, marketing analysis, and mobile applications.

11.6 Conclusion

Sketch recognition has the potential to make a powerful positive impact on education, and we believe that its application to new domains will continue to drive that change. To that end, we have presented a system and its algorithms for evaluating students' work when drawing rivers on a map. From our experiments, we show that the developed system is able to give accurate measures in cases with dissimilar location and/or shape to the intended river, and our users felt that they had gained a new familiarity with these rivers through the freehand drawing. In the future, we plan to extend these capabilities to more geographic features in order to help more students learn geography in a variety of contexts. Also, akin to how CogSketch [13, 40] is being developed to automatically detect various properties in sketches, we aim to apply machine learning techniques to help improve the automatic grading process and improve the feedback capabilities using AI.

Acknowledgements We would like to thank the Sketch Recognition Lab, the Computer Science and Engineering Department, and the Department of Geography at Texas A&M University for providing valuable feedback.

References

1. Alvarado, C., Davis, R.: Sketchread: a multi-domain sketch recognition engine. In: Proceedings of the 17th Annual ACM Symposium on User Interface Software and Technology, pp. 23–32. ACM (2004)
2. Andrienko, G.L., Andrienko, N.V.: Interactive maps for visual data exploration. Int. J. Geogr. Inf. Sci. **13**(4), 355–374 (1999)
3. Anthony, L., Brown, Q., Nias, J., Tate, B., Mohan, S.: Interaction and recognition challenges in interpreting children's touch and gesture input on mobile devices. In: Proceedings of the 2012 ACM International Conference on Interactive Tabletops and Surfaces, pp. 225–234. ACM (2012)
4. Atilola, O., McTigue, E.M., Hammond, T., Linsey, J.: Mechanix: evaluating the effectiveness of a sketch recognition truss tutorin program against other truss programs. In: 120th American Society for Engineering Education Annual Conference & Exposition (ASEE), ASEE, Atlanta, 23–26 June 2013, 15 pp.
5. Atilola, O., Valentine, S., Kim, H.H., Turner, D., McTigue, E., Hammond, T., Linsey, J.: Mechanix: a natural sketch interface tool for teaching truss analysis and free-body diagrams. Artif. Intell. Eng. Des. Anal. Manuf. **28**, 169–192 (2014). doi:10.1017/S0890060414000079. http://journals.cambridge.org/article_S0890060414000079
6. Auer, S., Bizer, C., Kobilarov, G., Lehmann, J., Cyganiak, R., Ives, Z.: Dbpedia: a nucleus for a web of open data. Springer, Berlin, Heidelberg (2007)
7. Belongie, S., Malik, J., Puzicha, J.: Shape context: a new descriptor for shape matching and object recognition. In: NIPS, vol. 2, p. 3 (2000)
8. Choi, H., Hammond, T.: Sketch recognition based on manifold learning. In: Proceedings of the 23rd National Conference on Artificial Intelligence, AAAI'08, vol. 3, pp. 1786–1787. AAAI Press (2008). http://dl.acm.org/citation.cfm?id=1620270.1620353
9. Cummings, D., Fymat, S., Hammond, T.: Sketch-based interface for interaction with unmanned air vehicles. In: CHI'12 Extended Abstracts on Human Factors in Computing Systems, CHI EA'12, pp. 1511–1516. ACM, New York (2012). doi:10.1145/2212776.2223664. http://doi.acm.org/10.1145/2212776.2223664
10. Cummmings, D., Fymat, S., Hammond, T.: Reddog: a smart sketch interface for autonomous aerial systems. In: Proceedings of the International Symposium on Sketch-Based Interfaces and Modeling, SBIM'12, pp. 21–28. Eurographics Association, Aire-la-Ville (2012). http://dl.acm.org/citation.cfm?id=2331067.2331071
11. Feng, G., Viard-Gaudin, C., Sun, Z.: On-line hand-drawn electric circuit diagram recognition using 2D dynamic programming. Pattern Recogn. **42**(12), 3215–3223 (2009)
12. Field, M., Valentine, S., Linsey, J., Hammond, T.: Sketch recognition algorithms for comparing complex and unpredictable shapes. In: Proceedings of the Twenty-Second International Joint Conference on Artificial Intelligence, IJCAI'11, vol. 3, pp. 2436–2441. AAAI Press (2011). doi:10.5591/978-1-57735-516-8/IJCAI11-406. http://dx.doi.org/10.5591/978-1-57735-516-8/IJCAI11-406
13. Forbus, K., Usher, J., Lovett, A., Lockwood, K., Wetzel, J.: Cogsketch: sketch understanding for cognitive science research and for education. Top. Cogn. Sci. **3**(4), 648–666 (2011)
14. Godwin, A., Stasko, J.: Drawing Data on Maps: Sketch-Based Spatiotemporal Visualization https://www.cc.gatech.edu/~john.stasko/papers/infovis15-poster-sketch.pdf. Accessed 15 Nov 2017

15. Gonzalez, H., Halevy, A.Y., Jensen, C.S., Langen, A., Madhavan, J., Shapley, R., Shen, W., Goldberg-Kidon, J.: Google fusion tables: web-centered data management and collaboration. In: Proceedings of the 2010 ACM SIGMOD International Conference on Management of Data, pp. 1061–1066. ACM (2010)
16. Hammond, T.: Ladder: a perceptually-based language to simplify sketch recognition user interface development. Ph.D. Doctoral Dissertation, Massachusetts Institute of Technology (MIT), Cambridge (2007). Advisor: Randall Davis, 495 pp.
17. Hammond, T., Davis, R.: Ladder, a sketching language for user interface developers. Comput. Graph. **4**(29), 518–532 (2005)
18. Hammond, T., Logsdon, D., Peschel, J., Johnston, J., Taele, P., Wolin, A., Paulson, B.: A sketch recognition interface that recognizes hundreds of shapes in course-of-action diagrams. In: CHI'10 Extended Abstracts on Human Factors in Computing Systems, CHI EA'10, pp. 4213–4218. ACM, New York (2010). doi:10.1145/1753846.1754128. http://doi.acm.org/10.1145/1753846.1754128
19. Hammond, T., Paulson, B.: Recognizing sketched multistroke primitives. ACM Trans. Interact. Intell. Syst. **1**(1), 4:1–4:34 (2011). doi:10.1145/2030365.2030369. http://doi.acm.org/10.1145/2030365.2030369
20. Hammond, T., Paulson, B.: Recognizing sketched multistroke primitives. ACM Trans. Interact. Intell. Syst. (TiiS) **1**(1), 4 (2011)
21. Hammond, T.A., Logsdon, D., Paulson, B., Johnston, J., Peschel, J.M., Wolin, A., Taele, P.: A sketch recognition system for recognizing free-hand course of action diagrams. In: IAAI (2010)
22. Johnston, J., Hammond, T.: Computing confidence values for geometric constraints for use in sketch recognition. In: Proceedings of the Seventh Sketch-Based Interfaces and Modeling Symposium, SBIM'10, pp. 71–78. Eurographics Association, Aire-la-Ville (2010). http://dl.acm.org/citation.cfm?id=1923363.1923376
23. Kara, L.B., Stahovich, T.F.: An image-based trainable symbol recognizer for sketch-based interfaces. In: AAAI Fall Symposium, pp. 99–105 (2004)
24. Kebodeaux, K., Field, M., Hammond, T.: Defining precise measurements with sketched annotations. In: Proceedings of the Eighth Eurographics Symposium on Sketch-Based Interfaces and Modeling, SBIM'11, pp. 79–86. ACM, New York (2011). doi:10.1145/2021164.2021179. http://doi.acm.org/10.1145/2021164.2021179
25. Kim, H.h., Taele, P., Valentine, S., McTigue, E., Hammond, T.: Kimchi: a sketch-based developmental skill classifier to enhance pen-driven educational interfaces for children. In: Proceedings of the International Symposium on Sketch-Based Interfaces and Modeling, pp. 33–42. ACM (2013)
26. LaViola Jr, J.J., Zeleznik, R.C.: Mathpad 2: a system for the creation and exploration of mathematical sketches. In: ACM SIGGRAPH 2007 Courses, p. 46. ACM (2007)
27. Miller, E.G., Matsakis, N.E., Viola, P., et al.: Learning from one example through shared densities on transforms. In: Proceedings IEEE Conference on Computer Vision and Pattern Recognition, 2000. vol. 1, pp. 464–471. IEEE (2000)
28. Oltmans, M.: Envisioning sketch recognition: a local feature based approach to recognizing informal sketches. Ph.D. thesis, Massachusetts Institute of Technology (2007)
29. Patterson, T., Kelso, N.V.: Free vector and raster map data. http://www.naturalearthdata.com/
30. Paulson, B., Hammond, T.: Paleosketch: accurate primitive sketch recognition and beautification. In: Proceedings of the 13th International Conference on Intelligent User Interfaces, pp. 1–10. ACM (2008)
31. Paulson, B., Rajan, P., Davalos, P., Gutierrez-Osuna, R., Hammond, T.: What!?! No Rubine Features?: using geometric-based features to produce normalized confidence values for sketch recognition. In: HCC Workshop: Sketch Tools for Diagramming (VL/HCC), p. 57–63. VL/HCC, Herrsching am Ammersee (2008)
32. Rubine, D.: Specifying Gestures by Example, vol. 25. ACM, New York (1991)
33. Rucklidge, W.: Efficient Visual Recognition Using the Hausdorff Distance, vol. 1173. Springer, Berlin (1996)

34. Sezgin, T.M., Stahovich, T., Davis, R.: Sketch based interfaces: early processing for sketch understanding. In: ACM SIGGRAPH 2006 Courses, p. 22. ACM (2006)
35. Sylverberg, T., Kristensson, P., Leifler, O., Berglund, E.: Drawing on paper maps: reliable on-line symbol recognition of handwritten symbols using a digital pen and a mobile phone. In: 2007 2nd International Conference on Pervasive Computing and Applications (2007)
36. Valentine, S., Vides, F., Lucchese, G., Turner, D., Kim, H.h., Li, W., Linsey, J., Hammond, T.: Mechanix: a sketch-based tutoring and grading system for free-body diagrams. AI Mag. **34**(1), 55 (2012)
37. Valentine, S., Vides, F., Lucchese, G., Turner, D., Kim, H.h., Li, W., Linsey, J., Hammond, T.: Mechanix: a sketch-based tutoring system for statics courses. In: IAAI (2012)
38. Vatavu, R.D., Anthony, L., Wobbrock, J.O.: Gestures as point clouds: a $*p* recognizer for user interface prototypes. In: Proceedings of the 14th ACM International Conference on Multimodal Interaction, pp. 273–280. ACM (2012)
39. Wobbrock, J.O., Wilson, A.D., Li, Y.: Gestures without libraries, toolkits or training: a $1 recognizer for user interface prototypes. In: Proceedings of the 20th Annual ACM Symposium on User Interface Software and Technology, pp. 159–168. ACM (2007)
40. Yin, P., Forbus, K.D., Usher, J.M., Sageman, B., Jee, B.D.: Sketch worksheets: a sketch-based educational software system. In: IAAI (2010)
41. Yu, B., Cai, S.: A domain-independent system for sketch recognition. In: Proceedings of the 1st International Conference on Computer Graphics and Interactive Techniques in Australasia and South East Asia, pp. 141–146. ACM (2003)

Part V
Increasing Liberal Arts Engagement and Learning with Sketch

Chapter 12
Insights from Exploration of Engaging Technologies to Teach Reading and Writing: Story Baker

Michel Pahud

Abstract To engage children in learning to write, we spent several years exploring tools designed to engage children in creating and viewing stories. Our central focus was the automatic generation of animations. Tools included a digital stylus for writing and sketching, and in some cases simple robots and tangible, digitally-recognized objects. In pilot studies, children found the prototypes engaging. In 2007, a decision not to develop new hardware was made, but at today's greatly reduced tablet cost and with more capable touch and pen technology, these experiments could inspire further research and development.

12.1 Introduction

Children learn to write at different ages and with different degrees of challenge. From 2004 to 2007 our team developed and tested over 30 prototype systems designed for children 5–7 years old who have challenges in learning to write. The intent was to create highly engaging interactive experiences for children who might have limited attention spans. We made use of animation, typing, a stylus for writing and sketching, machine-recognizable tangible objects for input, and simple robot activity for viewing the outcome of actions.

Khan Academy, Sprout Online, and other sites provide learning experiences for children with real-time feedback, often focused on STEM, memory and concentration, music, and art. Writing presents unique challenges. In particular, the goal of writing is to communicate, and writing tasks that are given to children are often artificial. After our project ended, games that include writing became available. Margaret Johnson, our general manager, started a company Sabi Games through Microsoft IP Ventures to pursue further early education efforts https://news.microsoft.com/2008/11/12/sabi-and-microsoft-launch-unique-interactive-drawing-and-education-game-for-children/. DC Comics' "Scribblenauts" [3]

M. Pahud (✉)
Microsoft Research, 14820 NE 36th Street, Building 99, Redmond, WA, 98052, USA
e-mail: mpahud@microsoft.com

T. Hammond et al. (eds.), *Frontiers in Pen and Touch*, Human–Computer Interaction Series, DOI 10.1007/978-3-319-64239-0_12

has a very large database of visual objects and actors that children can summon; only nouns and proper nouns are written.

We set out to make the computer an active partner by animating actions that a child enters as sentences, usually starting the animation when a end of sentence period is entered. We experimented with several ways for children to create simple end of sentence period. The easiest, for children who balk at a blank page, was to modify an existing story – swapping actors or adding a photo of oneself and an image of their dog to an adventure story, for example. Complete sentences result in visual animations that portray the actions described.

This requires a vocabulary comprising nouns with associated images, action verbs, and simple natural language parsing. With a limited vocabulary, we can guide children by displaying an array of available objects, with typing or writing narrowing the options. More significantly, we enabled children to add objects by inserting digital photos and sketches, making them available for use in subsequent sentences or stories. We also explored ways to enable children to add verbs. Future work could extend this, and include adjectives and adverbs.

Children's sketches are not usually polished. Some children might like to try specifying an unusual action, such as a pizza eating a dog, to see what happens. They might like to paste a head-shot photo on a story character. In consideration of such issues, we chose a non-realistic "South Park" cartoon style of animation with the concept of a mouth split for animated objects. Because highly predictable animations will not stay fresh, we developed techniques for injecting surprise and maintaining engagement. "Idle behaviors" are attached to objects – small actions through which actors appear alive and waiting to be told what to do next. Objects are given meta-data that allows a range of expression and results in less predictability; for example, a shy character flies or kicks differently than a bold one.

To encourage handwriting, one prototype rewarded small children who traced over a letter with a stylus by having images beginning with the letter appear in bubbles. Tracing a second letter popped bubbles of words that no longer fit, as the remaining images grew.

We created animated fonts through careful character-by-character analysis of how letters could stretch or distort in synchrony, allowing text beneath images to move rhythmically as a story was enacted, or emphasize words selectively. To make the fonts attractive to children, a design goal was to make them look organic and alive.

We went beyond the keyboard, stylus, and display. Cards with an image or word on one side and a scanned tag on the other could be used to construct sentences. Simple robots were used to act out story actions. We explored ways to combine visual information and robot movements to convincingly represent actions more complex than the robot could manage, such as falling down and getting up.

In addition to use in classroom contexts, we envisioned children sharing their stories online with friends and family, and a technology for remote story-telling. For example, a parent on a trip can create animated accounts of his days with photos and sketches, to be emailed to a parent at home and shared as 'bedtime stories.'

In pilot studies with children, the prototypes were well-received, but a rugged tablet was very expensive at the time. These concepts could be revisited today and extended with touch, depth cameras, and other technologies. They could be made available for younger children, and some features could find use in assistive or other settings seeking engaging captioning.

12.2 Story Baker: Will Converting Text to Animation Engage Children?

How will children react when typing 'The dragon flew over a beach' in a text field instantly produces a view of that activity? After a period was typed in our first prototype (Fig. 12.1), the dragon and beach appeared and the animation is launched (demo [2]). As the child types another sentence, the animation remains visible until the next period is typed (demo [2]). At that time, the animation of the new sentence replaces the previous one. To make the animations more interesting, Story Baker added background and foreground objects: For example, in Fig. 12.1b, the dragon flies behind the rock and the palm tree.

Each sentence typed creates a new animation and a new page. A series of sentences represent a story. The controls at the bottom allow navigation through the pages of the story, playing back the corresponding animations (demo [2]). At any time a child can augment the story by entering another sentence.

Our Natural Language Processing (NLP) software parses each sentence to extract the actor (noun; dragon in the example), action (verb; flew), direct object (noun; none in this example), and the location if any (noun; beach) [1]. The output is a tree containing the sentence structure. An interesting property is that different tenses, including passive tenses, have the same logical form. For example, 'The dragon

a b

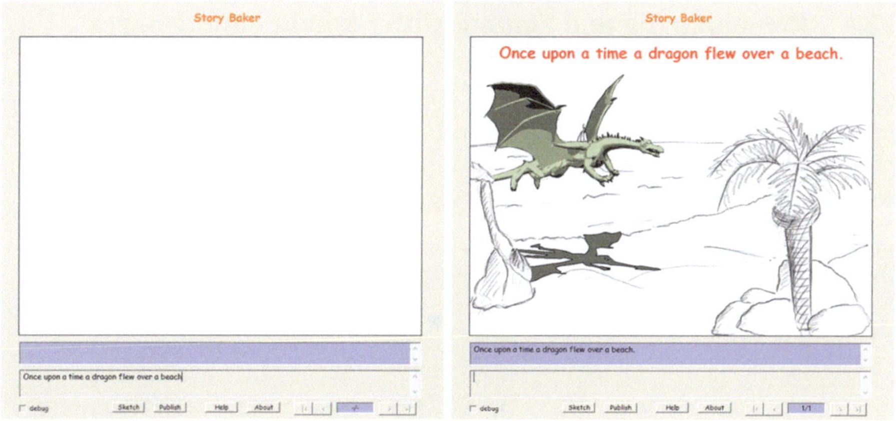

Fig. 12.1 (**a**) Before the child enters a sentence, the screen is blank; (**b**) After the period completing a sentence has been typed, the animation is launched

kicked the ball on the beach', 'The dragon is kicking the ball on the beach', 'The dragon kicks the ball on the beach', or 'The ball was kicked by the dragon on the beach' will generate the same logical form and animation. This allowed the prototype to understand a wide range of sentences and also allowed us to swap languages relatively easily – we experimented with English and French versions.

We had to address word strings that did not yield a valid sentence. The intent was to display whatever we could based on the information recognized in the sentence. For example, if the program could identify only an actor, the actor was shown looking confused.

12.3 Animation Style

We built and explored several animation engines. One goal was to enable children to insert images or use a pen to sketch actors as they created stories, and we soon saw that a 3D engine sets expectations of perfection that would inhibit such input. Realistic 2D animations also proved limiting: 'A man eats a ball' could be animated, but 'A ball eats a man' is a valid sentence that is difficult to animate when a ball has no mouth. Our solution: simple 'South Park Type Animation,' in which every object is a 2D image with a mouth defined as a horizontal split line. To make it visually more interesting, an optional body with an attach point pivot on the head can rotate and scale independently. To make 'flying' or 'jumping' visually understandable, a shadow below each actor reveals whether or not it is touching the ground. Figure 12.2 illustrates this with a dragon and a generic actor with placeholders for the head and body, and a shadow.

Because all nouns are given a mouth, a ball can eat a man. After an object is eaten, it reemerges in a whimsical way, to be available for subsequent use and to avoid scaring a child if what was eaten was a person, pet, or other valued item.

12.4 Personalizing and Scaling with Sketches and Images

Children inevitably want to add objects not in our set, including photos of family, friends, pets, and toys. We enabled them to sketch objects with a stylus, insert photos, and draw over photos, with the results incorporated into the available set.

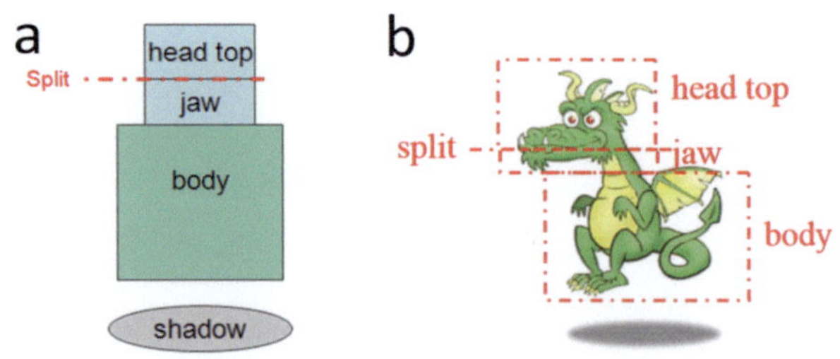

Fig. 12.2 (**a**) A generic actor with head, mouth, optional body, and shadow; (**b**) An instance of a dragon actor

Fig. 12.3 The pizza drawing (*above*) and its use in two sentences

Fig. 12.4 Actors with photographed and sketched faces

It was engaging for children to modify the system and create unique, personalized stories this way, rather than being restricted to pre-built content. Most sketches were rough, but they were easily incorporated into the South Park style animation.

The early example above illustrates the vocabulary extension, preceding the full South Park like animation style. Encountering 'The man ate a pizza in the town,' with no corresponding visual for pizza, the Natural Language Processing component detects that and the child is asked to draw it (demo [2]). From then on, the system associates that drawing with the word 'pizza' when it appears in the text (demo [2]) (Fig. 12.3).

We created simple templates of a headless man, woman, girl and boy on which a face could be drawn or a photo placed. One appeared in Fig. 12.3; two are shown in Fig. 12.4.

A photo could be adorned with glasses, moustache, tattoos, missing teeth, and so on, if desired. Figure 12.5 shows in sequence a man's body, after a face has been added, then when adorned with colored hair, and finally given a name ("Michel") and used in a story (demo [2]).

We created generic activities for verbs, described in more detail in the Appendix, and considered how new verbs might be added. In Fig. 12.6, a user sketches the verb 'cheers' (raising both arms) on a tablet with a stylus. (This prototype also included a physical robot, described later.) To define cheer, a user stroked on the tablet to raise first the right arm, then the left.

The general issue of enabling children to create verbs is an area for future exploration. It could make use of depth cameras and other new technologies.

Fig. 12.5 A sketchpad showing a body as face, hair and action are added

Fig. 12.6 A user defining the verb 'cheers' by sketching

12.5 Idle Behaviors and Metadata to Add Unpredictability and Surprise

Highly predictable behaviors will cease to engage children. To counter this, we borrowed from Sims game design by assigning metadata (or 'Digital DNA') to each object to reduce predictability and inject surprise. Discovering what different characters will do can be an incentive to keep writing stories. The metadata comprised physical characteristics (weight, strength, hardness), personality (Serious/Silly, Shy/Ongoing, Lazy/Hardworking, Grumpy/Happy, Dumb/Smart, Sleepy/Awake), and characteristics such as the horizontal split line coordinates for the mouth. They affected the visual appearance and auditory effects of actions specified by verbs. We also designed idle behaviors – small movements and sounds – that give the impression that an actor is 'alive' and waiting for new instructions. Idle behaviors also varied – not all actors fidgeted the same way. Manually setting up the metadata could be difficult for younger children, so the choice of character body (as seen in Fig. 12.5) was used to define the metadata. For example, when a child chooses a new character with ragged jeans or a clown suit, the character metadata is set to be more outgoing and silly than for a character given a body with a suit and tie.

The metadata constrain the resulting animation. The experience was designed to engage children even if they just riff a single sentence (in fact the first prototype to test the Southpark animation style had only one sentence). We knew that children would soon lose interest if they could predict the resulting animation upon swapping a noun. Generating a surprising animation when a word is swapped was important. For example, a child writes, "One day, a donut kicked a dragon in the backyard" and the kicked dragon follows a parabolic trajectory (Fig. 12.7). If the child modifies the sentence to be "One day, a donut kicked Michel in the backyard," (replacing "a dragon" with "Michel" who is more "silly" than the dragon) the animation shows Michel banging against the display glass, then becoming angry and arguing with the donut (Fig. 12.8). The behavior of actors can differ, based on the metadata. A "shy" strawberry will hesitate to kick and run back and forth before deciding to kick, where an "outgoing" house will go straight at the secondary actor. Because of

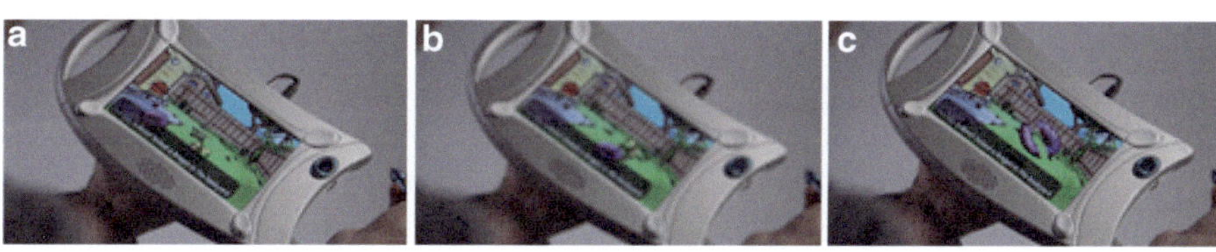

Fig. 12.7 Animation of "One day, a donut kicked a dragon in the backyard." (**a**) A donut approaches a dragon; (**b**) The dragon is kicked and follows a parabolic trajectory outside of the scene; (**c**) The donut celebrates

Fig. 12.8 Animation of "One day, a donut kicked Michel in the backyard." (**a**) A donut approaches Michel; (**b**) Michel is kicked and his head whimsically hits the glass of the device; (**c**) Michel with a red head angrily confronts the donut

the range of combinations, even the design team had no idea what would happen when actors were swapped. In addition to audio effects, we experimented with a laugh track to accompany comical events.

For a given sentence, the animation is always the same. This enables a child to show a story to other people without being surprised, and encourages children to continue creating or modifying sentences.

12.6 Remote Storytelling

A child could create a story to share with distant grandparents. We also prototyped the opposite scenario, in which a parent on a business trip could use photos and text to easily create an animated story (basically using Story Baker as an authoring tool) to send to the family, perhaps as a bedtime story. In the prototype, on the receiving side, a series of sentences would be projected on a wall to be read aloud by a parent or the child. When the speech recognition technology detected sentence completion, the corresponding animation was triggered. The animation could quickly be constructed by the traveler using selfies, pictures taken during the day, and other materials at hand. For example, a parent travelling to Paris for business could take a selfie with the Eiffel tower and make a quick cartoonish story to share with family at home.

12.7 Teaching Handwriting Through a Delightful Stylus Experience

Handwriting has significant educational advantages over typing [8, 9]. Some children take to it easily at a young age, others can use an incentive to work on it. We looked into recognition of free-form handwriting of kids learning to write to see if it could be used as input in story creation, but it is too inaccurate. For the youngest children, starting with a blank page is intimidating and yields many invalid sentences. Instead, we provided them with story templates that they could modify by swapping words, then seeing the new sentence that they had created acted out. To specify a word, they used a traditional letter-tracing task made more engaging through animation.

As the child traces, small animations (called ‘particles system’) appear to emerge from the stylus. In Fig. 12.9, where letters are grouped as in the familiar “Now I Know my A-B-C’s” song, as the child traces the letter ‘b’, images of ‘banana’, ‘belt’, ‘bee’, ‘book’, and ‘broccoli’ appear in bubbles (demo [2]). If the child next traces the letter ‘a’, the bubbles for ‘belt’, ‘bee’, ‘book’, and ‘broccoli’ will pop, leaving the visual ‘banana,’ which grows, its size inversely proportional to the number of letters remaining. The child can continue, or can pop the bubble with a quick stylus touch

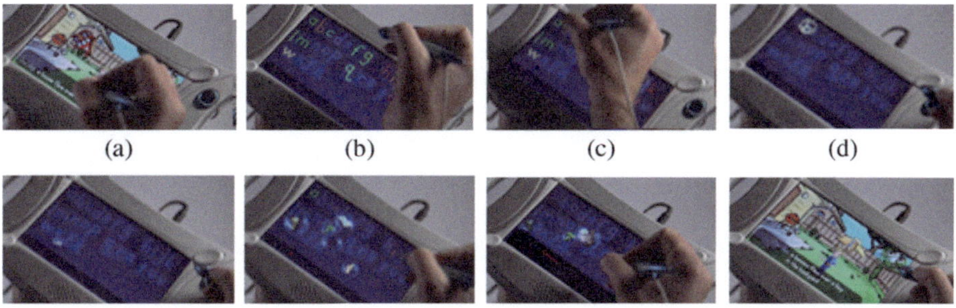

Fig. 12.9 (a) The animation plays before the child selects a word to replace; (b) The alphabet for tracing; (c) 'b' is traced, animates (d) and falls into place (e); (f) the visuals for 'banana', 'belt', 'bee', 'book', and 'broccoli' are visible in bubbles, with continuation letters highlighted; (g) a child pops the 'banana' bubble, the word is inserted and a new animation plays (h)

at any time to insert the word; if an entire word is traced, the bubble automatically pops and the word is inserted.

If a great many letter-matching choices are available for a given combination of letters, the bubbles could rotate over time to keep a manageable number visible at one time. Having to trace additional letters to constrain the set could motivate additional writing. When a hovering stylus can be sensed, bubbles could gently move aside to leave space for the child to trace the letter underneath.

12.8 Classroom Presenter

Because children having trouble learning to read and write may not be motivated by text in general, we considered how to make fonts themselves lively and engaging. Our goal was a "font with feelings," which we termed "Fontlings." Canned animations require extensive artistic work and do not scale to other font types. Previous work on parametric font modification [6] and kinetic typography [4, 5, 11] did not meet our needs. The former was designed for a different purpose; it was too CPU intensive for our prototype hardware and would be difficult for children to work with. The latter didn't convey organic, living sense that we sought for this experience. To achieve a simple, procedurally animated, lively font, we decided to adapt our South Park-like animation. The letter 'h' split above or below the horizontal line as with objects could look strange and perhaps not be recognizable. Instead, we created a join that connects the top and bottom part of each character, identifying a rotation point, as shown in Fig. 12.10 for the letter "f". Each part can move, rotate and/or scale independently while connected, yielding very interesting animations. Letters might have multiple joins; for example, for the letter "h," the split was set close to the base of the letter, so two joins connect the parts.

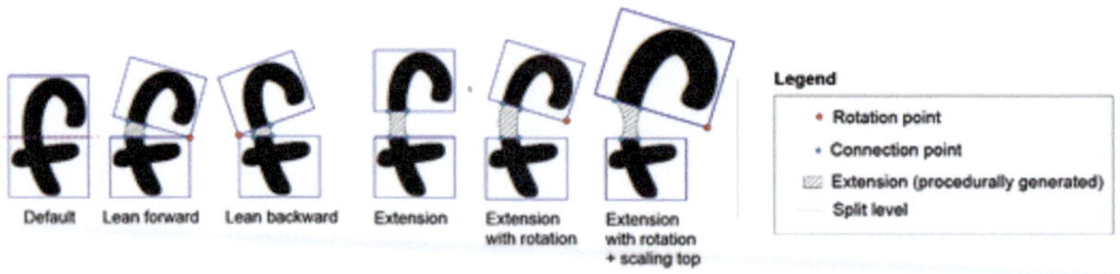

Fig. 12.10 A dynamic join connects parts of a character when animated

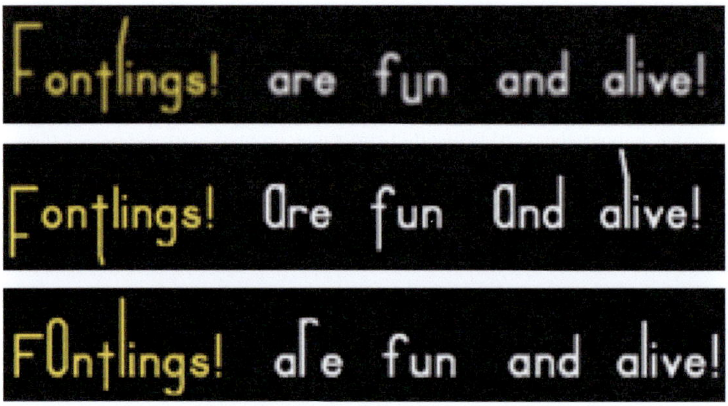

Fig. 12.11 Snapshots during idle behavior animation, illustrating the organic and lively impression

One interesting point to note with this approach is that a font type only need the split horizontal line height information for each character once for all. If someone wants to support other font types in the future, they just have to define a split line for each character for this new font type.

We implemented a fully working prototype, defined a scripting language, and wrote scripts to animate idle, happy, and misunderstood behaviors. The results impressed viewers (demo [2]). Especially appealing were a worm-like organic feel of idle behaviors and a rubber-band effect for misunderstood behavior. Figure 12.11 shows snapshots of idle behavior, although imagining the full effect is not easy. For more technical details, the data structure for Fontlings appears in the Appendix.

Within the context of Story Baker, Fontlings were used to connect images and text during interactions (Fig. 12.12). Touching the image of the dragon produces a dynamic link from the image to the word "dragon," and the font on the word becomes animated (Fig. 12.12a). Touching the word dragon animates it and produces a visual effect on the dragon image (Fig. 12.12b). The sentence exhibits idle behavior most of the time, but becomes fully playful and excited from time to time (Fig. 12.12c). This experience is enhanced with audio accompaniment. (Demo [2])

(a) (b) (c)

Fig. 12.12 Within the context of Story Baker, Fontlings were used to connect images and text during interactions. (**a**) Touching the dragon image produces a visual link to the word. (**b**) Touching the dragon word produces a visual link to the image. (**c**) The sentence occasionally animates spontaneously

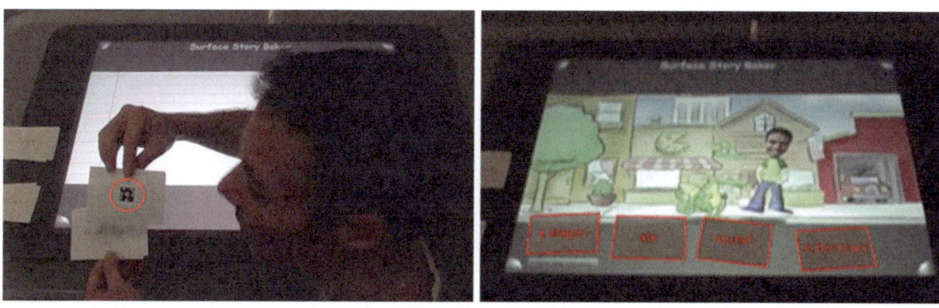

(a) (b)

Fig. 12.13 The "dragon ate Michel in the town". (**a**) A card with a tag (*circled*) that is recognized by the table. (**b**) Animation based on a sentence; cards are enhanced to be more visible

12.9 Incorporating Real-World Objects

We prototyped a simpler story construction interface that involved physical cards, each with a word or set of words, with a code recognizable by a Microsoft PixelSense [10] table on the back (demo [2]). The cards represented nouns, verbs, and locations. A child who could read the words could use them to construct a sentence to be animated, such as "the dragon ate Michel in the town." (Fig. 12.13.)

We experimented with the use of robot actors. With a robot positioned next to the tablet, when a story reached the word "robot," a virtual robot appeared on the screen and the specified action affected both the virtual and real robots. We also created an verb authoring with experience on a tablet with a stylus described earlier that optionally animated a robot when a new verb was authored (Fig. 12.6). We used a ROBOSAPIEN from WowWee [13] and built a robotic API to send infrared commands to the actual robot. Figure 12.14a shows a robot that was kicked by the virtual actor in response to the sentence, "Then Michel kicked the robot on the beach." We explored the use of text to instruct the robot to act on physical objects, within the limitations of what the robot can do. In Fig. 12.14b, the robot has responded to the sentence "Once upon a time, a robot grabbed a cup."

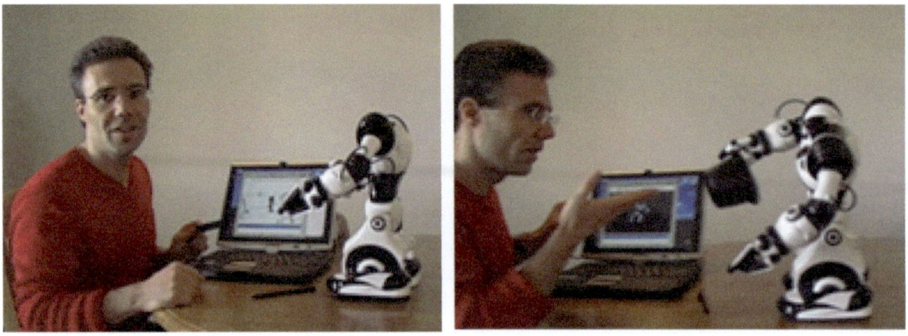

(a) (b)

Fig. 12.14 (**a**) A robot that was just kicked by a virtual actor. (**b**) A robot that just picked up a cup

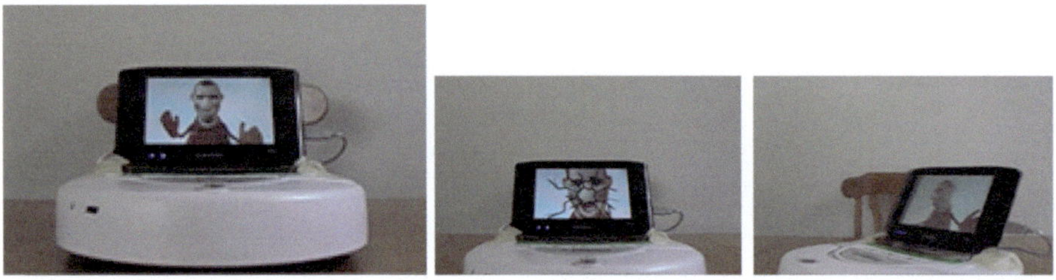

(a) (b) (c)

Fig. 12.15 "One day CoolBot was at home. CoolBot was very happy and started dancing. Suddenly he tripped and fell down. Finally, he stood up and started spinning". (**a**) CoolBot dances: the physical robot moves back and forth while the virtual character waves its arms;. (**b**) CoolBot falls. (**c**) CoolBot spins

We also positioned a display on an armless Roomba from iRobot to 'symbiotically' act out verbs, working to overcome some limitations (demo [2]). The Roomba had a UMPC (Ultra-Mobile PC) with a touch screen controller. To 'dance' the robot moved back and forth while the virtual character waved its arms (Fig. 12.15a). 'Fall down,' 'fly,' and 'eat' were more challenging. For 'fall down,' the robot moved forward very fast while the virtual actor screamed, then stopped with an image appearing that seemed to be the screen cracked by the character's head (Fig. 12.15b). For 'fly,' the virtual character could extend its arms like wings as the robot moved. To 'eat,' the robot could suddenly move forward as the display showed an object moving into the virtual character's mouth.

One story was "One day CoolBot was at home. CoolBot was very happy and started dancing. Suddenly he tripped and fell down. Finally, he stood up and started spinning." Figure 12.15 shows the robot 'dancing,' 'falling down,' and 'spinning.' In our simplified prototype, robot stories could not be authored, but the UMPC touch display would support it, perhaps by stylus tracing as described above.

Fig. 12.16 (**a**) Two robots, physical props (*trees*), a virtual ball, and a virtual shark hidden in the lake; (**b**) Acting out "Roby kicked a ball." (**c**) After acting out "A big shark upset kicked the ball very strongly toward Robo and made him fall down"

12.9.1 Robotic Playset

We also explored a robotics playset, which provides children with a starting point that constrains what they say while retaining a sense of freedom (demo [2]). The robots were programmatically controlled with infrared signals dynamically sent from the PC controlling the PixelSense table, synchronized with the animation displayed on the table (Fig. 12.16). For this prototype we provided the cards with words and tags to use in authoring stories.

12.10 Pilot Studies

12.10.1 Free-Form Study

In pilot studies of early versions of Story Baker, children typed free-form text using the prototype shown in Figs. 12.1, 12.3, and 12.5. These were undertaken to get feedback and guide or inspire design changes, and are described anecdotally based on review of videotapes. In general, children were fully engaged and did not seek a perfect experience. Rather than a defect, having to draw missing objects was seen as "teaching the computer"; personalizing the content was a positive experience.

12.10.2 Surprising Animation Study: Single Sentence

A subsequent study explored reactions to the South Park style of animation in which tracing was used to enter text, an early version of the prototype shown in Fig. 12.9 with the kind of surprising animations such as shown in Fig. 12.7 and 12.8. We had children adapt a sentence that they were given to see the effect of animation that varied based on digital DNA. As a child modified the text in a sentence with a simple word selection interface, others watched and made suggestions ("Do, 'the microscope kicks the house' "), and they laughed at unexpected animations. A parent reported that the children talked about it on their way home. With variation, even a single simple sentence could be highly engaging.

12.10.3 Extended Study Session 1: Modifying South Park Style Stories

In a more extended study, six children over two sessions with a mature version of Story Baker examined word entry by tracing with six stories (median length 9 sentences). The primary actor, secondary actor, action and location could be modified. Children could choose among realistic, space travel, and fairy tale story themes. The prototype had 13 actions,[1] 76 actors,[2] and 16 locations.[3] Each actor had a metadata file that defined physical and personality characteristics inspired by Snow White's dwarves.[4] The actors, actions and locations were inspired by a list of words familiar to children in the target age range.

12.10.4 Extended Study Session 2: Personalization by Inserting Pictures of Children, Family, Pets, and Toys

Personalization was introduced in the second session, a week after the first: parents were asked to provide pictures of their child, one or more parents or guardians, any siblings or pets, and preferred toys. We placed them in the system, using 'dad' and 'mom' for parents, with faces on a cartoon body. For children we used their names and put their faces on boy or girl cartoon bodies. We defined a split line for a mouth

[1]List of actions (verbs) initially on Story Baker prototype: chase, create, dance, eat, fall, fly, jump, kick, love, marry, meet, play, and see.

[2]List of actors (nouns) initially on Story Baker prototype: Alligator (an), apple (an), astronaut (an), bag, ball, banana, bee, beehive, belt, bicycle, book, boy, broccoli (some), cake, car, cat, chair, cheese (some), clock, clown, cookie, couch, cow, dinosaur, dog, donut, dragon, duck, fish, girl, Gus (actor name), hand, hippo, horse, house, invention (an), Jack (actor name), Jish (actor name), kangaroo, King, lasergun, lemon, letter, lion, mailbox, man, Michel (actor name), microscope, monkey, mouse, pancake, pants (a pair of), penguin, phone, pig, pizza, plane, prince, princess, queen, racecar, Red (actor name), robot, rock, rocket, shark, shoe, snake, strawberry, table, toy, tree, truck, turtle, underpants (a pair of), and woman.

[3]List of locations initially on Story Baker prototype: backyard, beach, bedroom, castle, classroom, den, factory, farm, forest, kitchen, moon, park, party, store, town, and zoo.

[4]Physical: Weight [0..9] (0 – Light, 9 – Heavy), Strength [0..9] (0 – Weak, 9 – Strong), Soft/Hard [0..9] (0 – Soft, 9 – Hard). Personality: Type Human, Animal, Vegetable, Mineral, Serious/Silly [0..9] (0 – Serious, 9 – Silly), Shy/Ongoing [0..9] (0 – Shy, 9 – Outgoing), Lazy/Hard worker [0..9] (0 – Lazy, 9 – Hard worker), Grumpy/Happy [0..9] (0 – Grumpy, 9 – Happy), Dumb/Smart [0..9] (0 – Dumb, 9 - Smart), Sleepy/Awake [0..9] (0 – Sleepy, 9 – Awake).

that was about the height of the actual mouth. For the pets and toys, we didn't attach the picture to a body, but defined a split line for the mouth.

Animating sentences in the virtual world engaged children, even or especially those making no sense in the real world. A child excitedly repeated "The pizza ate the fish," turning the device to show us his animation. Personalization was popular. An excited child said, "I got daddy", "I got mommy and daddy" and later "it was super fun" and "yeah!" when asked if she would do it again. One child laughed at length after making a rocket fly over his dad in the backyard. Surprised to see her pet in a bubble, one said "Oh <pet name>, cutie!" before creating a story in which her pet ate a fish in the zoo. Another said "Where is dad?" after tracing the letter "d," then "Here is daddy!" and giggling when seeing his face in a bubble. Other examples were "<child name> flew over dad in the castle." Asked about the experience, one said, "I like it very very very, I love it."

The decision not to continue was based on issues around producing a successful device given the hardware available in 2007. (Specifications are in the appendix.) The reactions to the prototypes was very positive, and a decade later, with costs down and capabilities up, these ideas merit reconsideration.

12.11 Conclusion

Over four years, we explored how to engage children in writing stories sentence by sentence, with each completed sentence instantly converted to an animation. We found that 'realistic' animations set high expectations and made many sentences difficult to animate. Our South Park type of animation yielded more whimsical and versatile animations and invited children to extend the vocabulary by sketching and annotating. We explored animating robot actors alongside the virtual world, and enlivening stylus tracing of letters with relevant animations. Our work is a step toward teaching literacy. Opportunities that seem worth exploring follow.

- Our prototypes limited sentences to two actors, one verb, and one location. This should be extended to more complex sentences, with more actors, adverbs and adjectives. For example, "The big dragon and small bird quickly ate the blue banana." For more advanced children, we could animate more complex sentences.
- We enabled new verb creation through the use of stylus strokes to 'push' limbs of a virtual robot (which also animated a real robot). Another approach might be acting in front of a depth camera. Variants of a verb for different digital DNA could be authored the system could prompt: "Please act out someone cheering shyly; now, act out someone cheering in a goofy way." The system could create variants for different levels of shyness, goofiness, and so on. Verb authoring with multi-touch and tracing trajectories could be explored.
- With the trend toward large pen-sensitive and multi-touch displays, with different devices working together, one could explore collaborative story creation. Multi-

ple children might co-author on a large display, or remotely via smartphones or tablets. Every child might contribute to a portion of the story, or small groups could make and share their stories. Children could draw sets of missing nouns on a large display.

Appendix

A.1 Prototype Tablet

In 2007, we made a prototype device with a 7" 800 × 480 pixel display (Fig. 12.17a) and 4 GB solid state memory, running Windows CE on an AMD Geode 500 MHz processor. It had a FinePoint digitizer with an active stylus tethered by a retractable power cord. The device was designed to survive 3 foot drops. Its overall size was 11 × 6.25 × 1.4 inches, weighing 0.8 kg. By design the device looked like a painter's palette to emphasize the stylus and creativity experience. The device also had a handle to make it easy to carry. The stylus was tethered to avoid losing it and to provide power without requiring a battery (Fig. 12.17a). A polished version of Story Baker was implemented on the device (Fig. 12.17b).

A.2 Sentence Structure

Figure 12.18 shows the sentence structure with some of the objects and actions. Sentences have a primary actor, action, a secondary actor, and optional background. We created an API to convert and animate a sentence (see Visual Sentence API documentation [12]).

The top row of Fig. 12.19 illustrates a generic eating animation. Beneath it is an instance of a dragon eating a banana: From left to right, the dragon faces the banana, its mouth opens, the banana enters its mouth, and the dragon chews. Chewing uses rotation and scaling, and can dynamically recolor the character; for example, the

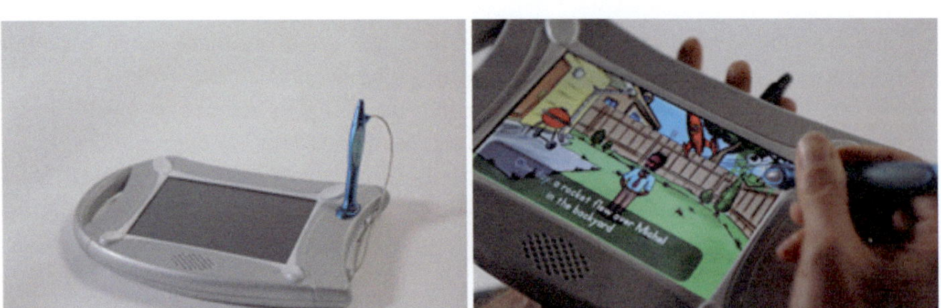

(a) (b)

Fig. 12.17 The prototype hardware (2007). (**a**) The prototype device. (**b**) Story Baker running on the device

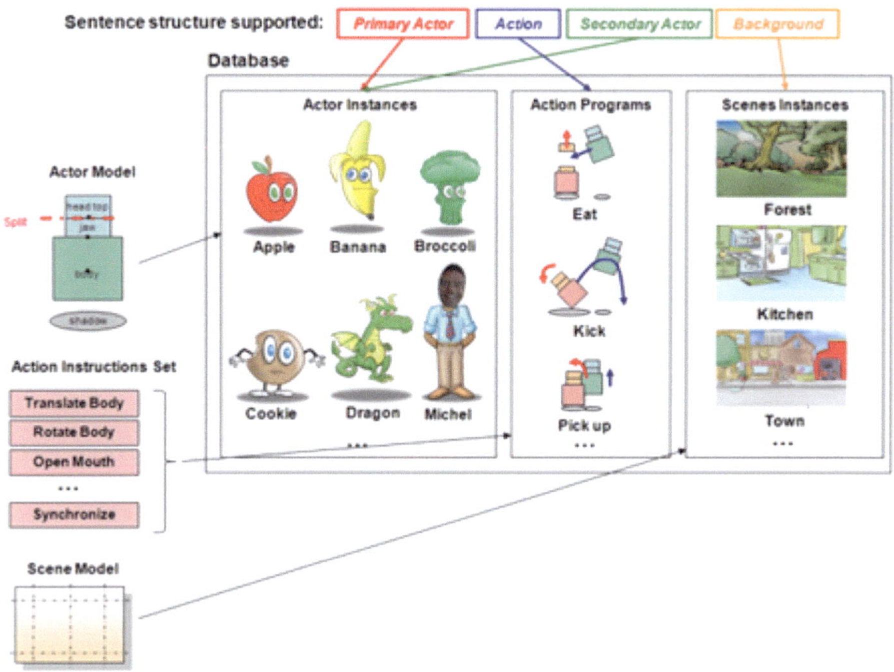

Fig. 12.18 Animation system overview

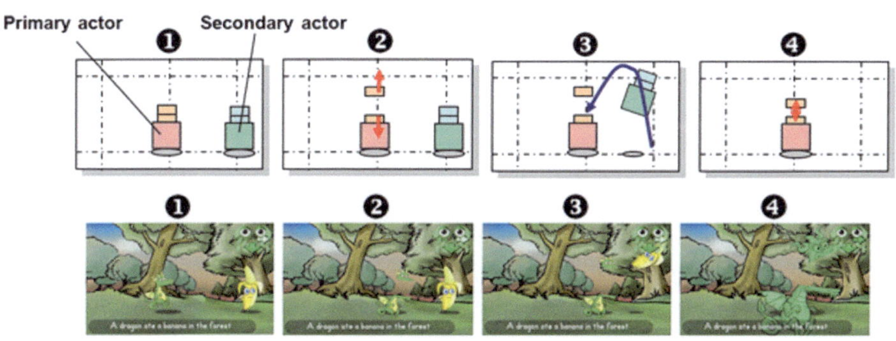

Fig. 12.19 Animating a dragon eating a banana in the forest. Top: Generic eating animation, Bottom: Animating a dragon eating a banana in the forest

head and/or body could turn green to depict disgust or overeating. After chewing, the primary actor ejects the food to avoid trauma if the secondary actor is a favorite character. The animation is accompanied with audio to make it convincing.

Figure 12.20 illustrates how our system transforms a sentence into an animation. The digital DNA, image, and audio of the primary and secondary actors are drawn

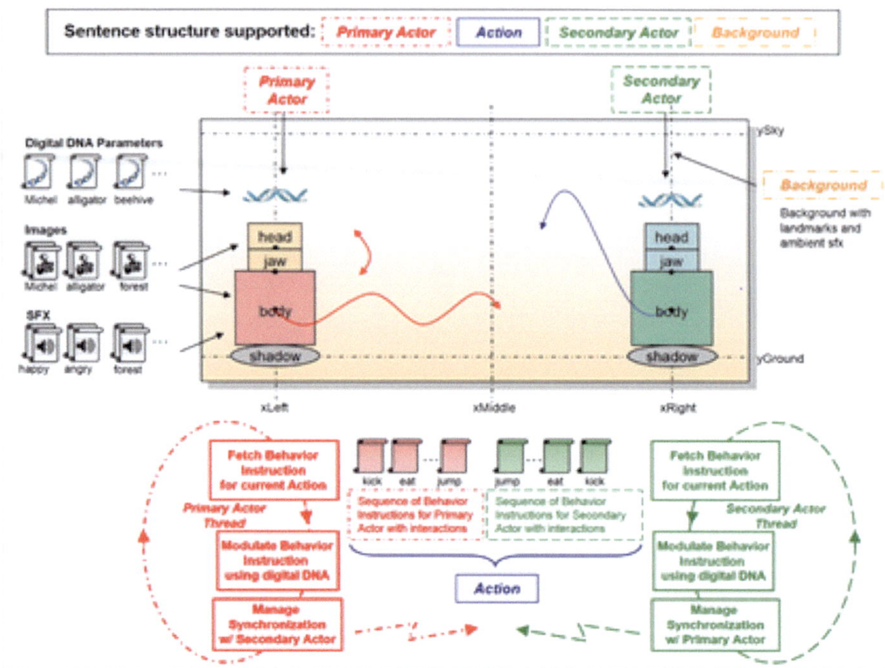

Fig. 12.20 Model for animating with digital DNA variations

from the KidWords dictionary. Behavior instructions are fetched for each actor and the resulting trajectories and sounds are modified to accord with their meta-data. A synchronization barrier ensures that the actors synchronize. For example, an actor that is kicked should not begin flying off before the kick is delivered. Since the primary actor will likely have different digital DNA than the secondary actor, they could behave quite differently (e.g., a "silly" primary actor might stumble around before kicking). Actions are defined as a sequence of behavior instructions, each of which indicates what the actor is to do, but not how: That is defined by the digital DNA. For instance, a digital instruction to move from point A to point B could be executed by striding, stumbling, or hopping. For more details, see this list of macro actions [7].

Figures 12.21 and 12.22 show generic animations for "kick" and "pick up" actions, respectively, without variants due to digital DNA.

A.3 Fontlings Data Structure

In the data structure for Fontlings (Fig. 12.23), each letter (or number/special character) has a split line and a rotation point for each part (top and bottom part). Each letter has position, angle (rotation), and scaling parameter (not shown) for

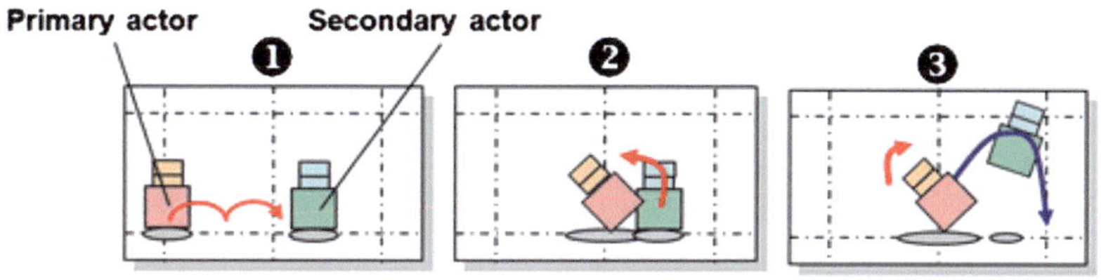

Fig. 12.21 Example of kicking

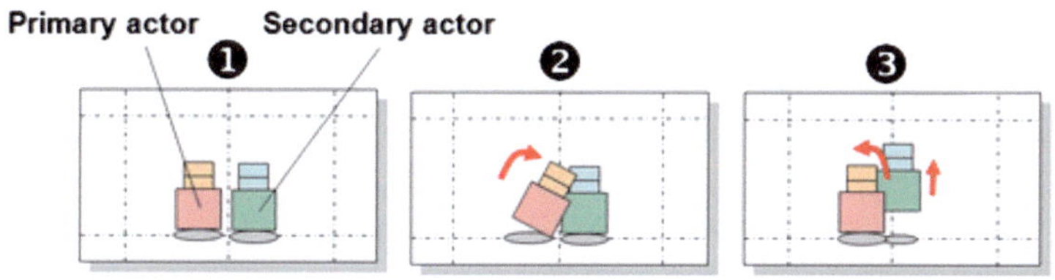

Fig. 12.22 Example of picking up

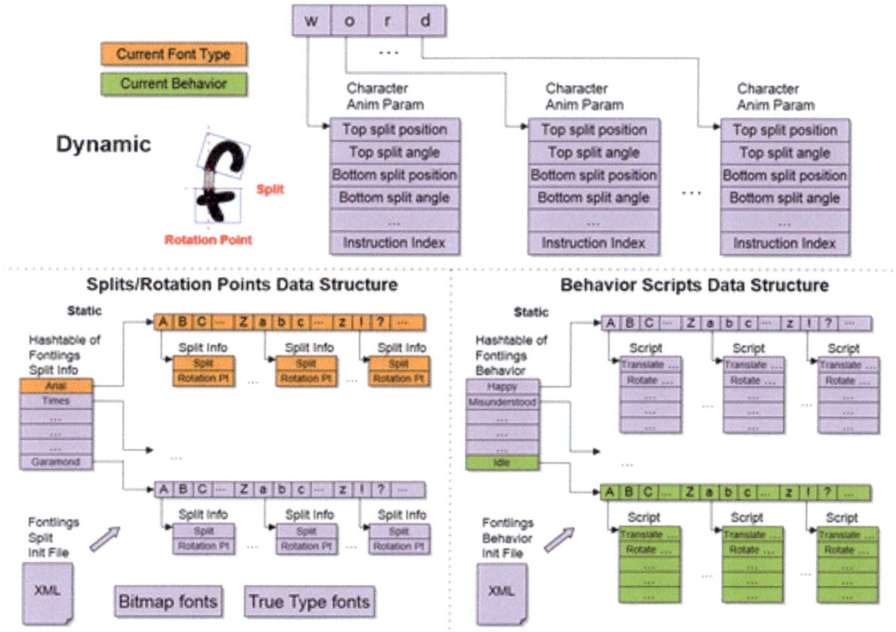

Fig. 12.23 Fontlings data structure

each part, and each letter animates independently (Top split position, Top split angle, Bottom split position, Bottom split angle, etc. in Fig. 12.23). An instruction index (Instruction Index in Fig. 12.23) tracks its position in the execution of the script. One letter or word could be in idle mode while another is happy. In this way, a specific word can highlight something while other words are alive but less dramatic. For each letter, scripts define behaviors (idle, happy, misunderstood, etc.). The Splits/Rotation Points Data Structure refers to a hash table with an entry key for each font type (Arial, Times, etc. in Fig. 12.23). It specifies rotation points and the height of the split line as a percent of the character's height. The Behavior Scripts Data Structure is a hash table with the behavior name as entry key (Happy, Misunderstood, Idle, etc. in Fig. 12.23). It contains a script for that behavior for each character. Figure 12.23 also highlights a specific example for the font type Arial (in orange color) and the behavior Idle (in green color).

The data structures are initialized by the xml file Fontlings Split Init File, which contains the height of the split line for each character of a given font, and Fontlings Behavior Init File, which contains the scripts to animate each character for each behavior (see bottom of Fig. 12.23).

Acknowledgements This work would not have been possible without Chuck Thacker, Craig Mundie, Margaret Johnson, Takako Aikawa, Lee Schwartz, Bill Buxton, Jonathan Grudin, Howard Phillips, Tom Steinke, Duncan, John D. Bransford, Diana Sharp, Heinz Schuller, Margaret Winsor, Allison Druin, Tim Kannapel, Arin Goldberg, Keith Daniels, Xiang Cao, Anoop Gupta, Chris Quirk, Bill Dolan, John Paquin, Neal Noble, Chad Essley, Lindsey Sparks, John Manferdelly, Xiaolong Li, the children participating in studies, and perhaps others I forgot to list.

References

1. Aikawa, T., Schwartz, L., Pahud, M.: Nlp Story Maker (2005)
2. Cptte demo page. https://aka.ms/mpahud-CPTTE2016-demos/
3. Dc Comics' "scribblenauts". www.scribblenauts.com
4. Lee, J.C., Forlizzi, J., Hudson, S.E.: The kinetic typography engine: an extensible system for animating expressive text. In: Proceedings of the 15th Annual ACM Symposium on User Interface Software and Technology, pp. 81–90. ACM (2002)
5. Lee, J., Kim, D., Wee, J., Jang, S., Ha, S., Jun, S.: Evaluating pre-defined kinetic typography effects to convey emotions. J. Korea Multimed. Soc. **17**(1), 77–93 (2014)
6. Loop, C., Blinn, J.: Resolution independent curve rendering using programmable graphics hardware. In: ACM Transactions on Graphics (TOG), vol. 24, pp. 1000–1009. ACM (2005)
7. Macro actions. https://aka.ms/mpahud-CPTTE2016-MacroActions/
8. Mueller, P.A., Oppenheimer, D.M.: The pen is mightier than the keyboard advantages of longhand over laptop note taking. Psychol. Sci. **25**(6), 1159–1168 (2014)
9. Oviatt, S.: The design of future educational interfaces. Routledge, New York (2013)
10. Pixelsense table. https://www.windowscentral.com/microsoft-surface-pixelsense-table
11. Strapparava, C., Valitutti, A., Stock, O.: Dances with words. In: IJCAI, pp. 1719–1724 (2007)
12. Visual sentence API documentation. https://aka.ms/mpahud-CPTTE2016-VisualSentenceAPI/
13. Wowwee. http://wowwee.com/products/robots

Chapter 13
A Multilingual Sketch-Based Sudoku Game with Real-Time Recognition

Caio D.D. Monteiro, Meenakshi Narayanan, Seth Polsley, and Tracy Hammond

Abstract Sudoku is one of the most popular puzzles of all time: easy to understand but still very challenging, Sudoku continues to captivate players all over the world through newspapers, puzzle books, and digital devices. The application introduced in this work is a multilingual, sketch-based version of the Sudoku game. Sketch input renders more flexibility to users and increases usability. Multilingual support paired with sketching allows the game to serve as an educational tool for those learning a new language, with the current implementation supporting Chinese and Hindi. The recognition algorithm proposed in this work, based on the Hausdorff metric, easily enables extending the application to support other languages. Preliminary results indicate an overall accuracy of over 93% when recognizing Chinese and Hindi numbers at the same time.

13.1 Introduction

Sudoku is a very well-known puzzle that appears in papers and magazines. It is a 9 by 9 grid filled with numbers from 1 to 9. The grid is partially filled when the game begins. Users have to complete the grid by filling in the missing fields. There can be no repeated number in any row, column or in the 3 by 3 sub-grids. It is naturally played with a pen and pencil. However, there have been many Sudoku applications available to play on the desktop and other mobile devices such as smartphones and tablets. Most of them use the keyboard to input numbers. Some of them have an online keypad from which numbers can be chosen, and played with a stylus. However, these designs do not facilitate natural user interaction.

In this paper we propose a sketch-based Sudoku game playable in multiple languages, currently supporting Chinese and Hindi. The game can be played on pen and touch surfaces very much like it is played on pen and paper, while it also can be

C.D.D. Monteiro (✉) • M. Narayanan • S. Polsley • T. Hammond
Sketch Recognition Lab, Department of Computer Science & Engineering, Texas A&M University, College Station, TX, 77843, USA
e-mail: caioduarte.diniz@tamu.edu; meen92@tamu.edu; spolsley@tamu.edu; hammond@tamu.edu

T. Hammond et al. (eds.), *Frontiers in Pen and Touch*, Human–Computer Interaction Series, DOI 10.1007/978-3-319-64239-0_13

used as an enjoyable and useful learning tool for beginners of a new language. The sketch-based input also encourages more natural interaction with the user, providing users the ability to modify fields through gestures or make partial strokes that the system will ignore so that they can keep notes without interfering with game play. These features further strengthen the analogy of the sketching surface to the feel of pen and paper.

Digital games can be appealing for many reasons, from the challenges they present to the degree of customization and freedom that they allow. This engagement factor makes digital games a powerful learning tool [11]. With that in mind, our proposed multilingual Sudoku game can serve as a support tool for learning the numbers of a new language because it keeps users engaged through the context of solving a puzzle. It combines the focus and attention required for completing a Sudoku board with the repetition of seeing sketched numbers and requires writing them as the input, exercising two important aspects of language learning. There is a large body of cognitive science research which supports the value of learning through problem solving [7, 9, 19, 20].

13.2 Related Work

Digital games have been used in education for quite some time, and the benefits have already been demonstrated [28]. One of the challenges faced by the usage of educational games is how to integrate them into more traditional learning schemas, Torrente et al. propose an educational games authoring tool that might ease this problem [27]. We elected to use a sketching interface to facilitate integration with player's current playing methods. The combination of sketch interfaces and educational games has also been discussed in recent years [13].

There have been many systems that recognize Chinese and other Asian characters using a geometric approach [17, 21–26]. These systems perform Chinese radical recognition using primitive recognizers [6, 14] and corner-finding [16, 29] for low-level interpretation, combined with LADDER [4, 5] for high-level recognition. While our system could have used LADDER for Chinese recognition because of the high geometric symmetry of Chinese numerals, our goal was to use one algorithm to support multiple languages.

Hindi numerals are generally drawn using a single stroke and could hence be classified well using a gesture recognition algorithm like Rubine's features [15]. In [1], Al-Omari proposes an Arabic numeral recognition system based on template matching, and [12] performs Hindi numeral recognition using neural networks. Since it deals with printed numbers instead of sketches, the pre-process stage is a little different than what it is commonly used in sketch recognition techniques. So, in order to provide multilingual support, we decided to use a simple template matching algorithm using a Hausdorff measure. Hausdorff is introduced in [8] and [18], and it is used extensively in many domains, including sketch recognition [3, 10] and image processing.

13.3 Methodology

Our system uses two separate applications, both developed using JAVA. The first application is used to collect templates from a certain number of people, which are used to train the system by assigning appropriate labels. These templates are then compared against user sketches in the recognition stage. The second application is the Sudoku game itself. Sketched user inputs for both applications were performed through a Wacom Bamboo Pen Tablet.

13.3.1 Training Data

This JAVA application was only created to ease the development and extension of the Sudoku game. It collects sketches from several test users in order to train the system in any new language. Figure 13.1 shows the interface for collecting training data.

We collected data from graduate students at Texas A&M University who were proficient in the languages of Chinese or Hindi. For each user, ten samples for each of the nine digits (1 through 9) were collected. In all, we have collected 270 samples, from three different users, for each language in the current implementation. For reference, Table 13.1 shows the digits collected for both Chinese and Hindi.

The application prints the digit to be written, along with the iteration number of that digit. A clear button is provided for the user to clear the canvas if he/she is

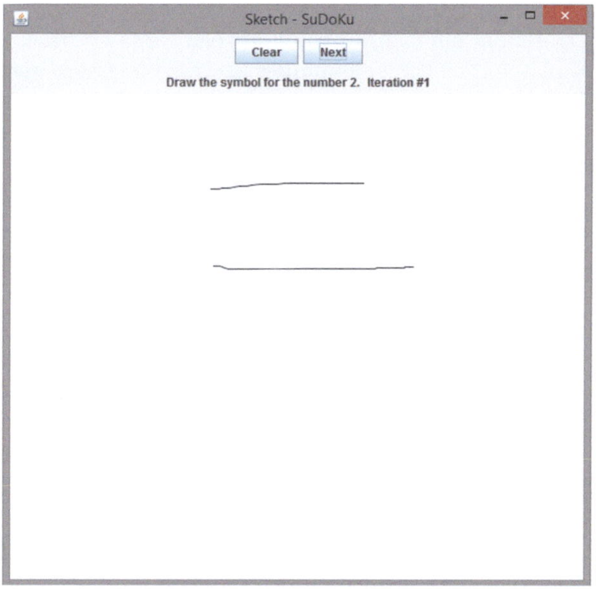

Fig. 13.1 The training app is a large canvas designed solely for the purpose of gathering a variety of user-drawn numbers

Table 13.1 Digits 1 through 9 in Chinese and Hindi

Number	Chinese	Hindi
1	一	१
2	二	२
3	三	३
4	四	४
5	五	५
6	六	६
7	七	७
8	八	८
9	九	९

not satisfied with the submission. After all the iterations are complete, an XML file is generated, which has coordinate information regarding the sketch drawn. There are tags to identify a number, iteration, the stroke number of that iteration, and the points (X and Y coordinates) for that stroke. This file is read at the start of the Sudoku application to normalize all the templates collected.

13.3.2 Sketch-Based Sudoku Application

The Sudoku game provides multilingual options in the form of radio buttons. Regardless of user selected language, sketches can be made in any of the languages supported by the system. So, if a Hindi user is playing in Chinese but forgot how a particular number is written, he can simply write it in Hindi and the system will display the recognized number in Chinese. Users interact with the game much as they would a pen and paper version of Sudoku, filling numbers in the grid according to the the rules of the game. Users can erase sketches using a scribble gesture, or an erase button that clears a cell on click.

The system also has a rough sketching mode that allows users to draw sketches without being recognized. Recognition begins after a certain timeout or as soon as the user starts drawing in another cell. Once a number is recognized, the user's sketch is replaced with the recognized digit in blue ink, differentiating it from the original numbers displayed in the puzzle. The Check button can be pressed whenever the user is done filling the grids or just want to assess the correctness of his current assignments, upon button press, all the user inputs are highlighted in green, if the number assigned to the cell is correct, or red, if the number assignment is incorrect. The game ends when there are no empty cells on the board and all the user inputs are highlighted in green. Figure 13.2 shows a game played in Chinese and Fig. 13.3 in Hindi.

Before running recognition, two conditions are checked. First, if the input sketch is smaller than a fixed threshold, then it is discarded. Second, if the sketch is

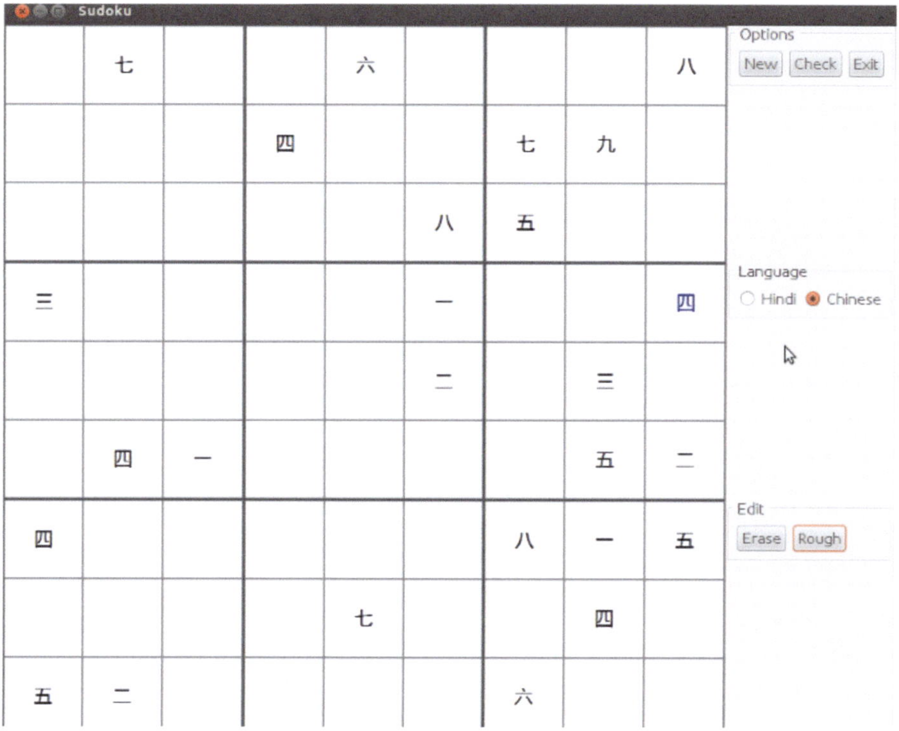

Fig. 13.2 Sudoku game played in Chinese

composed by a single stroke, a horizontal line test is performed to check if the input represents the Chinese numeral one. The second check was necessary because of the simplicity of the numeral concerned, since the system could confuse it with some other interface action. If neither condition is true then the system performs preprocessing and recognition.

Preprocessing is performed on the input sketch by resampling, scaling, and translating. These steps ensure input sketches can be matched to the normalized templates. Next, the one-sided Hausdorff distance is calculated according to $h(A, B) = \max_{a \in A} \min_{b \in B} |a - b|$. Hausdorff distance is a technique for comparing a sets of points; for each point of a model set, it measures how near they are from any point of another sketch set. We always compare the input sketch (A) to the template (B). After obtaining the distances to the template sketches, we use a k-NN classifier with a neighborhood size of 5 to improve selection accuracy. The label with the majority of votes when considering the 5 smallest Hausdorff distances is the recognized number.

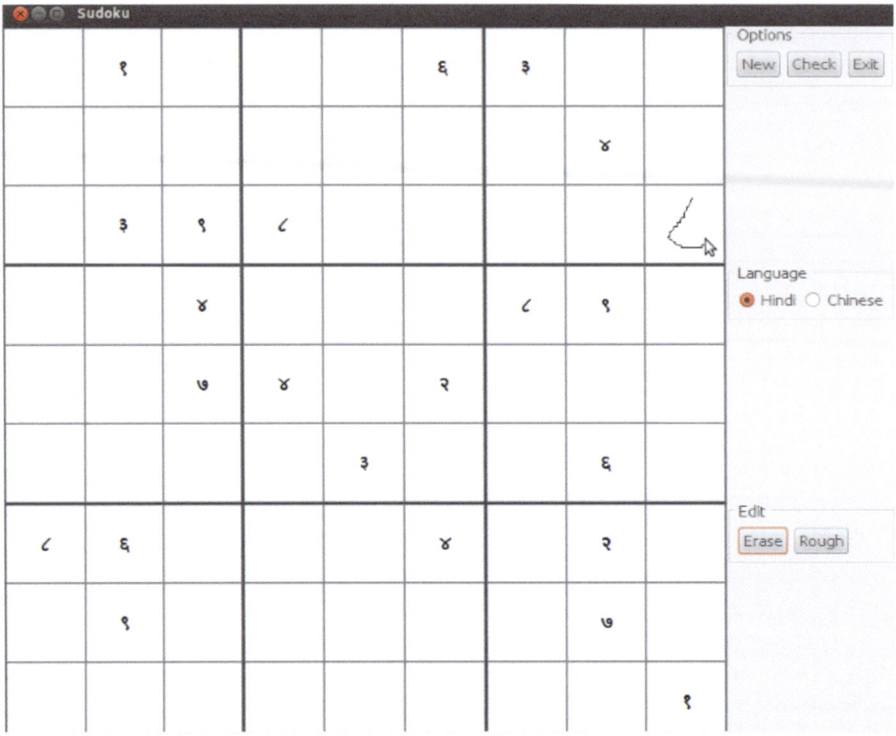

Fig. 13.3 Sudoku game being played in Hindi. Note that the user is currently drawing a symbol; this will be automatically recognized and replaced with the beautified number after a slight delay from when the pen is lifted

13.4 Evaluation

We use k-fold cross validation to evaluate our recognizer, with k = 10 [2]. We consider all-or-nothing accuracy, since that is all that concerns the user in this application, so accuracy is calculated as the number of correct recognitions divided by the total training set size. Each fold contained 54 sketches from our data set, 3 samples for each of the 18 different numbers, 9 Chinese and 9 Hindi. Each time, one of the 10 folds is used for testing and the other nine for training. This process is repeated until all the folds have been used for testing once.

Table 13.2 shows the confusion matrix for the system. As shown, the majority of the numbers were classified with an accuracy of over 90%, and 4 of the numbers were perfectly classified. By language, Chinese characters were recognized better than the Hindi numbers. Several digits in Hindi are complex and similar to each other; see Table 13.1 for a comparison of each number. We see this with the weakest recognition for a Hindi *3*, which is very similar to *2* in that script. As Table 13.2

Table 13.2 Confusion matrix for 10-fold cross validation across all supported characters, Chinese and Hindi

		Chinese									Hindi								
		1	2	3	4	5	6	7	8	9	1	2	3	4	5	6	7	8	9
	1	0.93	0	0	0	0	0	0	0	0	0	0	0	0	0	0.03	0.03	0	0
C	2	0	0.93	0.03	0	0	0	0	0	0	0	0	0	0	0	0	0	0.03	0
h	3	0	0	1.0	0	0	0	0	0	0	0	0	0	0	0	0	0	0	0
i	4	0	0	0	1.0	0	0	0	0	0	0	0	0	0	0	0	0	0	0
n	5	0	0	0	0.03	0.97	0	0	0	0	0	0	0	0	0	0	0	0	0
e	6	0	0	0	0	0	0.97	0	0	0.03	0	0	0	0	0	0	0	0	0
s	7	0	0	0	0	0	0	0.97	0	0	0	0	0	0	0	0	0.03	0	0
e	8	0.03	0	0	0	0	0	0	0.97	0	0	0	0	0	0	0	0	0	0
	9	0	0	0	0	0	0.07	0	0	0.9	0	0	0	0	0	0	0.03	0	0
	1	0	0	0	0	0	0.03	0	0	0	0.93	0	0.03	0	0	0	0	0	0
	2	0	0	0	0	0	0	0	0	0	0.03	0.93	0	0	0	0.03	0	0	0
H	3	0.03	0	0	0	0	0.07	0	0	0	0.03	0.1	0.73	0	0	0.03	0	0	0
i	4	0	0	0	0	0	0	0	0	0	0	0	0	1.0	0	0	0	0	0
n	5	0	0	0	0	0	0	0	0	0	0	0	0	0	0.83	0.07	0.03	0.07	0
d	6	0	0	0	0	0	0	0	0	0	0	0	0	0	0	0.93	0	0.03	0.03
i	7	0	0	0	0	0	0	0	0	0	0	0	0	0	0	0	1.0	0	0
	8	0	0	0	0	0	0.03	0.07	0	0	0	0	0	0	0	0	0.07	0.83	0
	9	0	0	0	0	0	0	0	0	0	0.03	0	0.03	0	0	0	0	0	0.93

shows, the recognizer favored *2* on classification for that digit over *3*, so *2* retains a high recognition accuracy. The overall classifier accuracy is 93%.

13.5 Discussion

While the results found on the confusion matrix are mostly positive, there is still room for improvement. In general, Hindi numbers seems to be harder to classify using our current template matching technique. These numbers are very curvy and share many similarities with one another, making template matching sensitive to noise. Template matching is still a desirable approach since we wanted a flexible system that could easily support enrolling new languages, but some additional language-specific rules may need to be added as in the case of the Chinese numeral one.

In regards to the user interface of the application, there are also improvements to be considered. For the application to run successfully on smaller mobile devices, it should allow the user to expand the cell size to draw the sketch comfortably. Too small cell sizes can restrict the user from drawing with ease. However, any cell expansion should not hinder the view of the rest of the Sudoku grid.

As mentioned before, the application can be extended to support more languages, but it should be possible to come up with on-the-fly language recognition. In this mode, the game board numbers would change between the available languages according to the user-sketched language. This adds another layer of learning and difficulty, which could be integrated with further optimizations to include different difficulty levels.

13.6 Conclusion

In this work, we presented a multilingual, sketch-based Sudoku game. Sudoku is a puzzle that is famous all over the globe. Our application can provide a nice variation to those already familiar with the game by providing them with the option of playing in other languages through a natural sketch interface. Multilingual support also allows players to use it as a learning tool to familiarize themselves with writing other languages.

Acknowledgements We thank the Texas A&M University Sketch Recognition Lab, especially Paul Taele, for his guidance, and all the volunteers that helped us train and test the supported languages. Author Caio D. D. Monteiro also gratefully acknowledges the scholarship from CAPES #11886/13-7.

References

1. Al-Omari, F.: Handwritten Indian numeral recognition system using template matching approaches. In: ACS/IEEE International Conference on Computer Systems and Applications, pp. 83–88. IEEE (2001)
2. Bengio, Y., Grandvalet, Y.: No unbiased estimator of the variance of k-fold cross-validation. J. Mach. Learn. Res. **5**, 1089–1105 (2004)
3. Field, M., Valentine, S., Linsey, J., Hammond, T.: Sketch recognition algorithms for comparing complex and unpredictable shapes. In: IJCAI, pp. 2436–2441 (2011)
4. Hammond, T.: Ladder: a perceptually-based language to simplify sketch recognition user interface development. Ph.D. Doctoral Dissertation, Massachusetts Institute of Technology (MIT), Cambridge (2007). Advisor: Randall Davis. 495 pp.
5. Hammond, T., Davis, R.: Ladder, a sketching language for user interface developers. Comput. Graph. **29**(4), 518–532 (2005)
6. Hammond, T., Paulson, B.: Recognizing sketched multistroke primitives. ACM Trans. Interact. Intell. Syst. **1**(1), 4:1–4:34 (2011). doi:10.1145/2030365.2030369. http://doi.acm.org/10.1145/2030365.2030369
7. Harlow, H.F., Harlow, M.K., Meyer, D.R.: Learning motivated by a manipulation drive. J. Exp. Psychol. **40**(2), 228 (1950)
8. Huttenlocher, D.P., Klanderman, G., Rucklidge, W.J., et al.: Comparing images using the Hausdorff distance. IEEE Trans. Pattern Anal. Mach. Intell. **15**(9), 850–863 (1993)
9. Iba, G.A.: Learning by discovering macros in puzzle solving. In: IJCAI, pp. 640–642 (1985)
10. Kara, L.B., Stahovich, T.F.: An image-based, trainable symbol recognizer for hand-drawn sketches. Comput. Graph. **29**(4), 501–517 (2005)
11. Mitchell, A., Savill-Smith, C.: The use of computer and video games for learning: a review of the literature (2004)
12. Pandey, A., Kumar, A., Kumar, R., Tiwari, A.: Handwritten devanagri number recognition using majority voting scheme. Int. J. Comput. Sci. Inform. Technol. Secur. (IJCSITS)(2), 2249–9555. ISSN (2012)
13. Paulson, B., Eoff, B., Wolin, A., Johnston, J., Hammond, T.: Sketch-based educational games: drawing kids away from traditional interfaces. In: Proceedings of the 7th International Conference on Interaction Design and Children, pp. 133–136. ACM (2008)
14. Paulson, B., Hammond, T.: Paleosketch: accurate primitive sketch recognition and beautification. In: Proceedings of the 13th International Conference on Intelligent User Interfaces, IUI'08, pp. 1–10. ACM, New York (2008). doi:10.1145/1378773.1378775. http://doi.acm.org/10.1145/1378773.1378775
15. Rubine, D.: Specifying Gestures by Example, vol. 25. ACM, New York (1991)
16. Sezgin, T.M., Stahovich, T., Davis, R.: Sketch based interfaces: early processing for sketch understanding. In: ACM SIGGRAPH 2006 Courses, p. 22. ACM (2006)
17. Shahzad, N., Paulson, B., Hammond, T.: Urdu Qaeda: recognition system for isolated urdu characters. In: Proceedings of the Workshop on Sketch Recognition at the 14th International Conference of Intelligent User Interfaces (IUI). ACM, Sanibel (2009)
18. Shapiro, M.D., Blaschko, M.B.: On Hausdorff Distance Measures, vol. 1003. Computer Vision Laboratory University of Massachusetts Amherst, Amherst (2004)
19. Sweller, J.: Cognitive load during problem solving: effects on learning. Cogn. Sci. **12**(2), 257–285 (1988)
20. Sweller, J.: Cognitive technology: some procedures for facilitating learning and problem solving in mathematics and science. J. Educ. Psychol. **81**(4), 457 (1989)
21. Taele, P., Hammond, T.: Chinese characters as sketch diagrams using a geometric-based approach. In: Proceedings of the 2008 IEEE Symposium on Visual Languages and Human-Centric Computing Workshop on Sketch Tools for Diagramming, pp. 74–82 (2008)
22. Taele, P., Hammond, T.: Using a geometric-based sketch recognition approach to sketch Chinese radicals. In: AAAI, pp. 1832–1833 (2008)

23. Taele, P., Hammond, T.: Hashigo: a next-generation sketch interactive system for Japanese kanji. In: Proceedings of the Twenty-First Innovative Applications of Artificial Intelligence Conference (IAAI), pp. 153–158. AAAI, Pasadena (2009)
24. Taele, P., Hammond, T.: Enhancing instruction of written east asian languages with sketch recognition-based intelligent language workbook interfaces. In: Hammond, T., Valentine, S., Adler, A., Payton M. (eds.) The Impact of Pen and Touch Technology on Education, 1st edn., pp. 119–126. Springer, Cham (2015)
25. Taele, P., Hammond, T.: An intelligent sketch-based educational interface for learning complex written east asian phonetic symbols. In: Hammond, T., Valentine, S., Adler, A. (eds.) Revolutionizing Education with Digital Ink: The Impact of Pen and Touch Technology on Education, pp. 129–140. Springer International Publishing, Cham (2016). doi:10.1007/978-3-319-31193-7_9. http://dx.doi.org/10.1007/978-3-319-31193-7_9
26. Taele, P., Hammond, T.A.: A geometric-based sketch recognition approach for handwritten mandarin phonetic symbols I. In: 2008 International Workshop on Visual Languages and Computing (VLC) at the 14th International Conference on Distributed Multimedia Systems (DMS). Knowledge Systems Institute, Boston (2008). 6 pp.
27. Torrente, J., Del Blanco, Á., Marchiori, E.J., Moreno-Ger, P., Fernández-Manjón, B.: < e-adventure>: introducing educational games in the learning process. In: Education Engineering (EDUCON), 2010 IEEE, pp. 1121–1126. IEEE (2010)
28. Virvou, M., Katsionis, G., Manos, K.: Combining software games with education: evaluation of its educational effectiveness. J. Educ. Technol. Soc. **8**(2), 54–65 (2005)
29. Wolin, A., Eoff, B., Hammond, T.: Shortstraw: a simple and effective corner finder for polylines. In: Proceedings of the Fifth Eurographics Conference on Sketch-Based Interfaces and Modeling, SBM'08, pp. 33–40. Eurographics Association, Aire-la-Ville (2008). doi:10.2312/SBM/SBM08/033-040. http://dx.doi.org/10.2312/SBM/SBM08/033-040

Part VI
Let's Collaborate

Chapter 14
NuSys: Collaborative Insight Extraction on Shared Workspaces

Philipp Eichmann

Abstract Gaining insights – the act of acquiring a deep and intuitive understanding of a thing or concept – is often achieved through discussions in small work groups. Students frequently conduct informal study group sessions where scholarly material is revisited, and questions and ideas are bounced back and forth as a way to gain and present new insights. NuSys is a software framework that enhances how groups develop ideas when a shared workspace on an interactive whiteboard (IWB) and smaller devices driven by pen & touch gestures replace a physical whiteboard as the collaborative focal point. The core component of our system provides a fluid interface to an unbounded two-dimensional workspace, where study material can be seamlessly imported, laid out, grouped, annotated, linked and tagged collaboratively on multiple clients in real-time and in a free-form fashion.

14.1 Overview

The goal of extracting insights from multiple diverse sources is as old as civilization, as can be seen in such precomputer examples as the Library of Alexandria [3], Paul Otlet's UDC [6], and Vannevar Bush's Memex [7]. For the design of NuSys, we define Insight Extraction as an iterative interactive process consisting of the following three phases that provide an "open" spatial hypermedia system that interoperates with Office apps and the web: *Information Gathering:* When beginning to investigate a topic or hypothesis, relevant information in today's digital environment is gathered from such sources as PDF files, HTML documents found on the Internet, images, videos, formatted files and presentations such as Microsoft Office and Open Office documents, spreadsheets, emails, as well as structured data in the form of databases. *Sensemaking:* Concurrent with the gathering of relevant material, information workers analyze, interpret, and relate their document contents. Activities include writing notes, building relationships, and structuring document groups in order to gain new insights, and to support or refute hypotheses based on

P. Eichmann (✉)
Computer Science Department, Brown University, 115 Waterman Street, 4th floor, Providence, RI, 02912, USA
e-mail: philipp_eichmann@brown.edu

T. Hammond et al. (eds.), *Frontiers in Pen and Touch*, Human–Computer Interaction Series, DOI 10.1007/978-3-319-64239-0_14

evidence found in the gathered data. *Export:* At some point during the Sensemaking phase, information workers usually present interim findings to colleagues and supervisors or compile a summary/handout of their results that is used for further brainstorming and reviewing.

While many research projects and commercially available software [1, 4, 5, 8, 9] provide support for some or all three phases, they often fail to allow for seamless transitions between them. For instance, when importing content from a web browser or Word document, the user is often forced to perform multiple consecutive copy-paste operations or use the hard disk as temporary storage. Similarly, in order to view, edit and annotate different media types, users must switch back and forth between different applications, thereby disrupting the flow of their work. In addition, few existing solutions exploit the benefits of pen & touch and do not provide a fluid interface [2] that scales from tablet devices to laptops and large interactive whiteboards (IWBs). Existing approaches also make it unnecessarily tedious to create structures such as groups or add metadata to content, discouraging information workers from appropriately organizing their material as they progress in their task, thereby limiting the system's ability to assist users by providing search and visualization features. Furthermore, there is often a lack of robust multi-user experience due to the applications' limited collaborative features.

Inspired by WorkTop [10], we put our focus on the Information Gathering and Sensemaking phase, and address the issues outlined above with an interconnected collection of software components described below.

14.2 NuSys

14.2.1 NuSys App

The NuSys App is a Windows 10 Universal App that provides a workspace to view, lay out, organize, annotate and share heterogeneous materials such as Office documents, multimedia files, and entire websites or fragments thereof, with a focus on pen & touch interaction. A workspace is an unbounded, zoomable, and pannable 2D canvas that can be shared among multiple users; changes to it are reflected among all connected clients in real-time, and locking mechanism prevent synchronization issues. It facilitates explicit and implicit (spatially-contiguous objects) data linking, allowing for whiteboard content to be transformed to rich visual information networks that illustrate emergent patterns and help gain perspective. Following a "Post-It on a whiteboard" metaphor, a workspace allows users to import content as "nodes" that can be arranged, grouped, annotated, linked, and tagged in a free-form fashion for subsequent retrieval. In contrast to other information worker software, nodes and user-generated relationships such as links or groups are first class citizens, meaning that they can be linked to any other item of the same status and can carry metadata. NuSys App provides viewers and editors for first class citizens, allowing, for instance, for images, PDFs and videos to be displayed directly in the workspace, instead of having to open them in a different application. Furthermore, NuSys App incorporates ink, handwriting recognition and speech-to-text support, allowing for textual content to be added without a keyboard, as well as various intuitive gestures that can be used to navigate and augment the workspace.

14.2.2 Microsoft Word and Powerpoint Add-Ins

Our Word and PowerPoint add-ins provide extended clipboard functionality that allows information workers to select multiple individual pieces of content and batch-export them to an opened workspace, thereby minimizing context switches. Once a selection is exported to a workspace within NuSys App, a bi-directional link between the region within the original Office document and the created node on the workspace is established, enabling a user to jump from the node in the workspace to its original selection within an Office document.

14.2.3 Google Chrome Plugin

Selecting fragments of content on websites, such as text, lists, tables and images, and copying them to note-taking applications or word processors is a common task for information workers. However, web browsers only offer crude support for selections due to the restrictive underlying HTML/CSS model, making it difficult and sometimes even impossible for users to select and copy content. Our plugin for Google Chrome allows for selection gestures to be drawn directly onto websites and maps selection gestures to underlying HTML content, in contrast to other clipping tools, which merely take an image-based snapshot. It thereby enables users to more intuitively mark and extract regions of websites while preserving textual and semantic information, which, analogous to the Word and Powerpoint Add-ins, can be exported to a workspace in NuSys App. In particular, URL's in selections remain live when exported to NuSys.

14.3 Conclusion and Future Work

NuSys supports collaborative insight extraction by facilitating real-time viewing, organizing and editing of content on a shared, unbounded workspace and by providing plugins to third-party applications that lower the technical barrier when importing content to a workspace. For future research, a direction we are particularly interested in is to extend the NuSys App's functionality by enabling users to add structured metadata to content and suggest tags for imported media using an improved topic modeling approach. Using this information, we are then able to autogenerate visualizations such as timelines, infer relationships between nodes and improve data-dependent features such as search. In the current version of NuSys, there is no support yet for the Export phase. For a future release, we plan to integrate algorithms that interpret the user-generated structures on the workspace and the contents' metadata in order to create informative exportables such as linear presentations, branching narratives, and meeting summaries. We also will conduct user studies of NuSys' effectiveness.

Acknowledgements The author thanks Andy van Dam, Bob Zeleznik, and Rosemary Simpson for their valuable feedback and discussions, and all undergraduate researchers who have contributed to this project.

References

1. Bernstein, M.: Can we talk about spatial hypertext. In: Proceedings of the 22nd ACM Conference on Hypertext and Hypermedia, pp. 103–112. ACM (2011)
2. Elmqvist, N., Moere, A.V., Jetter, H.C., Cernea, D., Reiterer, H., Jankun-Kelly, T.: Fluid interaction for information visualization. Inf. Vis. **10**(4), 327–340 (2011)
3. Erskine, A.: Culture and power in ptolemaic Egypt: the museum and library of Alexandria. Greece and Rome (second series) **42**(1), 38–48 (1995)
4. Microsoft: Onenote. http://www.onenote.com. Retrieved on 10 Jan 2016
5. Microsoft: Sway. http://www.sway.com. Retrieved on 10 Jan 2016
6. Rayward, W.B.: Visions of Xanadu: Paul Otlet (1868–1944) and hypertext. J. Am. Soc. Inf. Sci. (1986–1998) **45**(4), 235 (1994)
7. Simpson, R., Renear, A., Mylonas, E., van Dam, A.: 50 years after "as we may think": the brown/mit Vannevar bush symposium. Interactions **3**(2), 47–67 (1996)
8. Storyboard: Storyboard. http://www.storyboardthat.com. Retrieved on 10 Jan 2016
9. Wright, W., Schroh, D., Proulx, P., Skaburskis, A., Cort, B.: The sandbox for analysis: concepts and methods. In: Proceedings of the SIGCHI Conference on Human Factors in Computing Systems, pp. 801–810. ACM (2006)
10. Zeleznik, Bob: Worktop. http://cs.brown.edu/research/ptc/worktop/. Retrieved on 10 Jan 2016

Chapter 15
Collaborative Spaces: One Note in the Middle School Classroom

Sara Mata

Abstract Collaboration is an essential component of dynamic learning environments. It promotes a sense of community and allows individuals to work together as they generate new ideas. In the middle school classroom, it is especially important to build a community in which students can openly share their insights and work with one another to critically think about ideas and issues they are wrestling with. In my English classroom, students participate in small book clubs year round. As a certified reading specialist and English teacher, I believe that reading literature should be a shared experience and that students should collaborate and explore ideas together as they read, write, speak, and listen to one another. This research paper summarizes the process of developing physical and digital collaborative spaces in a middle school English classroom while using Microsoft OneNote software, which incorporates digital ink and tablet technology. Students completed a round of book clubs in three weeks as they shared a space in the classroom where they could read together, discuss their ideas as they wrestled with complex concepts, such as character motivation, author's purpose, and the theme of the book they read. In addition, students utilized OneNote's collaboration space to help them work together as they read their book. The use of digital pen and tablet technology also enabled me as their teacher to give them instant feedback as I evaluated their reading progress all along. As a result, students were able to build innovative, collaborative spaces that enabled them to work together, provide feedback to one another, and have dynamic dialogues about literature.

15.1 Problem Statement and Context

Whitfield School is an independent, coeducational college preparatory day school that caters to students in grades 6–12. It is located in suburban Saint Louis County. Whitfield School's mission is to "cultivate ethical, confident, and successful students in a community of innovation, collaboration, and trust." I teach sixth and seventh

S. Mata (✉)
Whitfield School, 175 South Mason Road, Creve Coeur, MO, 63141, USA
e-mail: sara.mata@whitfieldschool.org

T. Hammond et al. (eds.), *Frontiers in Pen and Touch*, Human–Computer Interaction Series, DOI 10.1007/978-3-319-64239-0_15

grade English to a group of fifty-eight students. Roughly 25% of those students have diagnosed learning disabilities, mostly related to reading.

As I made observations of my students throughout the first several weeks of this school year's semester, I observed that they were not highly engaged in the process of reading literature and discussion of novels. As I collected data about their reading habits, I found that a large number of students did not enjoy reading and only read novels when they were required for school assignments. One of the reasons why this was the case is due to a lack of selection of books by reading whole-class novels. In addition, students with processing delays and/or diagnosed reading disabilities were not involved in collaborative groups that enabled them to process their novels as they were reading them. These valuable observations led me to implement highly collaborative book clubs that would not only allow students to work together as they read their books but would also engage them in profound discussions about the text. In an article published in the Journal of Adolescent and Adult Literacy [3], Lloyd claims the importance of engaging in meaningful discussion during literature circles. She specifically emphasizes that reading should be a shared experience and that teacher-led discussions with teacher-centered discussion questions won't promote the motivation and reading stamina of students. Furthermore, Lloyd indicates that many educators turn opportunities for rich literature conversations and turn them into empty assignments by telling students which questions and which topics to explore and discuss from a book. Additionally, OneNote has repeatedly been shown to be a successful platform for engaging k–12 students [1, 2, 4–7].

In order to promote authentic, student – centered discussion amongst my students, I implemented the use of OneNote's collaboration space for each of the book clubs so that students could gather and share ideas, resources, and weekly annotations/notes from their readings and Socratic Seminar discussions.

The Essential Questions for my English courses are:

- How does literature reflect a culture's values and beliefs?
- How do I effectively communicate who I am and what I believe?
- How do our values and beliefs shape who we are as individuals?
- How are individuals transformed through their relationships with others?
- How do I effectively communicate who I am and what I believe?
- What transforms personal identity?
- What distinguishes personal identity from group identity?
- How do societal identifiers such as race, culture, language, gender, and social class influence an individual's idea of self?

Although students did not directly respond to these questions during their book club discussions, the themes they discussed tied flawlessly into many of these questions.

15.2 Method Employed

As a reading specialist, I conducted a diagnostic baseline reading assessment to learn more about my students as readers. From the information gained from this data, I gathered that a large number of my students had strong decoding skills but could not recall information that they had read. However, when discussing the passage instead of simply answering my questions, students could demonstrate their understanding of the passage. Having the opportunity to discuss what they had read gave them an opportunity to process it more easily instead of answering the questions I had prepared for them based on my assessment. In addition, I gave students a reading interest survey to learn more about them, their interests, and their feelings towards reading. Based on this survey, I learned that many of my students were only reading books that were assigned to them and were not highly engaged in the process of reading. Most of the students that felt this way were students whose reading level is lower than the rest of the class and have a diagnosed reading disability.

Based on this information, I decided that it would be best for my students to have authentic reading experiences that would promote active dialogue, critical thinking, and collaboration. In my sixth grade classes, I offered two different novels for book club: *Runaway*, by Wendelin Van Draanen, and *Hatchet*, by Gary Paulsen. In seventh grade, I offered three different novels: *Touching Spirit Bear*, by Ben Mikaelsen, *Part of Me*, by Kimberly Willis Holt, and *Where the Red Fern Grows*, by Wilson Rawls. These books are different not only in their themes and plots, but also in reading level. This allowed for differentiation as needed based on students' reading needs. I made sure that students knew the challenges of each book so that they could make a selection that would be appropriate for them. I conferred with students about their book club choice to ensure that they were picking a novel that would challenge them but not be too difficult.

I set up a collaboration space on OneNote for each of the book club books to ensure that they would have all the resources they needed to successfully run their book club. I included items such as KWL templates, book club schedule templates, discussion norms/skills, Socratic Seminar rubrics, annotation rubrics, etc. I conferred with each book club (Fig. 15.1) to discuss group expectations as well as expectations for OneNote's collaboration space. Some of these conversations emphasized the idea of being open to learning from others and also being open to having others collaborate with us and provide us with feedback. We discussed the idea of collaboration as a whole and how our book club would enable us to become better at working with one another. I provided students with the option of choosing a spot in the classroom that would help them collaborate as a book club (some students chose the main seminar table, other students chose the carpeted area by the classroom library, etc.). Regardless of whether their book club space was digital (on OneNote) or physically in the classroom, students understood that the goal of book club would be to collaborate as they engaged in dynamic discussion or analysis of the text each week.

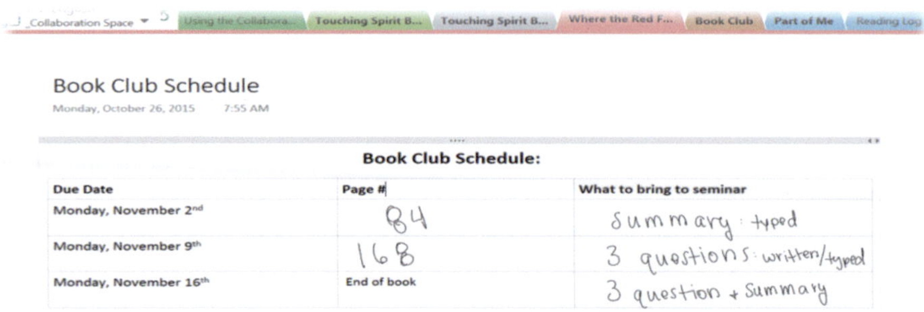
Book Club Schedule

Monday, October 26, 2015 7:55 AM

Book Club Schedule:

Due Date	Page #	What to bring to seminar
Monday, November 2nd	84	summary: typed
Monday, November 9th	168	3 questions: written/typed
Monday, November 16th	End of book	3 question + summary

Fig. 15.1 Book club schedule template

Fig. 15.2 Conferring with book club groups

Figure 15.1 includes an example of a resource I included on the collaboration space. This Book Club Schedule resource helped students stay on track and organized. In addition, it held them accountable for different assignments they would need to bring to book club in order to discuss. Students were supposed to each bring their annotations from that week as a resource for discussion as well.

Figure 15.2 illustrates the conferring that took place individually and in small groups with students to not only discuss their book club selections but also to discuss book club norms as well as individual student progress/feedback as book clubs were taking place.

Figure 15.3 shows a group of sixth grade students reading and discussing the novel *Runaway* by Wendelin Van Draanen. Just as this group did, other book club

Fig. 15.3 Sixth grade during a Socratic Seminar

groups found spaces in the classroom that would enable them to collaborate together as they discussed the novel.

As one can see on Fig. 15.4, the use of the collaboration space while having a Socratic Seminar allows students to take notes that they all have access to at the same time. This way, they are able to keep track of the details being shared throughout the discussion. The note – taker is able to use digital ink while taking notes, which helps him/her to better process the information being shared by synthesizing it and only writing out the main details of the discussion versus copying down everything being discussed word for word. Figure 15.5 illustrates an example of collaboration as students brainstorm to determine strong details for a written constructed response. Students used digital ink to add their responses to the collaboration space as they discussed these together in the classroom.

Students were also able to have meaningful conversations about their book annotations. Figure 15.6 illustrates student feedback for peer-based indicators on a 4 point rubric that focused on Signposts, questions, summarizing, characterization, and significant quotes/passages. Students utilized digital ink to provide meaningful feedback on the margins of the rubric. In addition, since multiple students gave feedback to one another, they selected different colors of digital ink in order to distinguish themselves from one another.

The use of collaboration space has also enabled my students to post their individual written responses for each other to read and learn from. For example, students wrote summaries that included key details from their novels and posted these to the collaboration space. They read one another's summaries in order to compare and contrast what plot details their book club members considered

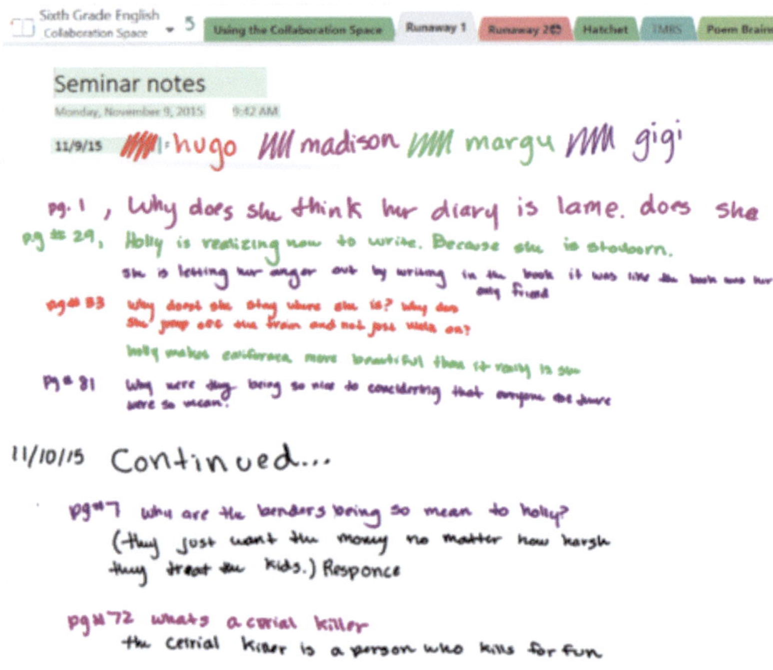

Fig. 15.4 Socratic Seminar notes on collaboration space

important/unimportant to include in their individual summaries. As one can see on Figs. 15.7 and 15.8, students even wrote comments to one another praising each other's work and also giving them ideas for key details they might consider adding to their summaries in order to make them more complete. Students partnered up to share their feedback with one another and to collaborate on ways to make their summaries stronger.

Since the use of collaboration space has been successful in my classroom, I invited students to have the opportunity to annotate digitally by using a "double-entry annotations" instead of having to annotate inside a book. The idea of digital annotations was something that I began to explore this school year that I had not previously considered in the past. Double-entry annotations are organized in a T-chart, where the left space of the T is dedicated to writing down the quote or passage and the right space of the T is dedicated to writing down the annotation based on that quote. Students began to post their annotations on the collaboration space so that members of their book club could keep up with one another's ideas about the book as they would independently read the book throughout the week. Figure 15.9 shows an example of a student's double entry annotations.

Another form of collaboration that takes place in my classroom occurs one-on-one between my students and me as I provide them feedback on their class work. Although I use a conferencing table to meet individually with students, the dialogue

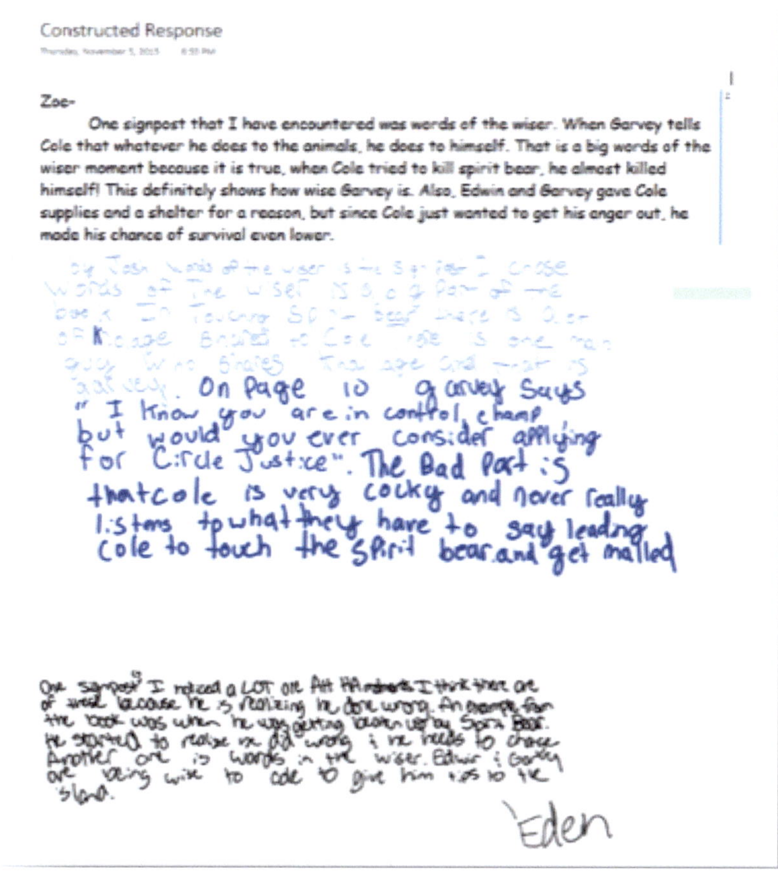

Fig. 15.5 Constructed response brainstorming collaboration

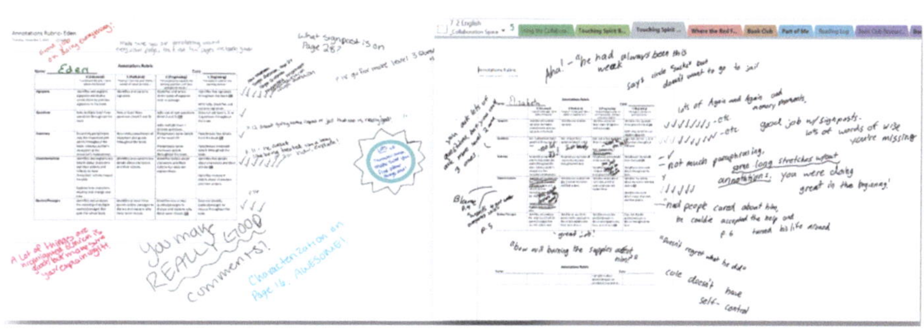

Fig. 15.6 Annotations rubric with feedback

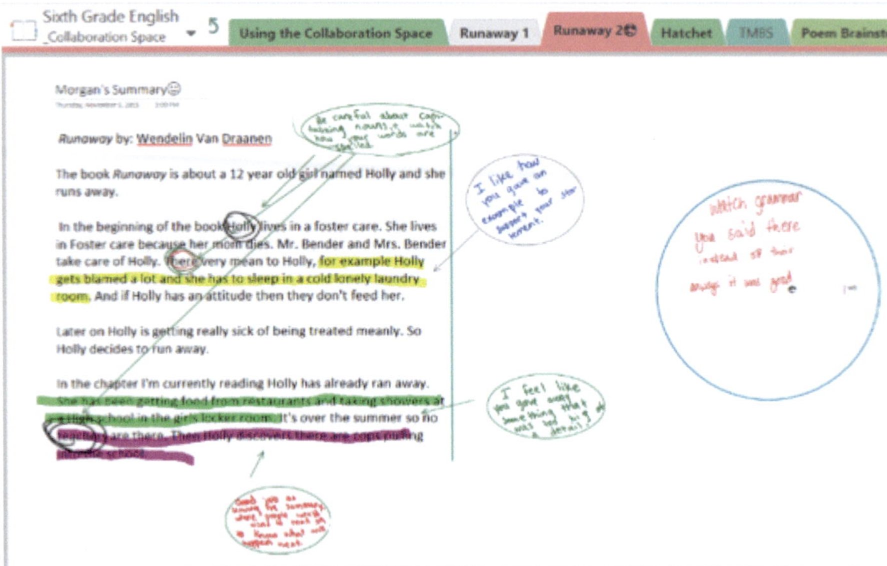

Fig. 15.7 Peer feedback on book summary

Fig. 15.8 Student providing feedback with digital ink

about their growth does not stop with those conferring meetings. In addition to these, students and I collaborate on OneNote. I am able to provide them with individual and private feedback that only they can see on their individual OneNote notebooks. Figure 15.10 illustrates a Socratic Seminar rubric in which I provide a

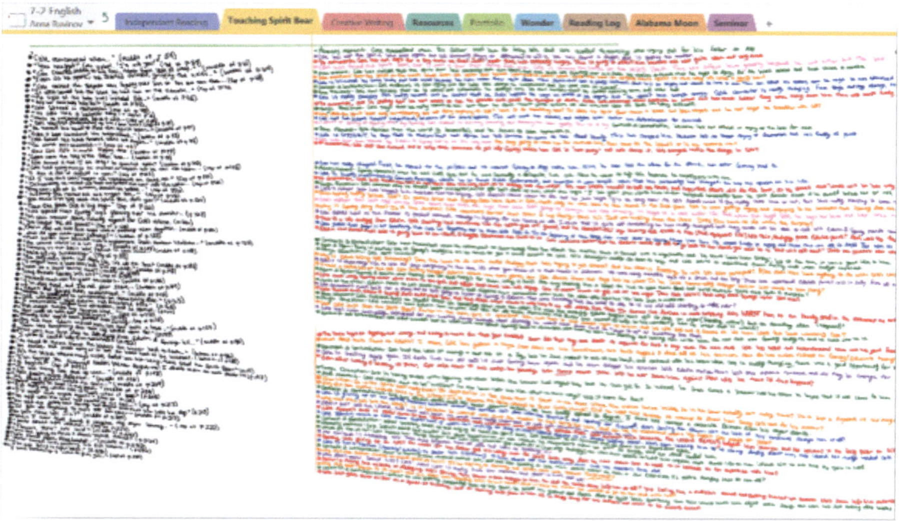

Fig. 15.9 Digital double-entry annotations on OneNote's collaboration space

specific student with feedback on her contributions to the discussion. The digital ink allows me to be flexible with my feedback around the rubric and it also allows me to directly write onto my tablet while walking around the classroom as I listen to student discussions.

15.3 Results and Evaluation

By promoting a sense of collaboration in the English classroom, students have gained a deeper appreciation for literature and engaged in rich discussions. Struggling readers have been able to demonstrate their understanding of novels through conversations with their peers and students have received helpful feedback from one another as they have worked on written assignments and projects together. The use of digital ink has enabled students the flexibility to collaborate with peers on the same page and comment on each other's ideas.

Based on data collected from students' reading logs as well as their Goodreads' accounts, on average, my students now read anywhere from 1,200 to 1,500 min per month, while before, students were only reading the assigned number of minutes given to them (600). In addition, the average number of novels my students have read in a four month period of time is anywhere from 6–8. This is a higher number of books that many of the students read overall the previous school year.

Based on the quality comments that many of the students make during Socratic Seminars, it is evident that they understand the material that they are reading, as they are providing strong text evidence and making inferences that demonstrate strong comprehension of the book.

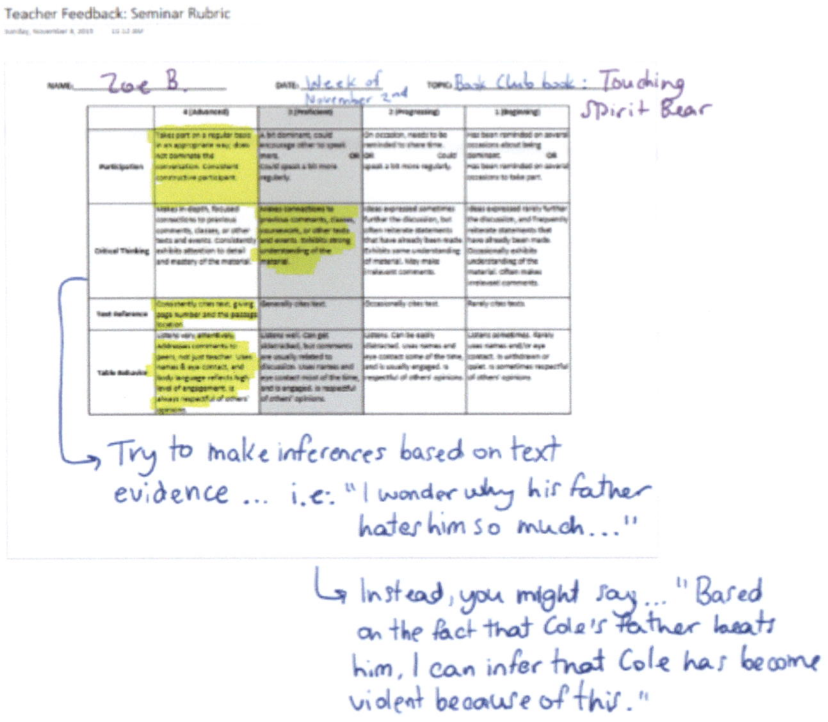

Fig. 15.10 Teacher feedback on seminar

15.4 Future Work

After having established highly collaborative book clubs in my classroom, I am looking forward to continuing to implement more book clubs throughout the school year. I would also like to explore the idea of collaboration for other reading, writing, or researching classroom projects by sharing OneNote's collaboration space.

In the coming months, I also hope to collaborate with English language learning (ELL) students in Shenzhen, China, to implement book clubs that my students would be part of with them. This opportunity would give students a more global perspective about the issues and themes in the literature that they read. My goal would be that students can collaborate via OneNote Collaboration space and also via Skype. Although they would not be sharing a physical space in the same classroom, the virtual possibilities for collaboration would be endless.

I look forward to providing my students with authentic learning experiences revolving around reading, which will enable them to collaborate by engaging in profound discussions and working together on group projects.

Acknowledgements I am very thankful for all the help and guidance that Matt DiGiulio, director of technology at Whitfield School, has provided me with. Matt has not only taught me how to use software on my tablet, but he has also provided me with meaningful feedback on best ways to use technology to support my classroom instruction. In addition, I would like to thank Mark Payton, former director of technology at Whitfield School, for his ongoing support and for his efforts in helping Whitfield School have one-to-one cutting-edge technology. Their help has made it possible for me to provide my students with rich learning experiences.

References

1. Armstrong, C., Llorin, J.: Using Shared Microsoft OneNote "Binders" to Create a School Environment for Sharing, Continual Formative Assessment, & Information Organization, pp. 241–247. Springer International Publishing, Cham (2015). doi:10.1007/978-3-319-15594-4_24. http://dx.doi.org/10.1007/978-3-319-15594-4_24
2. Cristy, J., Tront, J.G.: VText: A Tablet Based E-Textbook Framework, pp. 241–247. Springer International Publishing, Cham (2015). doi:10.1007/978-3-319-15594-4_15. http://dx.doi.org/10.1007/978-3-319-15594-4_15
3. Lloyd, S.L.: Using comprehension strategies as a springboard for student talk. J. Adolesc. Adult. Lit. **48**(2), 114–124 (2004)
4. Palou, E., Ramirez-Apud, Z., Ramirez-Corona, N., Lopez-Malo, A.: Analysis of Student Perspectives on Using Tablet PCs in Junior and Senior Level Chemical Engineering Courses, pp. 231–240. Springer International Publishing, Cham (2016). doi:10.1007/978-3-319-31193-7_21. http://link.springer.com/chapter/10.1007/978-3-319-31193-7_21
5. Ploesser, A.L.: Student Producers: OneNote, Camtasia Studio, and the Authentic Project, pp. 231–240. Springer International Publishing, Cham (2016). doi:10.1007/978-3-319-31193-7_16. http://dx.doi.org/10.1007/978-3-319-31193-7_16
6. Tront, J.G.: Facilitating Pedagogical Practices Through a Large-Scale Tablet PC Deployment, pp. 241–247. Springer International Publishing, Cham (2015). doi:10.1007/978-3-319-15594-4_36. http://dx.doi.org/10.1007/978-3-319-15594-4_36
7. Williams, L.: The Integratino of Inking Touch, and Flipping within the Mathematics Middle School Classroom, pp. 231–240. Springer International Publishing, Cham (2016). doi:10.1007/978-3-319-31193-7_23. http://link.springer.com/chapter/10.1007/978-3-319-31193-7_23

Part VII
Domain Independent Educational Tools

Chapter 16
Design Considerations for a Large Display Touch Application in Informal Education

Lucy Van Kleunen, Trent Green, Miranda Chao, Tiffany Citra, and Carlene Niguidula

Abstract Large touch screens are an increasingly common platform for education. Effective design for this form factor will directly influence the effectiveness of these screens as educational tools. The sheer size of a large touch display can bring up issues related to distal access of touch targets and limited field of vision. The visibility of the screen at a distance and the ability for large groups to gather in front of such a display also mean that designers must create applications that are effective teaching tools for proximal visitors who only see part of the screen as well as attractive for distal visitors who see the screen from afar. We discuss our approach to these design challenges through the iterative design and qualitative evaluation of a touch screen exhibit about the life of Alfred Nobel hosted on an 82-inch touch screen. From this early investigation into the efficacy of large displays as informal educational tools, we present practical guidelines for designers working with this form factor.

16.1 Introduction

Large touch screen displays are becoming increasingly common tools for informal educational settings such as museums and public kiosks. Large screens provide a unique opportunity for facilitating engagement due to their immersive effect and their potential as performative tools, in which either one user drives the experience while others watch or all users passively watch. Effective user interfaces for these displays must be able to convey a body of content in multi-user scenarios and for multiple age groups. Upright screens such as 82-inch displays present unique design challenges related to usability. In this paper, we discuss the iterative design and early qualitative evaluation of an upright large-display museum exhibit. We provide

L. Van Kleunen (✉) • T. Green • M. Chao • T. Citra • C. Niguidula
Graphics Lab, Computer Science Department, Brown University, 115 Waterman, Providence, RI, 02912, USA
e-mail: lucy_van_kleunen@brown.edu; trentdgreen@hotmail.com; miranda.c173@gmail.com; tiffany_citra@alumni.brown.edu; niguidulac13@gmail.com

T. Hammond et al. (eds.), *Frontiers in Pen and Touch*, Human–Computer Interaction Series, DOI 10.1007/978-3-319-64239-0_16

practical guidelines for museum kiosk designers working with a large touch display form-factor in an informal educational setting.

Large displays have been an active area of research in recent decades as they are becoming more affordable and common. Ill-designed applications present difficulties due to inaccessibility of icons and menus and do not take advantage of the opportunity to design with a user's periphery in mind [3]. We expand on prior research by examining our own strategies for dealing with these challenges in the context of informal education.

16.2 Previous Work

Previous work has observed common gesture patterns used by visitors at touch tables in order to motivate more intuitive designs for tabletop games or interactive science visualizations at small touch screens [4–6, 8]. Other work has focused on group interaction at museum kiosks. An algorithm was introduced by Block et al. [2] for defining group clusters at a touch kiosk and identifying specific classes of exhibit attendees. In Peltonen et al. [9], the social configurations of visitors was observed at an upright touch screen located in a large urban center. Novel interaction techniques have also been developed that address issues of inaccessibility of icons and content due to screen size in the context of productivity applications [1, 7, 10]. Most of the literature in this area has either focused on kiosk exhibit design for tabletop displays and small screens, group social dynamics at touch kiosks, or novel interaction techniques for large touch displays. The goal of our investigations, which are in the early stages, is to leverage these discussions to derive effective design principles for large displays in an informal educational environment.

16.3 Methodology

16.3.1 The Nobel Will Interface

The Nobel Will is a touchscreen kiosk experience that presents narrative and visual information about Alfred Nobel, the founder of the Nobel Prize. The experience uses a digital scan of Nobel's handwritten Will as its anchor. Users drag an orange highlighter over the document to discover key words and phrases, which appear traced in red ink under the highlighter. These key words connect to informational bubbles on the right side of the window via orange curved links. Tapping on these bubbles brings up a larger pop-up with additional information. From the Will interface, as shown in Fig. 16.1, visitors can link to two types of related experiences: collections and interactive tours. These experiences are accessible via buttons in the pop-ups, which present options specific to the content of the pop-up, as well as from a "task bar" on the bottom of the Will interface, which offers immediate accessibility to any collection or interactive tour.

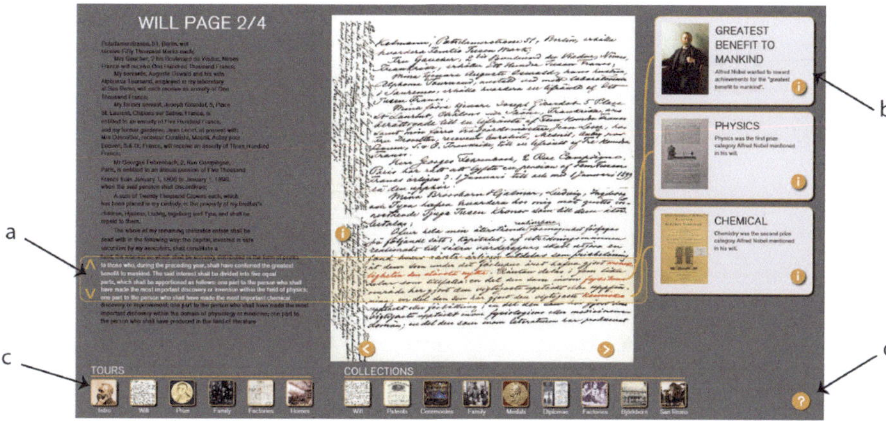

Fig. 16.1 Main Nobel Will interface with features indicated: (*a*) *orange* highlighter, (*b*) small pop-ups, (*c*) task bar, and (*d*) help icon

Collections are groups of high-resolution images. The collections interface displays thumbnail previews of the images and their titles in a region that is scrollable via a panning gesture. Tapping a thumbnail brings up a deep-zoom viewer of the image that fills the entire screen. Descriptive information is available via a sidebar, and images can be explored using pinch-zoom and pan gestures. The six interactive tours combine audio narration, panning and zooming images, and text and ink annotations. One is an introductory tour, which introduces Alfred Nobel and the history of his Will and the Nobel Prize. The other five deal with specific historical themes such as Nobel's family life, his inventions, and the establishment of the Nobel Foundation. Each tour lasts three to five minutes when played without interruption. Tours can be paused at any point during playback by explicitly pressing the "pause" button or by touching the screen. Unlike in a video, when an interactive tour is paused, all elements on screen are live. Visual elements on screen are deep-zoom images that can be explored with pinch-zoom and pan gestures. If an image has additional descriptive information, then a sidebar slides out when it is touched as shown in Fig. 16.2; about half of images in each tour had this sidebar information.

Also of note is the instructional flow of the exhibit. When users first enter the exhibit, they see a video screen-saver of the rotating Will document with the instruction "Touch to Explore." The screen goes back to this screen-saver every ninety seconds. Once the screen is touched, users see an instructional window with general instructions for exhibit use, as shown in Fig. 16.3. Once closed, this window can be re-accessed by tapping on a question mark icon in the bottom right of the screen. This constant accessibility of the instructions was partially due to the consideration that during high-volume use of the screen many people might approach the exhibit before it had re-set.

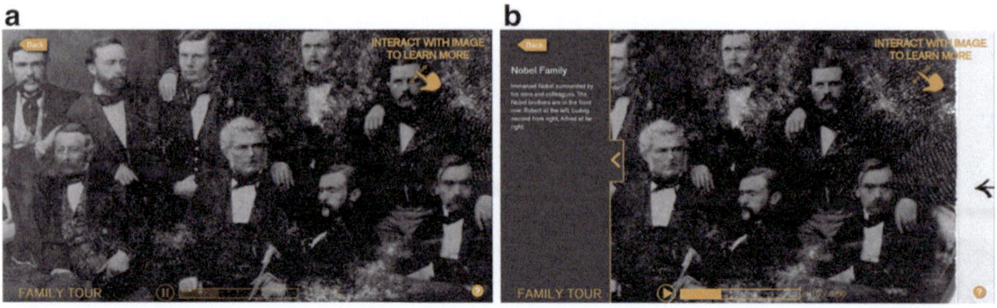

Fig. 16.2 The tour interface (**a**) during playback with no visible sidebar and (**b**) once the image has been touched and the tour has paused, with information relevant to the image displayed in a sidebar on the *left* side of the screen

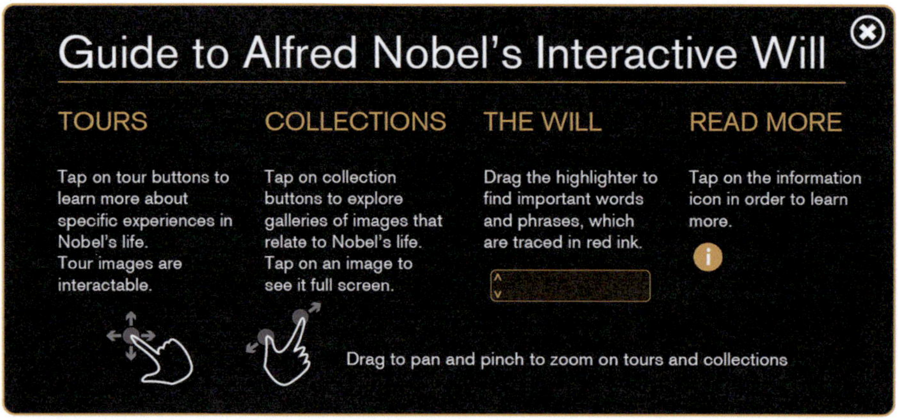

Fig. 16.3 Instructional window, accessible from the question mark icon

16.3.2 Iterative Design and Implementation

In collaboration with representatives from the Nobel Museum, Nobel Media, and Microsoft, we designed and implemented The Nobel Will exhibit over a period of four months. Design sessions took place remotely via Skype and in two in-person design meetings. Initial design decisions made during these meetings were based on the goals of (1) effectively communicating a complicated personal and societal narrative using a historical document as the semantic anchor, and (2) facilitating engagement with the material in a novel way through use of a large touch display. Design decisions also took into account previous practical knowledge by collaborators on designing interfaces for smaller touch kiosk displays. Two outside designers as well as two UI researchers took part in aesthetic design of the exhibit. Key evaluation considerations included aesthetic design, content presentation, and usability. In addition, the design of the exhibit had to integrate with the look-and-feel of elements of the already existing Nobel Museum exhibition. In some cases, as we will discuss, there were trade-offs between exhibit aesthetics, content presentation, and interface usability.

16.4 Usability Studies

16.4.1 Laboratory Testing

We completed formative evaluations of the usability of the exhibit in our laboratory in two separate rounds. In the first round, evaluations happened with family members and friends we invited to experience the exhibit, with participants ranging in age from 10 to 70. They were observed in order to determine common sources of confusion and their reactions and observations were recorded. We made multiple small adjustments to the UI after these sessions to disclose interaction opportunities more explicitly.

A second round of more systematic testing took place over the course of two weekends. Fifty testers, who were primarily undergraduate students in the age range of 17 to 22, evaluated the exhibit in hour-long slots in groups of 1 to 4 for a total of 21 h of testing. Each session was observed by at least two researchers who took notes on the interactions between testers and their interactions with the displays, including points of confusion and excitement. In order to mimic a museum environment, testers were given no introduction to the system by the observing researchers. After each session, the testers were asked qualitative questions about the exhibit either face-to-face or, for 29 participants, via an online survey. Some questions were open-ended, such as "What features of the exhibit did you find unintuitive?" While others were more specific, such as "Did you notice the question mark icon?" And "Did you know that tours could be stopped and interacted with?" Again, changes were made to the UI as a result of our observations and analysis of tester responses.

16.4.1.1 Reservations

Studies have shown that evaluation of kiosk exhibits in non-natural environments can lead to abnormal levels of engagement and more conventional forms of user interaction than in a museum setting [2]. The testing periods were also the first time that the newly created software system was used by many users at a time for an extended period. The student volunteers who tested the exhibit also functioned as application testers and therefore spent some of their time looking for "bugs" and trying to create system "crashes" rather than interacting in a standard fashion. The extent to which individual students appeared to see their role as system testers as opposed to usability testers varied. Due to time constraints, we were unable to recruit a set of volunteers that more appropriately represented a museum audience. In particular, we were unable to test the exhibit with younger volunteers in elementary and middle school age ranges. Due to these reservations, it is difficult to make concrete judgements about the efficacy of the exhibit in terms of engagement and presentation of exhibit content. However, practical revisions to design due to usability of the interface are worth discussion in the context of designing for a large screen, as our conclusions took into account activities common to any potential user of the exhibit.

16.4.2 In-Situ Initial Observations

The exhibit was observed for one day of normal use after installation, during which it was used by approximately 150 visitors. Users stayed, on average, for about 4–5 min each in front of the exhibit, although dwell time ranged from 10 s to an hour. One observation of note was that few visitors paused the tours and interacted with them during playback, and instead stood back and watched the tours as if they were videos. Often this behavior was observed when visitors were at the kiosk in large groups, perhaps because they did not want to be rude and interrupt the viewing experience of others. An insignificant amount of qualitative data was gathered to make a direct comparison with our laboratory studies except to say that in-situ use was, as expected, less regular and more influenced by social interaction.

16.5 Design Guidelines

A number of our design decisions were made after evaluation of usability studies. These evaluations focused on how easy or difficult it was for users to follow common paths of interaction. For example, if a user was confused, how easy was it for them to find the button that would bring up the instructional window? How much of this ease or difficulty resulted from the size of the screen and how could design decisions be made to help in increasing noticeability of the question mark icon? Through this process of iterative design and evaluation of usability, we have determined three practical design guidelines for creating digital exhibits for a large screen form factor.

16.5.1 Consider Physical Accessibility of Touch Targets

In designing for a large display, we recommend placing essential icons and toolbars on the bottom half and center of the screen and considering alternative mechanisms for interaction that do not involve specifically tapping a touch target. The Nobel Will is hosted on an 82-inch screen mounted upright at 77.5 inches. Adult visitors of average height can easily reach the top of the screen, however in order to ensure that content was also accessible for shorter visitors, children, and visitors using wheelchairs, vertical accessibility of buttons was considered. In many traditional design paradigms, menu items and help icons are in toolbars at the top of the screen or aligned vertically on the left or right. In the Nobel Will, these buttons are on the bottom of the screen instead.

Horizontal asymmetry of buttons can also present issues of accessibility. There were two ways to close pop-ups, such as the informational window shown in Fig. 16.3, by tapping the X icon in the upper right or tapping anywhere else on the screen outside of the pop-up. However, users had difficulty reaching the icon when standing at the screen in groups because they had to reach across the front of the person standing to their right. Users did not realize that they could close the pop-up simply by tapping the area outside of it. Two close buttons, one on each of the top corners, was aesthetically unpleasing and therefore avoided. In another iteration, the X icon was absent entirely from both the informational pop-ups and the instructional pop-up. After entering the exhibit, users saw the instructional window with no X icon. Although some users expressed confusion about how to close the pop-up, all groups figured out within a few seconds that they could close it by tapping the area outside. After this initial "training" they would close any other pop-up on the interface quickly by tapping outside of it. In order to avoid the initial confusion caused by not including an explicit "close" button, the final iteration included the X icon on the upper right. However, the concept of "training" walk-up users to use more convenient, yet less familiar, interaction techniques is an interesting area of research that deserves more in-depth exploration given that large upright touch screens might call for non-traditional touch interaction techniques.

16.5.2 Consider Immediate Field of Vision

We observed that many users of the Nobel Will would step back during passive use to see images and text from farther away, perhaps because the visuals were more decipherable when seen in context of the whole screen. However even at a distance of a few feet, the corners of large screen are not in a viewer's immediate field of vision. Therefore, we had to consider making icons large enough to be noticeable in a user's peripheral vision.

One of our goals in designing the interactive tour interface was to remind viewers that they were not watching a video, and that they could instead stop the tour to interact with and learn about the images on screen. In an initial iteration, users learned about this feature in the instructional pop-up. However, many would still select the tours and then watch them in their entirety, like videos, without touching the screen. In survey responses, viewers mentioned that they forgot that the tours were interactive without an explicit reminder during tour playback. In order to encourage interaction during tours, we added an alert that showed up on screen whenever one of the visible images had descriptive content (visible in Fig. 16.2). It was placed in the upper right due to the concern that more central placement would obstruct visual material on screen and therefore disrupt the experience of passively watching a tour. While most viewers would step back to view tours, many did not step back far enough to notice this alert appear on the upper right of the screen. This problem worsened when users watched tours in groups, as those on the left side of the group would have their view of the upper right corner of the screen obstructed. In the final iteration, the alert was made bigger in order to make it more visible. This decision represented a trade-off between content presentation on the one hand, since relevant context was sometimes obscured, and usability on the other.

16.5.3 Consider Obstruction and Viewing Distance

For smaller kiosk displays or table top displays, it is often essential that a user be directly in front of or leaning over the screen in order to have a meaningful experience. However, in the case of an 82-inch display, much of the screen can still be visible to passers-by when there are one or more people already interacting with the display. We also observed that the screen can be viewed comfortably from a distance of up to 25 feet. During a day of in situ observation, many visitors who spent time at the exhibit never touched the screen at all. Designing for a large upright display means considering the experience of the observer, who is not touching the screen but can see some or all of the on-screen content. This must be done with the additional consideration that the observer can be viewing the screen from afar or only partially. Interactive tours, for example, are interesting for both those directly touching the screen and a crowd of observers since interactive tours or full-screen imagery can be appreciated passively from afar as well as interactively. We recommend creating large display experiences that are still engaging while a user is only viewing part of the screen or viewing the screen from afar.

It is worth noting that based on our observations we chose to make touch targets, such as icons, much larger than the necessary size for registering a finger-press. This was due to considerations of accessibility, field of vision, and obstruction. Large icons are simply easier for visitors to find visually, reach, and then successfully press when navigating a large screen in the presence of a group.

16.6 Future Work

The Nobel Will exhibit records visitor interaction in the form of text file interaction logs. After three months of in-situ use, we will be able to evaluate these logs in order to gain quantitative data about exhibit use, such as how often particular buttons were pressed, how often tours were paused, and how often the exhibit re-set. This data will be useful for gaining a basic understanding of use patterns. Unfortunately, very little additional qualitative data will be recorded from this period of up-time.

We have begun a collaboration with the Massachusetts Historical Society on an exhibit that will also include interactive tours. We hope to test the next exhibit with a more representative audience, especially elementary-school aged children who are often the target audience for informal education. Children may be more intimidated by the large screen size or have more significant distal access problems due to shorter height. On the other hand, we wonder if children may have a different relationship with technology than their parents, which would lead to contrasts in walk-up use. We will also continue experiments in order to determine more concretely the optimal icon-to-screen ratio for a large screen as well as investigate the possibilities of "training" walk-up users in new modes of touch interaction.

16.7 Conclusion

We found that a large screen can work well as a platform for an experience that does not require a complete view of the screen at all times. However, due to design challenges with distal access of icons and limited view of the screen at short proximity, some applications such as interactive games or visualizations might be experienced more comfortably on a smaller screen. Walk-up users cannot be expected to learn novel interaction techniques in order to combat these accessibility issues, but designers can address them by adjusting size and placement of icons. We also found that large screens work well for experiences that combine passive and active use. Due to their visibility at a distance and the ability for groups to form around these displays, they are natural platforms for semi-passive experiences such as interactive tours. However, large screens may prove intimidating for some to use publicly because they feel observed. Larger screens also raise problems of engagement because users may be unfamiliar with the concept of large touch screens and therefore interact less often with the display than they would with more familiar smaller touchscreens or table tops. In order to use large touch displays as effective tools for informal education, designers must consider these constraints.

Acknowledgements We would like to thank our sponsors at Microsoft, our collaborators at Microsoft Research, especially Roy Zimmerman and Amy Sorokas, our collaborators at the Nobel Museum, especially Fredrik Skog and Ulf Larsson, our collaborators at Nobel Media, especially Magnus Gylje, as well as Jonas Lindkvist for initial mock-ups, Jessica Herron and Anne Spalter for consultations on User Interface aesthetics, Andries van Dam for advising and general consultation, Rosemary M. Simpson for content design consultation, and Tanay Padhi and Jessica Fu for implementation assistance.

References

1. Bezerianos, A., Balakrishnan, R.: The vacuum: facilitating the manipulation of distant objects. In: Proceedings of the SIGCHI Conference on Human Factors in Computing Systems, pp. 361–370. ACM (2005)
2. Block F., Hammerman, J., Horn, M., Spiegel, A., Christiansen, J., Phillips, B., Shen, C.: Fluid grouping: quantifying group engagement around interactive tabletop exhibits in the wild. In: Proceedings of the 33rd Annual ACM Conference on Human Factors in Computing Systems, pp. 867–876. ACM (2015)
3. Czerwinski M., Robertson,G., Meyers, B., Smith, G., Robbins, D., Tan, D.: Large display research overview. In: CHI'06 Extended Abstracts on Human Factors in Computing Systems, pp. 69–74. ACM (2006)
4. Hinrichs, U., Carpendale, S.: Gestures in the wild: studying multi-touch gesture sequences on interactive tabletop exhibits. In: Proceedings of the SIGCHI Conference on Human Factors in Computing Systems, pp. 3023–3032. ACM (2011)
5. Horn M., Leong, Z.A., Block, F., Diamond, J., Evans, E.M., Phillips, B., Shen, C.: Of bats and apes: an interactive tabletop game for natural history museums. In: Proceedings of the SIGCHI Conference on Human Factors in Computing Systems, pp. 2059–2068. ACM (2012)
6. Hornecker, E.: I don't understand it either, but it is coolİ-visitor interactions with a multi-touch table in a museum. In: 3rd IEEE International Workshop on, Horizontal Interactive Human Computer Systems, TABLETOP 2008, pp. 113–120. IEEE (2008)
7. Khan, A., Fitzmaurice, G., Almeida, D., Burtnyk, N., Kurtenbach, G.: A remote control interface for large displays. In: Proceedings of the 17th Annual ACM Symposium on User Interface Software and Technology, UIST '04, pp. 127–136. ACM, New York (2004). doi:10.1145/1029632.1029655. http://doi.acm.org/10.1145/1029632.1029655
8. Ma J., Liao, I., Ma, K.L., Frazier, J.: Living liquid: design and evaluation of an exploratory visualization tool for museum visitors. In: IEEE Transactions on Visualization and Computer Graphics, pp. 2799–2808. IEEE (2012)
9. Peltonen P., Kurvinen, E., Salovaara, A., Jacucci, G., Ilmonen, T., Evans, J., Saarikko, P.: It's mine, don't touch!: interactions at a large multi-touch display in a city centre. In: Proceedings of the SIGCHI Conference on Human Factors in Computing Systems, pp. 1285–1294. ACM (2008)
10. Wu, M., Balakrishnan, R.: Multi-finger and whole hand gestural interaction techniques for multi-user tabletop displays. In: Proceedings of the 16th Annual ACM Symposium on User Interface Software and Technology, UIST '03, pp. 193–202. ACM, New York (2003). doi:10.1145/964696.964718. http://doi.acm.org/10.1145/964696.964718

Chapter 17
DataSketch: A Tool to Turn Student Sketches into Data-Driven Visualizations

Michelle Hoda Wilkerson

Abstract Highly visual, interactive, data-driven displays are a ubiquitous part of life. We see them on industrial interfaces, at museums, and on daily news websites. Although they are popular and visually appealing, data visualizations require specific skills and knowledge to be understood. Most K-12 curricula in the U. S. only expose learners to conventional representations such as line graphs or tables rather than more complex visualizations of this sort. This paper describes DataSketch: a tool designed to allow young learners to develop data visualization literacy by creating their own data-driven digital ink visualizations. Informed by theory and empirical research in the learning sciences, I argue that leveraging a sketch-based paradigm offers a powerful way to develop youths' data visualization literacy by leveraging the familiarity and flexibility of drawing as an expressive medium.

17.1 Introduction

Data visualizations are an important new class of tools for understanding the world [3]. But while they are popular and powerful, they are not always transparent. Consider Fig. 17.1, which shows a screenshot from an interactive visualization featured on the Bloomberg news website [10]. The visualization builds slowly, adding year after year of monthly average temperatures. After the animation is finished playing, users can scroll over each year's data points (in the figure below, we have highlighted 1983) to evaluate them more closely. To make sense of this visualization, users must understand how to control the visualization (hovering, replaying); how the basic conventions of graphing are used and extended to highlight the passage of time (animation), increase in temperature (color) and breaking of records (bolded lines); and how these data serve as indicators of climate (by highlighting difference in average temperature relative to the twentieth century average).

M.H. Wilkerson (✉)
Graduate School of Education, University of California-Berkeley, 4645 Tolman Hall, Berkeley, CA, 94720, USA
e-mail: mwilkers@berkeley.edu

T. Hammond et al. (eds.), *Frontiers in Pen and Touch*, Human–Computer Interaction Series, DOI 10.1007/978-3-319-64239-0_17

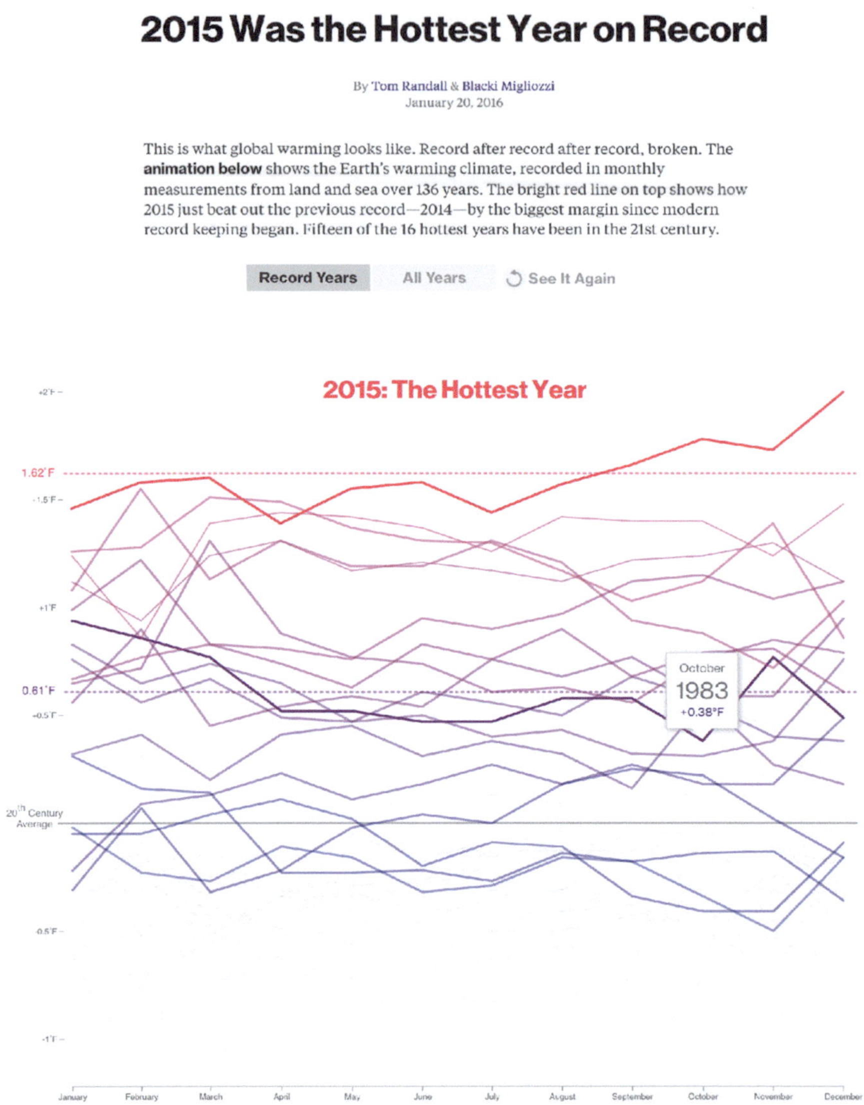

Fig. 17.1 A visualization of average global temperatures. Visualizations like these require one to leverage knowledge of the domain and to make sense of novel representational patterns

Data visualizations are very different from the sorts of representations studied in the current K-12 curriculum. Because of their complexity, there are some concerns about whether leveraging visualizations in K-12 will help, or simply complicate, the math and science curriculum [13]. But, the ubiquity of data visualization suggests it is not a question of whether they are appropriate, but rather how educators can prepare students to work with them.

A number of researchers and educators have recognized, and worked to address, the need for learners to develop data visualization literacy. There are of tools that allow learners to explore new representations of data flexibly such as Tinkerplots [2] or Tabletop [4]. Others are exploring how to engage learners in the design of highly visual data representations such as infographics [6], or in collecting their own data that can then be represented by students themselves or with the aid of computational tools [7]. These efforts have established the feasibility and promise of engaging even elementary students in data representation and analysis activities. Thus far, however, few tools have engaged students with the blended data-driven, narrative and computational aspects of data visualization. In this short paper, I describe the design of DataSketch: a visualization toolkit that leverages digital ink as a foundation for introducing young learners to data visualization. I describe the theoretical and empirical inspirations for the project, and the design of the DataSketch tool.

17.2 Theoretical Framework

The DataSketch project is informed by two major theoretical perspectives in education and the learning sciences: (1) That learners possess implicit, but powerful, ways of making sense of representations even before formal instruction; (2) That developing competence with representations involves developing a flexible collection or 'conglomerate' of skills and knowledge rather than becoming familiar with a limited set of representational forms. Here, I review each of these motivating perspectives.

By middle school, learners can create, critique, and interpret unconventional representations. diSessa [1] and colleagues call this metarepresentational competence, and suggest it has strong roots in childhood activities such as drawing [11]. For example, learners are familiar with using conventions such as right-to-left to represent time, and with coordinating visual properties such as length, size, and direction of objects with quantitative measures such as position and speed. These competencies are further developed when learners work together to create and refine shared representations; and can contribute to understanding professional forms of data representation.

Rather than understanding specific types of representations as unified concepts, research is showing that understanding representation requires a flexible network of knowledge and skills. Kosslyn [5] described how making sense of diagrams involves deconstructing it into a number of familiar subcomponents such as axes, points, and shapes. Similarly, Superfine and colleagues [12] argue that when young learners work to translate across different mathematical representations, they activate a number of related clusters of knowledge and skills. This suggest that representation – and data visualization – should be as a flexible set of mapping and interpretation practices rather than a set of specific forms to be learned.

These two lines of research suggest that learners already have knowledge and skills that can inform their development of data visualization literacy, and that this knowledge can be developed as a loosely connected network. Furthermore, it suggests that this set of existing knowledge and skills has basis in drawing as an activity.

17.3 How Young People Already Represent Data

To exemplify the knowledge and skills many learners already have, consider the following empirical data. Figure 17.2 shows a seventh grade student's invented representation of the growth a collection of plants with a distribution of heights, for which the shortest was 20 mm and the tallest was 80 mm, collected during a classroom study [14]. The participant drew the shortest and tallest plants (indicated by a vertical axis), and used a group of smaller plants labeled "in between" to represent what he thought was the mean height of the collection of plants. He included dotted lines to indicate how he wanted these objects to change to indicate growth, and a sun to indicate a causal relationship he expected would affect growth.

This representation illustrates how the student employed a loosely connected collection of skills, and an inclination to make use of dynamic elements of visualization. He superimposed an axis over iconic representations of flowers. A specific property of the flowers – their height – is used as a quantitative indicator. And, the student illustrated a dynamic growth over time, or animated within the figure. Research on students' representations of data and quantitative situations suggests that this type of juxtaposition of iconic, decorative elements and systematic, symbolic indicators is not uncommon [1, 8, 9, 11]. While such

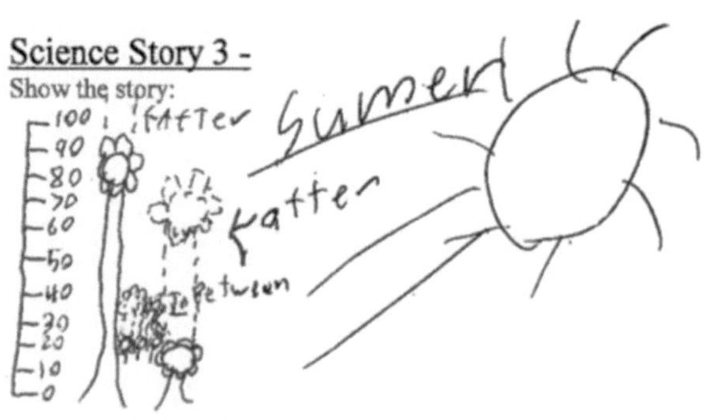

Fig. 17.2 A student-generated sketch to describe the growth of plants over the summer

research often focuses on students' progression from aesthetic and idiosyncratic concerns to the development of more symbolic and generalizable representations, it also recognizes the role that context and specificity play in how learners develop, make sense of, and refine such representations. Additionally, modern visualizations increasingly preserve such idiosyncratic features rather than minimizing them.

17.4 DataSketch: Data-Driven Visualizations with Interactive Ink

The DataSketch project is an NSF-funded effort to educate young learners about interactive data visualization, including enabling them to create their own digital sketches and program them to respond to live or archival data sources. A major component of this effort is DataSketch, a tool inspired by interactive visual data analysis software such as TinkerPlots [2] and TableTop [4]. DataSketch complements these tools by using students' own sketches rather than conventional display formats. The goal is to create an entry point to data visualization, by bridging the familiarity and ownership learners experience through drawing with the capability to directly link to data.

With DataSketch, users begin by sketching directly onto the screen of the application. These sketches are decomposed into independent vector objects; users are also able to select and define a group as one object if they would like. The user loads data as a comma separated file: with rows for each parameter of interest, and columns for each measurement over time.

Consider again Fig. 17.2 from the last section. Using DataSketch, this student would be able to create the same drawing as before by sketching directly on the screen. Each time they pick up their finger or stylus, the new sketches are converted into vector objects on the screen. Multi-part objects, such as the cluster of flowers indicating the mean, can be grouped through an available menu item (Fig. 17.3).

In Fig. 17.3, we have loaded a comma-separated values (CSV) file with time series time series data indicating the heights of the tallest and shortest plants in a collection, along with the collection's average height, and measures of humidity and light intensity. Each of these parameters can be programmatically linked to any sketched object. Imagine, for example, that a user wishes to link the sun object to the measure of light intensity they have available in their data set. They would click on the sun, and a menu would become available for the user to select which property of the drawn object – position, width, height, rotation, scale, color or transparency – can be linked to changes in any given parameter. The user can then select an object parameter. In the screenshot below, "scale" was selected, which will make the sun object grow and shrink as its corresponding light intensity parameter grows and shrinks. To calibrate the link, the user indicates max and min values (in this case, a multiple of the sun's current scale) that should be mapped to the max and min values of the corresponding data parameter. In the example below, other mappings have

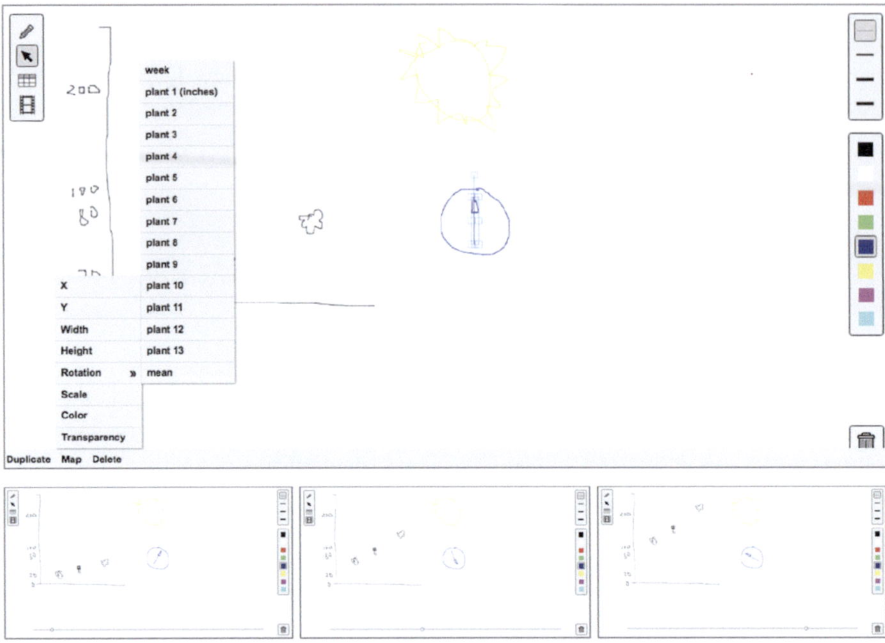

Fig. 17.3 A reproduction of Fig. 17.2 within a working, early prototype of DataSketch. Users link data to properties of each sketched object (*top*). The sketch is then "played" to observe changes in data over time (*bottom*)

also been made: between the height of flowers and their y-position, and between humidity and an overlay image with small blue droplets.

Once such mappings are made, the user can click the play button to watch the sketch animate to reflect changes in data over time. These changes in the visualization are driven only by the data; the user simply connects a single drawing to these driving data parameters. In this way, they can visually explore hypotheses about relationships between factors using the visualization, and communicate relationships that are known to exist. DataSketch is actively under development, and a free, open source prototype is available for use at https://github.com/CalCoRE/DataSketch.

DataSketch is designed to leverage and strengthen the flexible network of knowledge and skills about representation reviewed in the last section. Students start with a visual interpretation of the problem as they see it – a drawing that communicates the relationships that they think are most important about a situation. This drawing is decomposed into constituent objects in a way similar to the way graphs can be decomposed into points and axes, and each subcomponent can be assigned as set of rules for how it works – the same way a point's position in space indicates a pair of quantities or a bar's height indicates a single quantity, so can any objects, position, color, height, rotation, or other visual features represent quantities in the students' constructed representation. The goal is to connect learners' existing commonsense knowledge about data representation to the types of rules and mappings that define formal representational systems.

17.5 Future Work

Research is ongoing to explore whether and in what ways DataSketch can help learners to develop data visualization literacy. But any single tool – DataSketch, spreadsheet software, or even an activity book – is only one part of a curricular system that requires aligned goals and support from a classroom teacher, curricula, and the student community. Therefore, this project investigates how students make sense of and use the DataSketch tool alongside broader questions about how they reason about data visualizations more generally, through interviews and planned classroom studies. In currently ongoing interviews, learners are asked to make sense of a set of interactive, idiosyncratic data visualizations. Analyses focus on how learners talk about the underlying relationships that govern interactive visualizations, what strategies they use to decipher unfamiliar or difficult visualizations, and what visualization techniques and quantitative relationships are most salient to them. We also consult with learners as design consultants to help us understand what refinements might make DataSketch more usable. In future classroom studies, we plan to work with teachers to incorporate DataSketch as one of a number of tools (including spreadsheet applications, online data analysis tools, and traditional charts and tables) used for data-centric exploratory projects.

Our ultimate goal is for students to better understand how DataSketch and other visualization tools can be used to deepen and enrich their own explorations of data related to consequential topics such as climate change, human impact on Earth systems, and biodiversity; or, personally interesting topics such as local community issues. Our goal is for students to see data visualizations as a way for them to explore and argue about data in ways that help them understand the world around them.

Acknowledgements This work is funded by the National Science Foundation Grant Number IIS-1350282. Thanks to four anonymous reviewers for their feedback on previous versions of this draft. Special thanks to Vasiliki Laina and Kimberle Koile.

References

1. diSessa, A.A.: Metarepresentation: native competence and targets for instruction. Cogn. Instr. **22**(3), 293–331 (2004)
2. Finzer, B., Erickson, T., Blinker, J.: Fathom. In: Computer Software. Oakland, CA, USA: Key Curriculum Press (2001)
3. Friendly, M.: A brief history of data visualization. In: Chen, C.H., Härdle, W., Unwin, A. (eds.) Handbook of Data Visualization, pp. 15–56. Springer, Berlin/Heidelberg (2008)

4. Hancock, C.: The medium and the curriculum: reflections on transparent tools and tacit mathematics. In: diSessa, A.A., Hoyles, C., Noss, R., Edwards, L.D. (eds.) Computers and Exploratory Learning, pp. 221–240. Springer, Berlin/Heidelberg (1995)
5. Kosslyn, S.: Understanding charts and graphs. Appl. Cogn. Psychol. **3**(3), 185–225 (1989)
6. Lamb, G.R., Polman, J., Newman, A., Smith, C.: Science news infographics: teaching students to gather, interpret, and present information graphically. Sci. Teach. **81**(3), 25–30 (2014)
7. Lehrer, R., Kim, M.J., Schauble, L.: Supporting the development of conceptions of statistics by engaging students in measuring and modeling variability. Int. J. Comput. Math. Learn. **12**(3), 293–331 (2007)
8. Lehrer, R., Schauble, L.: Modeling natural variation through distribution. Am. Educ. Res. J. **41**(3), 635–679 (2004)
9. Nemirovsky, R., Tierney, C.: Children creating ways to represent changing situations: on the development of homogeneous spaces. Educ. Stud. Math. **45**(1), 67–102 (2001)
10. Randall, T., Migliozzi, B.: 2015 was the hottest year on record (2016). http://www.bloomberg.com/graphics/hottest-year-on-record/
11. Sherin, B.: How students invent representations of motion: a genetic account. J. Math. Behav. **19**(4), 399–441 (2000)
12. Superfine, A., Canty, R., Marshall, A.: Translation between external representation systems in mathematics: all-or-none or skill conglomerate? J. Math. Behav. **24**(4), 217–236 (2009)
13. Tversky, B.: Prolegomenon to scientific visualizations. In: Gilbert, J.K. (ed.) Visualization in Science Education, pp. 29–42. Springer, Dordrecht (2005)
14. Wilkerson, M.H., Maldonis, J.: Toward a data expression toolkit: identifying the elements of dynamic representational competence in young learners. In: Proceedings of Constructionism (2012)

Chapter 18
A Survey Based Study on Personalizing Student Learning Using MyEduDecks Application

Samuel Lawrence Cohen, Parv Shrivastava, Nicholas Wilke, Aileen Owens, Matt Antantis, Prateek Jukalkar, Ashumi Rokadia, and Vinay Pedapati

Abstract MyEduDecks is a pen-based digital flashcards application that has been in development since 2013 at South Fayette High School. MyEduDecks is an extension of the Tablet Flashcard Application that was originally developed at Carnegie Mellon University (CMU). The MyEduDecks application was created to understand how K-12 students use and benefit from pen-based technologies. It was our goal to personalize student learning and improve peer-to-peer communication. The application can be used to create custom study decks for students and give them the ability to use their stylus to input answers and rely on automated grading (based on ink recognition) or self-grading to progress through their studies. The use of the stylus allows students to enter free hand input using text, numbers and sketches. It is particularly appropriate for subject areas where the answers are multi-dimensional (e.g.,: equations). This paper presents our application, its utility in a K-12 school environment, and preliminary data on the use of application at South Fayette School District.

18.1 Introduction

18.1.1 Problem Statement and Context

Paper based Flashcards have played a major role in helping students review and comprehend concepts in short knowledge segments [5]. New technologies have enabled flashcards to be presented as mobile or web applications. The benefits of electronic flashcards are the ease of access and potential for sharing.

S.L. Cohen (✉) • P. Shrivastava • N. Wilke • M. Antantis • P. Jukalkar • A. Rokadia • V. Pedapati
South Fayette School District, 3680 Old Oakdale Road, McDonald, PA, 15057, USA
e-mail: samuel.lawrence.cohen@gmail.com; parvshrivastava@gmail.com; nick.wilke26@gmail.com; mattantantis@gmail.com; psjukalkar@gmail.com; ashumi33@icloud.com; vinaycp2001@gmail.com

A. Owens
South Fayette High School, 3640 Old Oakdale Road, McDonald, PA, 15057, USA
e-mail: amowens@southfayette.org

T. Hammond et al. (eds.), *Frontiers in Pen and Touch*, Human–Computer Interaction Series, DOI 10.1007/978-3-319-64239-0_18

However, the current flashcard applications do not give students the ability to use a stylus in order to answer questions using sketches or complex diagrams as answers, a task that is difficult or impossible to do using a traditional keyboard input. Furthermore, creating decks with complex figures and sharing them with students and enabling collaborative learning in class has not been adequately addressed by existing flashcard applications. Our context is K-12 institutions where flashcards can be effectively integrated into instructions and young students can be trained to demonstrate and communicate using a flexible pen input modality. MyEduDecks also intends to bridge the gap between students who have well-developed typing skills and those who do not, by using pen-based input. Pen-input brings an exciting opportunity to build applications that are easy to use, and allow developers and researchers to understand how sketch based inputs can be integrated into class instructions and measure their effectiveness [1–3, 6, 9, 10]. Our research and development was motivated by the early work on pen-based applications by Carnegie Mellon University researchers [4]. In this paper we describe our ongoing development of MyEduDecks, its use in pilot studies at South Fayette School District, and the preliminary data we collected from student and teacher surveys.

18.1.2 MyEduDecks Application

MyEduDecks is a pen-based flash card application that is in development at South Fayette School district since 2013. The project is an extension to what was started as a Tablet Flashcard application project at CMU [4]. Each year, high school and college student groups with different skills work on the project in designing, developing, testing and improving its features. Former members of the team who are now in college work as part of the development team remotely. This project has created a pipeline of high school students who gain valuable experience in designing and developing applications that can one day become an application that is shared with many other K-12 institutions. The MyEduDecks application was created to serve the needs of K-12 students and teachers so that a pen-based computer interface could serve as a replacement to the traditional pencil and paper. The application can also connect students in their learning through sharing and creating decks together [11]. While most electronic flashcards [7] are created using a keyboard or some existing templates, MyEduDecks allow the additional benefit of creating cards using free-hand drawing. The questions and answers can be easily expressed using diagrams and complex equations. For example, it is difficult to create Math flashcards that contain fractions using existing electronic flashcard applications.

MyEduDecks uses Microsoft Ink API as its recognition engine to provide automated grading on handwriting that are accurately recognizable (e.g., text, numbers). The ink API recently received an update with the release of Windows 10. The updated API provides the application with better recognition of words written by the user and improves physical to digital accuracy. Teachers can create classes, add students, and allow quick creation of cards or decks with pen-input and provide access to decks for the students within minutes. In addition, teachers have the ability to assign and monitor student activity during the assignment using dashboards. The teacher can track student completion rates and correctness data for individual students or for the entire class. The students' completion and correctness data is then able to be compared between individual students, as well as the entire class set. We are currently building features that give students the ability to create peer-to-peer networks to enhance their learning and increase the opportunity to collaborate with their classmates. Figure 18.1 shows how students were engaged with the MyEduDecks program during our pilots in a class with teacher supervision. Figure 18.2 shows some screenshots of the application that allows easy creation of the pen-based card decks. If the answer can be recognized (e.g,. a number or word) the program automatically grades it. If the answer is a drawing, students are currently asked to self-grade the answer when the correct answer is shown.

Fig. 18.1 Students using the MyEduDecks platform on their computers during a beta test

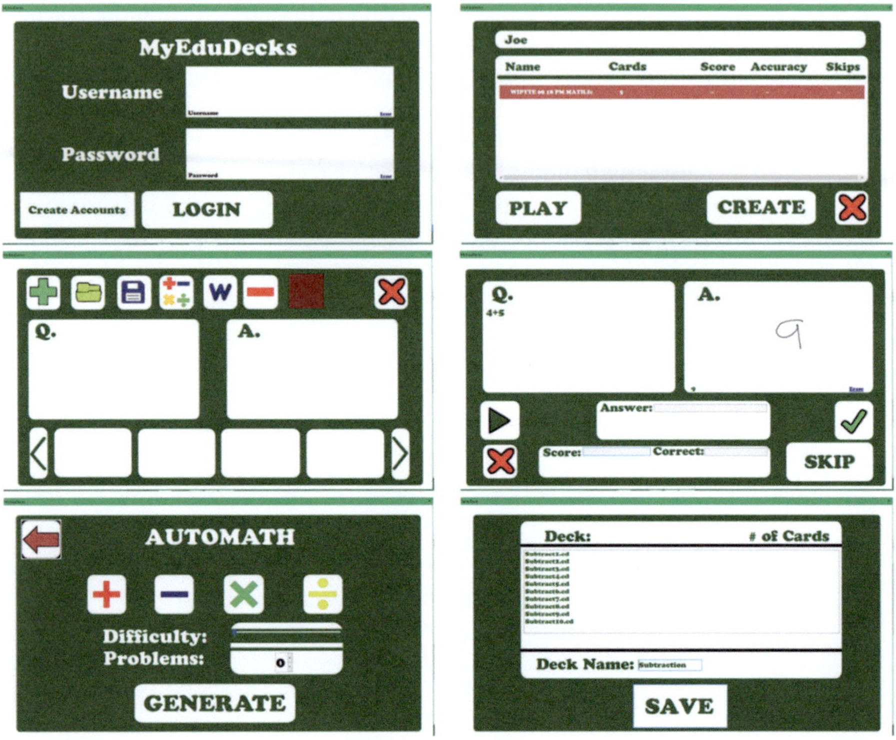

Fig. 18.2 Various screens in the MyEduDecks application

18.1.3 Comparison to Current Applications

Quizlet, a similar application, is an online based program that allows users to create a deck of flashcards by typing the question and answer. Quizlet is widely used in schools around the world because it is easy-to-use and user-friendly. However, Quizlet does not allow students to use any type of pen input, and therefore, does not support pen-based text recognition. The omission of this feature is one of its main flaws, since using graphical imagery to express ideas can offer increased opportunities for engineering and design. Research shows that students learn better when they write rather than type their answers [8]. The MyEduDecks application provides similar online features including student connectivity, but instead the application is focused more on pen recognition and easy input methods. Most existing flash card applications, including MyEduDecks, provide methods for creating, deploying and sharing decks, but MyEduDecks provides the added benefit of pen modality for creating and distributing cards with complex diagrams. Furthermore, it can use a formatted word document to import a suite of questions. Teachers who use MyEduDecks have the added flexibility of creating their decks in MS Word and then importing them to MyEduDecks. It is our goal to continually improve the application based on research data and release it to a wider group users soon. The application fills a void in K-12 education where pen-based flash-card programs can make a significant difference in enabling learning and improving collaboration. Furthermore, it may encourage teachers and students to use more pen-based applications that have shown to improve retention, filling a void in the pen-based applications market.

18.2 Prior Work

The MyEduDecks application began as a summer project four years ago at South Fayette High School, initially called "Flashcards." The project was motivated by the Table Flashcard projects that were done at CMU since 2007 [4]. The application was later renamed MyEduDecks to give the program a new identity in K-12 and to separate it from traditional, pen and paper, flashcards. An initial prototype of the app was built two years ago, and was tested in different classrooms. However, the previous versions had bugs and slight nuances in the code, as well as outdated graphics. The program has been since rewritten to patch the code and ensure smooth performance. The team has also completely overhauled the graphics and designed the interface to be cleaner and more efficient. Previous versions were tested in second and third grade classrooms. Elementary school teachers who beta tested this product in the past viewed the application as helpful in their lessons and indicated they would be willing to incorporate the finished version of the program into their curriculum. The results and suggestions of the educators were incorporated into the latest version of this application. This ongoing research evaluates opportunities for personalized learning and peer-to-peer connections using MyEduDecks for students at different grade levels and cognitive stages of development.

18.3 System Details

As shown in Fig. 18.2, the user starts off at the login screen when the user opens up the application. The Students can input their username and password through the digital ink and recognition. When the user logs in, they arrive at the dashboard where it displays their unfinished decks and the scores of their finished decks. There are three ways to create decks: the Automath feature, the Word feature, and the handwriting feature. In the Automath feature, students can request random math problems generated by the system. In the Word feature, a deck can be uploaded as a Word document. The word document template is provided to the user. In the handwriting mode, teachers or students can create cards by writing the question and the answer. Students are encouraged to use creating decks using handwriting and share with other students. Once they have created a deck, they can name the deck and save it with a name that helps find the deck easily.

18.3.1 Pen Recognition Software (Ink API)

The Microsoft Ink API was used to detect handwriting and convert the ink strokes into strings that can be used to autograde student answers that are recognizable. If the student feels that the system was incorrect in recognizing the answer, they can override and self-grade the problem. The ink API recently got an update in Windows 10, and its recognition capabilities have been improved. The API can save the strokes to a panel while the user writes, and read the answer in real time to tell the user what their writing has been interpreted as. The recognition can detect many forms of handwriting, and is sophisticated enough to handle an elementary school child's handwriting. There are some errors in the recognition, however, the real time recognition allows the user to rectify their mistakes once they happen. The writing capabilities are very smooth, and have little input lag, providing an immersive writing experience that mimics handwriting.

18.4 The Pilot Project

MyEduDecks enables students and teachers to create cards which can be answered with a digital ink pen. The Microsoft Ink API interprets the digital ink into a readable text format which the program compares to the correct answer. Teachers can create decks by writing them manually or importing them from a Word document, with a predetermined bulleted format as shown in Fig. 18.3. Another way to create decks is the "Automath" feature (Fig. 18.2), which automatically generates math problems after specifying the operation and the number of cards. Teachers can check their students' progress on any assignment to view in real time, as the students complete their assigned decks. Students can see what the Ink API interprets through a label to avoid unintended mistakes. Furthermore, students can upload custom decks that they have created in a shared cloud which will eventually allow classmates and teachers to take advantage of the collective resources. The database is coded in Microsoft Azure, which provides enough free storage for the beta test of the application, but could be upgraded in future versions as necessary. The ability to collaborate among students is the main component in creating a peer-to-peer learning network. The final design and coding of this feature is currently under development.

The application was tested in classroom by Mr. Scott Philipp, a South Fayette seventh grade science and social studies teacher. A pretest was administered to three seventh grade classes. Phillip was asked to provide insight as to how the application could aid his teaching and his personal thoughts on the ongoing development of the product. Philipp was also asked to explain how MyEduDecks might meet his teaching needs. He responded, "This year for the first year we have been using OneNote Notebooks to personalize learning. We have just launched our second

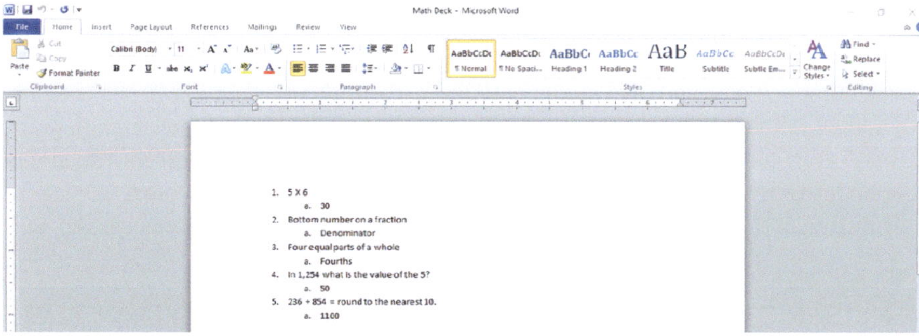

Fig. 18.3 Example of formatted word document

notebook and are learning more about the power of personalized learning. For us, self-assessing is the next step of personalized learning, and there is the potential for this through MyEduDecks." He also explained that there are some students in his classes who like to pre-learn the material before he teaches it, and for those students, this application could significantly help enhance their learning. Philipp observed his students' reactions to the pre-tests, and later reported that the class was excited and enthusiastic about trying a new type of study tool. The pretest that was administered yielded encouraging results and important input for areas of improvement. A beta test was then initiated in January of 2016, to compare the results of the pre-test to the post-test. Seventy-seven students from Philipp's three classes participated in this test. One of the members of the team directed the classes on how to use MyEduDecks. The students were directed to create their own decks from a Microsoft Word document, the results of which would be sent to Philipp. The students were also assigned to: (1) share three MyEduDecks quizzes created in a Word document with other students outside of school, (2) create and play an "Automath" deck, and (3) play a deck created by using the Word feature. Philipp received the results from the tests as the students completed their assigned decks and collected the data. The students completed a post-survey after the beta test, which included input regarding their vision for using MyEduDecks to improve learning and to identify improvement to the program.

Mr. Scott Phillip is currently considering possible options for decks and how he could further implement the program into his classroom most effectively. Since Philipp teaches science and social studies, he stated that the application would most likely be useful in testing vocabulary knowledge as that topic benefits greatly from a pen-based component. The team will continue to work with Philipp over the next year, using his input and information collected from his students, to improve the product.

18.5 Results

Conducting tests that involve students in their learning environment is challenging. We made the survey to deploy to approximately sixty students in three seventh grade classes. Our study was designed as a way to measure student perceptions about new learning technologies. The survey questions were designed around the use of MyEduDecks in their learning. Table 18.1 displays the post-test survey results.

18.6 Evaluation

The survey outcomes indicate that significant percentage of students reacted positively to the use of MyEduDecks. This is a first step in setting up a more detailed experiment that involves pre- and post-test scores and how they compare. The preliminary data gathered shows the students' interest and ideas on how MyEduDecks can best be used to enhance their learning experience. The analysis of beta test results shows the majority of the participants agree that the MyEduDecks application is a useful tool in a classroom environment. The results from the survey and the interview are discussed below.

Independent learning is a technology enabled method that students use effectively in and outside of schools to increase the speed and quality of learning. Based on the response from the students in the table above, seventy-eight percent find learning independently is important to them. MyEduDecks promotes independent learning by providing learning resources created through by teachers and peers. This collaborative tool provides the ability to learn independently and is supported by sixty-nine percent of the survey results depicted in the table above. The mark of a useful education application is the ability to make learning more efficient and effective. MyEduDecks promotes active learning, ease in creating new decks, and an instant gratification grading system. As seen in the table above, eighty-four percent of students agreed with the idea that MyEduDecks could enable students to personalize their learning of concepts through design and creation of unique cards and decks that aligns with their learning needs. It will allow students to focus on subjects where they have experienced challenges in the past and benefit them with additional practice. Independent and personalized learning are major components of the MyEduDecks program.

Increased self-confidence correlates to improved learning and increased information retention. A majority of students believed that MyEduDecks would help them succeed in learning content more effectively. The psychological benefit of confidence, along with the increased information retention with digital ink, may increase the efficiency and effectiveness of studying with MyEduDecks. The ability to quickly make personalized decks together with the use of pen technology aid in improving the native skill of writing, improves information recollection, and the effectiveness of the studying with MyEduDecks over typed flashcards. The

Table 18.1 Post survey test results

Question	Responses	
	Agree(%)	Disagree(%)
Learning independently is important to me	93.5%	6.5%
Based on what I known about MyEduDecks, I think using MyEduDecks could enhance my ability to learn independently	77.92%	22.07%
Based on what I know about MyEduDecks, I think using MyEduDecks could enhance my ability to work collaboratively with teammates	68.84%	31.16%
I like to have the ability to pre-learn concepts before they are introduced in class	88.32%	11.69%
Based on what I know about MyEduDecks, I believe that using MyEduDecks could enhance my ability to pre-learn concepts before my teachers teach the material	76.62%	23.58%
I like to be in control of my own learning	93.51%	6.49%
Based on what I know about MyEduDecks, I believe that using MyEduDecks could enhance my ability to be in control of my own learning	80.26%	19.74%
I like to personalize my own learning	88.32%	11.68%
Based on what I know about using MyEduDecks, I believe I could personalize the way I learn by designing and creating unique cards or decks that align to the concepts I most need and want to learn	84.42%	15.58%
I much prefer to create and design my own assessments or study cards than to have the teachers design them for me	53.24%	46.76%
Based on what I know about MyEduDecks, I think I would enjoy taking assignments more using the online MyEduDecks system rather than using the traditional paper and pencil flashcard system	57.14%	42.86%
I enjoy creating my own study guides and help guides	66.23%	33.77%
Based on what I know about MyEduDecks, I believe I could create my own personalized assessments to help me learn concepts that I find difficult	77.63%	22.37%
I believe I would like answering questions on MyEduDecks using digital ink and a stylus rather than simply typing answers on the cards	44.35%	55.84%
I would like to be able to add my own graphics and pictures on the MyEduDecks flashcards system rather than using plain white paper-based flashcards with print	84.41%	15.58%
I can think of important reasons to include audio when creating MyEduDecks flashcard quizzes	77.92%	22.28%
I think using the MyEduDecks flashcards would help me learn things that were hard because I could test myself and practice what I don't know	99.32%	11.70%
I think being able to add picture on the flashcards would be very helpful	93.43%	6.56%
I think hearing a person's voice reading, or having sounds on the flashcards would be very helpful	85.52%	14.47%
If I could make my own online flashcard deck using MyEduDecks, I believe I could make my own questions that could help improve my learning	80.52%	19.47%

use of pen also allows students whose typing skills are not up to par, progress as fast as those who have excellent typing skills by using their native method of communication.

One of the most surprising data points found from this pilot test is student aversion to ink, that against previous data gathered that showed that students preferred ink over typing. This outlier in the data most likely originates from the type of questions asked with the test. Students in seventh grade are generally proficient in typing, and the ink is more useful for expressing complex ideas such as formulas and graphics, rather than simple text. The use of these ideas would likely increase ink approval and match the data that was gathered previously. The students prefer to pre-learn concepts before they are introduced in class as seen in the table above, where eighty-eight percent agreed. As seen in the table above, seventy-seven percent of the surveyed students believe that MyEduDecks will enhance their ability to pre-learn concepts before they are taught by the teacher.

As seen in the table above, seventy-eight percent of the surveyed students find inclusion of audio while creating MyEduDecks flashcard quizzes important to them. The integrated audio with the visual in the play decks will enhance the understanding and grasp of the concepts. Based on the recommendations from the students, we will provide an audio icon to be able to listen to the question and introduce differentiated sounds for the correct and incorrect responses. Additionally, the option to add audio while creating a flashcard will be provided. The most important part of implementing an educational tool in the classroom is teacher approval to ensure that the students gain full advantages of the program from the teacher. The teacher who tested the program responded positively, declaring that MyEduDecks would be useful for students to learn concepts independently and in a peer-to-peer scenario. The value of the program is inherent in its ability to instant grading of cards, assign decks to students, as well as allow them to create their own.

18.7 Future Work

Collaboration is an integral part in modern learning, and MyEduDecks is an effective tool that allows sharing decks with one another in order to learn collaboratively. One of the future goals of MyEduDecks is to enable students in a class to share decks with one another, and in the current iteration, MyEduDecks can be used for peer-to-peer cooperation through offline means. With all of the new data that have been collected with the pre-test survey and beta test, the MyEduDecks application will continue to be enhanced using suggestions and preferences of the students. A new development team aided by the advice from previous members will continue to develop and improve the application in the months to come. Several key features that will be added and improved are: an upgraded Word function making it easier for teachers to create decks from Word documents and a feature to add images and audio to the cards based on strong response from the students and teachers surveyed. We will improve the usability of the program with refined autocorrect feature, ability to add graphics and pictures on the MyEduDecks digital flashcards system, and ability to add audio to the flashcards. Additional features that will be added to the MyEduDecks based on the participant responses include: flashcard sharing for enhancing collaborative learning, ability to include audio with the visuals to enhance understanding and grasping of the concepts, a feature to invite other students to share decks and play together, and support for foreign languages. Finally, a feature will be added to allow peer-to-peer sharing of decks to allow students to prepare and send decks to one another electronically rather than the current offline sharing method.

Acknowledgements The MyEduDecks application could never have been completed without the aid of several individuals as well as support from the South Fayette School District. The original MyEduDecks application was developed as a University project under Dr. Ananda Gunawardena, who was a professor at Carnegie Mellon University and oversaw the application from its inception until year three, and the team appreciates his advice. The team also appreciates the willingness of Mr. Scott Phillip, a 7th grade history teacher at South Fayette Middle School, to give class time as well as some personal time for training out of his day in order to test the MyEduDecks application. His recommendations and suggestions proved extremely helpful in guiding future development of the application. The team appreciates the time the seventy-eight students took in order to provide us with valuable feedback as well as an excellent data set for future research. Lastly, the South Fayette School district, under Dr. Bille Rondinelli, has given the team continued support through the last four years and has provided the funding for trips and conferences like CPTTE. Without any of these individuals the team could not have reached this level of accomplishment and opportunity that the MyEduDecks application provides.

References

1. Adamchik, V.: Pen-Based Problem-Solving Environment for Computational Science, pp. 365–372. Springer, Cham (2015). doi:10.1007/978-3-319-15594-4_38. http://dx.doi.org/10.1007/978-3-319-15594-4_38
2. Caceffo, R., Rocha, d.H.V., Azevedo, R.: A Ubiquitous, Pen-Based and Touch Classroom Response System Supported by Learning Styles, pp. 373–380. Springer, Cham (2015). doi:10.1007/978-3-319-15594-4_39. http://dx.doi.org/10.1007/978-3-319-15594-4_39
3. Chien, M.E., Jheng, C.M., Lin, N.M., Tang, H.H., Taele, P., Tseng, W.S., Chen, M.Y.: iCAN: a tablet-based pedagogical system for improving communication skills of children with autism. Int. J. Hum. Comput. Stud. **73**, 79–90 (2015). http://dx.doi.org/10.1016/j.ijhcs.2014.06.001, http://www.sciencedirect.com/science/article/pii/S107158191400086X
4. Jeong, Y., Gunawardena, A., Koedinger, K.R.: Designing a pen-based flashcard application to support classroom learning environment. In: CHI '10 Extended Abstracts on Human Factors in Computing Systems, CHI EA'10, pp. 4695–4698. ACM, New York (2010). doi:10.1145/1753846.1754214. http://doi.acm.org/10.1145/1753846.1754214
5. Kornell, N.: Optimising learning using flashcards: spacing is more effective than cramming. Appl. Cogn. Psychol. **23**(9), 1297–1317 (2009). doi:10.1002/acp.1537. http://dx.doi.org/10.1002/acp.1537
6. Kowalski, V.F., Kowalski, E.S., Colling, J.T., Cuba, G.J.V., Gardner, Q.T., Greivel, G., Palou, E., Ruskell, T.: Using InkSurvey with Pen-Enabled Mobile Devices for Real-Time Formative Assessment II. Indications of Effectiveness in Diverse Educational Environments, pp. 307–314. Springer, Cham (2015). doi:10.1007/978-3-319-15594-4_32. http://dx.doi.org/10.1007/978-3-319-15594-4_32

7. Li, J.T., Fan, Y.N.: Using iPad to Make E-Flashcards and Sharing the E-Cards Through Doceri, pp. 249–257. Springer, Cham (2015). doi:10.1007/978-3-319-15594-4_25. http://dx.doi.org/10.1007/978-3-319-15594-4_25
8. Longcamp, M., Zerbato-Poudou, M.T., Velay, J.L.: The influence of writing practice on letter recognition in preschool children: a comparison between handwriting and typing. Acta Psychol. **119**(1), 67–79 (2005). http://doi.org/10.1016/j.actpsy.2004.10.019, http://www.sciencedirect.com/science/article/pii/S0001691804001167
9. Raymond, D., Liew, J., Hammond, A.T.: A Vision for Education: Transforming How Formal Systems are Taught Within Mass Lectures by Using Pen Technology to Create a Personalized Learning Environment, pp. 355–363. Springer, Cham (2015). doi:10.1007/978-3-319-15594-4_37. http://dx.doi.org/10.1007/978-3-319-15594-4_37
10. Read, C.J., Horton, M.: Studying Digital Ink Technologies with Children—Methods and Insights for the Research Community, pp. 23–33. Springer, Cham (2015). doi:10.1007/978-3-319-15594-4_3. http://dx.doi.org/10.1007/978-3-319-15594-4_3
11. Roshan, S.: The flipped classroom: touch enabled, academically proven. In: The Impact of Pen and Touch Technology on Education, pp. 215–222. Springer, Cham (2015)

Part VIII
Panels and Keynotes

Chapter 19
CPTTE Education Panel: Education in Transition

Jonathan Grudin, Calvin Armstrong, Robert Baker, Eric Hamilton, and Mark Payton

Abstract At the 2016 CPTTE Conference on Pen and Touch Technology on Education, five expert researchers and educators spoke at a panel in front of the audience of CPTTE attendees where they discussed past influences and future directions. The members of the panel were Eric Hamilton, Rob Baker, Jonathan Grudin, Mark Payton, and Cal Armstrong. The discussion also included comments and questions from Ken Forbus (2017 CPTTE conference chair) and Andy van Dam (2016 CPTTE conference chair). Topics included the practice, the difficulties, and the added value of what leads to success in device-per-student deployments in K-12 educational settings. A philosophical discussion emerged including arguments for and against domain-specific and domain-independent solutions, as well as future directions to better preparing students for success.

J. Grudin (✉)
Microsoft, Microsoft Way, Redmond, WA, 99052, USA
e-mail: jgrudin@microsoft.com

C. Armstrong
Appleby College, 540 Lakeshore Road West, Oakville, ON, Canada
e-mail: carmstrong@appleby.on.ca

R. Baker
Cincinnati Country Day School, 6905 Given Road, Cincinnati, OH, 45243, USA
e-mail: bakerr@countryday.net

E. Hamilton
Graduate School of Education and Psychology, Pepperdine University,
24255 Pacific Coast Highway, Malibu, CA, 90263, USA
e-mail: eric.hamilton@pepperdine.edu

M. Payton
Peraza Dermatology Group, 32 Broad Street, Claremont, NH, USA
Whitfield School, 175 South Mason Road, Saint Louis, MO, 63141, USA
e-mail: mark.payton@outlook.com

T. Hammond et al. (eds.), *Frontiers in Pen and Touch*, Human–Computer Interaction Series, DOI 10.1007/978-3-319-64239-0_19

19.1 Introduction

At the 2016 CPTTE Conference on Pen and Touch Technology on Education, five expert researchers and educators spoke at a panel in front of the audience of CPTTE attendees where they discussed past influences and future directions. The members of the panel were Eric Hamilton, Rob Baker, Jonathan Grudin, Mark Payton, and Cal Armstrong. The discussion also included comments and questions from Ken Forbus (2017s CPTTE conference chair) and Andy van Dam (2016 CPTTE conference chair) (Fig. 19.1).

19.2 Panel Transcript

Jonathan Grudin [3]: We are ready to start the Education in Transition panel. I am very pleased and honored to be introducing and moderating these four people. I want to start by saying that they changed my research focus these past four years, and had a big impact on Microsoft. I will start with that story. In 2013, I was asked to do a short-term project in K-12 education. I went to a prior year's conference, which was hosted at Pepperdine University by Eric (Hamilton). Three presentations were given by Cal Armstrong, Mark Payton and a colleague of his, and Rob Baker. They showed how technology could be used in K-12 in ways that went beyond what we were aware of at Microsoft; beyond even the model schools to which we took people. One common refrain of these gentlemen was that Microsoft should be listening more to them than

Fig. 19.1 Members of the education panel, from left to right are [1] Eric Hamilton, [2] Rob Baker, [3] Jonathan Grudin, [4] Mark Payton, and [5] Cal Armstrong. Their bios are found at the conclusion of this chapter

we were. I said that I could not promise that Microsoft would do anything, but if they would join me for lunch and tell me what exactly we should be paying attention to, I would be glad to communicate it back. I did that and was so impressed with what they said that I continued working in K12 education after my project ended. They have had a big impact on what we are building at Microsoft. I owe a lot to these four gentlemen, and so does the field. Because they were out in front of the curve three years ago, we thought it would be interesting to hear their thoughts now on where we are going, what others can learn from what they did, and what they have done the past three years. So, first I will have them introduce themselves and describe their backgrounds, how they got into this, and what they have been doing. And then we will take some more forward-looking questions. Would you like to start, Cal?

Cal Armstrong [5]: I am Cal Armstrong. I am a three-quarters math teacher and three-quarters technology integrator.

Jonathan Grudin [3]: That does not add up to 1.

Cal Armstrong [5]: It does, however, reflect the reality of the situation. I have been at Appleby College since 2001. Appleby is a great 7–12 school just outside of Toronto, Canada. They went one-to-one laptop in 1997, and then one-to-one pen-based tablet in 2004. It is important to note that, until we ran the OneNote project in 2012, the one-to-one tablet program was a marketing success; but, it was not a pedagogical success. And I think that is an important outcome that my school has been very honest in saying. Before OneNote, it was a great experiment, and the students and teachers enjoyed the one-to-one devices, but they really did not affect pedagogy in a meaningful and systemic way until we introduced our version of what became OneNote Class Notebook. It was only that tool that made the incredible push towards what students and faculty experienced in the classroom. Yes, we have amazing teachers doing amazing things with the one-to-one program, but it was not until we actively began digitizing content using pen based ink in a clean and collaborative way that there was really any effect system-wide across the classroom. And for us, success is: if we walk into our school and throw a rock to hit any random classroom, when we get to the classroom we see that something has changed as a result of that pen being active in OneNote.

Mark Payton [4]: Mark Payton, here. I am most recently from Whitfield school in St. Louis, Missouri. Sadly, I am not currently working in education. I am open to change that. My degree is in early childhood education, so I started as a preschool teacher, but I was out of education for quite a while.

Back in 1997 or 1998, I started working at Vermont Academy in Southern Vermont, a relatively small, very low-tech independent boarding school. We remained low tech intentionally for a little while given the student population we had and the state of the art at that time. But when the Tablet PC was released in 2002, we made the argument that this was finally the right device for us to make a change to the culture of our school and introduce a one-to-one program. We did that; it was ubiquitous, but it was ubiquitous for the wrong reason. It was very top-down mandate from the headmaster, “You will do this, you will use these things." With the consequence that the success rate was what you would expect with a program like that. Everybody used it; some people used it amazingly well. We started with the Compaq, then HP TC1100, which I will probably show later for any who haven't seen it.

I moved from there to Amman, Jordan, or just outside of Amman, Jordan, to King's Academy, a new boarding school, that the King of Jordan had started and started a one-to-one program there, also based on Tablet PCs. This program was significantly more successful for a number of reasons. One is that I had learned a number of lessons about working with faculty by that point and would not support a dictatorial approach to it. By working with faculty, I began learning what they needed to get out of the technology, and approaching it not from top down but from bottom up, if you will, though that is still not quite the right terminology. That program became fairly successful and, to my knowledge, remains in place today although I haven't been there for quite a while.

Whitfield School, however, was quite a different story. I went into a school that had a currently implemented failed one-to-one program. By everybody's acknowledgement, it had just been an abysmal failure. I went there with the expectation that I would be heading the BYOD[1] program. But the demand for identical one-to-one devices was still there on the part of teachers. I recognized that they wanted a standardized environment where everybody could learn together and support one another on. But the teachers wanted and needed an environment that was supported by the school, rather than dictated by the school. Tablet PCs seemed to be the preferred devices by the teachers, which was definitely what I also preferred, and thus, presented that as one of several options. I showed the benefits and weaknesses of each of the several platforms that we looked at to the teachers and administration. There was near unanimity that digital ink was something that had phenomenal possibilities within education.

About the same time we developed a similar approach to OneNote that let us work collaboratively in the classrooms. And between that and DyKnow, it became quite a smashing success, attributable in large part to the qualities of the teachers and the fact that our approach had always been: "What are your educational needs?" to which they would reply, "We want to use technology." And we would respond by collaborating with them, analyzing what they needed, talking about it—my co-partner and I were both educators—and then pointing them in the right direction, encouraging them in the right direction, working together to develop the right direction, even if that meant no technology at all, because technology wasn't to be an end in itself, but a tool for the support of education.

That program continues now and now that I left Whitfield, we have one teacher there. Sarah, raise your hand. If you have any questions, feel free to talk to Sarah about how it is going now. That is the first program that, I think, is actually going to be self-sustainable. The state of the art for so much of this time required an advocate—a staunch support person and somebody to really

[1] Bring your own device.

steer and manage the program. And I am glad to see that the state of the art has gotten to that point. We will discuss that a little bit further, but that's a program I am very proud of and Whitfield is very proud of it and rightly so, because of the quality of the hardware and software and the teachers involved, and that is an important collaboration.

Robert Baker [2]: I am Robert Baker. I am the Director of Technology at Cincinnati Country Day School. And, very eerily similar to Cal's statement, we were the first school in the nation to go one-to-one back in 1996; tablets didn't exist then, obviously. Six years into that journey—we were a Toshiba shop at that time—Toshiba brought us a basically handmade Tablet PC at one of the laptop conferences we were running at that time. The limitations of one-to-one devices in grades 5–12, and the ability to enable every teacher to also use the same device were obvious and painful to us. Seeing the tablets really energized and excited us. We switched immediately in 2003. So we have been one-to-one with Tablet PCs, with inkable devices, for 13 years now.

And I am here with you. I love WIPTTE. I have been on the committee in years past. And I am your end user. I love being here and hearing about the giants of the industry: the people doing the research and working in the background. But we are the end user. We are the ones with boots on the ground dealing with end users. My user base starts with 10 year olds, just to put this in perspective. We think we are doing some really amazing things; we know we are. One of the main reasons we believe so, is because of the environment—because **devices do matter**.

My stated mission is to create the most powerful teaching and learning environment on the planet. And again, that could have meant that computers could have been bad for learning, but we found this not to be true. Part of our mission is sharing our methods and results. That is why we run tablet conferences. We have had educators come from all over the world to see what we do. I have been blessed to travel all over the world to try to share what we do.

Like Cal said, ink was a big deal back in 2007. When Office 2007 came out, you heard from the OneNote team—it was a three year process, and when we finally saw it, it fundamentally changed what we wanted to do. We started to hack around—we had no idea what it was nor what a OneNote TOC2 file was. We didn't know what the heck those things did. But we were able eventually, with Nate Johnston—he is back there in the crowd—to make an alpha version of shared OneNote notebooks. So, in 2007 we started that and it was so interesting and powerful that it spread very rapidly through the school, much like it does with any school.

I will keep it brief and stop, but I have the honor of keynoting the closing keynote. And that will be a huge honor, assuming anyone will be remaining after lunch. I do have a lot of samples and like Cal, we felt like we had the best-kept secret education for the last 13 years. And we have done everything we can to try to encourage Microsoft and anybody who would give us a bigger megaphone to share what's possible when you have the right kind of environment, because digital ink really can transform what you can do and can transform pedagogy.

Unfortunately, even now one-to-one programs, to a large extent, are for publicity, so you can put it on your website, because your competitors are one-to-one, or the vision and the constraints or the expectations of what a one-to-one environment can do are low. Generally the bar is already so low and many programs do not even meet that low bar. But when you have the right kind of device, you have a pen and touch and wireless projection, and everyone has the same functionality, then amazing things happen. Great educators do great things in that kind of environment.

That's why I am here to share, and to make sure that this group knows that cool things are happening on the boots on the ground. I would like to make sure that you know, that this really is spreading. The numbers are increasing, even for our current class' notebook that we had a really rough sketchy version for. Cal's school still has the best version; theirs is better than the one everyone else has access to. I am a little jealous, but we will get there. It is really out there; the numbers are huge. We are now talking about three million notebooks worldwide, with almost half a million teachers using these notebooks. And, it is a big deal.

But the increase in K12 users is only part of it. We love what you guys are doing with them. I have had great conversations with a bunch of you about finding ways to streamline the formative assessment process for educators and being able to focus on the process for young children up through 12th grade and in post-secondary. The process is such a big part of the assessment. In a nutshell, you get more of what you subsidize. I think a lot of us up here are just workflow specialists. We're trying to make stuff as easy as possible so more people will try and do it and find interesting ways to make it powerful and engaging.

Eric Hamilton [2]: Rob, I will come to your keynote. I appreciate, that at the WIPTTE 2013 conference, Rob called digital ink, "the best-kept secret". I don't know if he had coined the expression before, but the expression at least vindicated for a lot of people that gnawing sense of what the heck is going on? and why are more people not engaged in this kind of technology? It at least affirmed that the question was being shared, because I do consider digital inking to be one of the best-kept secrets there is in education.

My story—in terms of digital ink—began maybe before even any of the others who are up here. Back in 1984 I was teaching a class for the Chicago public schools for their so-called gifted program. I was teaching students at Northwestern University in the Math Department. These students were all very advanced-some of them were hyper-advanced and some were only top 5 percentile. I craved the opportunity to be able to see what they were doing at their workstations. Well, they did not have computer workstations back in 1984, but I started thinking about the idea of being able to see everything that kids were doing at their respective desks, so that I could interact with them and not deal with them at a product level. When I would ask them for a solution to a problem, I did not want to know what their answer was. Rather, I wanted to see what their work process was. And, although you could walk around the class, I craved something that would be more automatized and that would give me a line of sight, and that would also get kids to engage when other kids were moving ahead of them. So I started developing some software through the following years and eventually developed some screen sharing collaboration software that required and relied on digital ink. This was all about digital ink and things you could do in mathematics, but that you could not do with the keyboard.

Before there were tablet computers, there was, as you may be aware, a big push in the late 1980s and early 1990s to develop digital pens that work with glass. We would spend $1,000 a pop for micro-touch surfaces. We patented some software on screen sharing that required digital ink. It was very, very exciting, except there was no market for it. It was too darn expensive. So it was very exciting to see the Tablet PC operating system arrive in 2002–2003 and then to see the advent of tablet computers.

A lot of work that I have done in digital ink has involved undergraduates, where we use the kind of technologies by which a teacher can see what everyone is doing using digital ink and can intervene with them and interact with them in real time, so that's been very powerful. Microsoft has supported some of that work. We have also looked at ways that digital ink can provide a portal to some of the most elusive and powerful goals of education. We do a lot with getting teachers and educators and students to collaborate creating interactive digital media using digital ink especially in mathematics.

I worked at NSF in the US federal government for a number of years. I signed out a fortune over several years to some of the smartest and best people in the US in research and teacher professional development: How do we get teachers to be better? To be more effective? And there are some really, really smart people in this country who could lead that research.

What we discovered though, and what I also discovered after I left the federal government was that **none of us is as smart as all of us**—that old expression. We found out that **when teachers could collaborate**, and they had the input tools and representational tools using digital ink, and they were using all the other representational affordances: **magical, creative, imaginative things happened**—things that never happened when they were under the tutelage of the smart professor.

Our work started focusing quite a bit on using digital ink to free the imagination of teachers to create either digital versions of what they were already doing, in which case the digital versions were often and usually better, or to create the entirely new ways of looking at the kind of work that they were doing. I am not into the *We don't do things differently; we do different things.* I think, you do things differently and you do the same things differently. I do not know if you see that dichotomy, but effectively we saw digital ink as transformative in many ways.

Most recently, I have been working on a project as a visiting professor at the University of Namibia under a Fulbright Research Fellowship from the State Department. And we are in an exciting process; I think this is more exciting than what they are trying to do in Estonia, which has been a very powerful small country example of going digital. The Republic of Namibia is actually looking at and building towards the strategy by which the educators of that country digitize their entire curriculum, not for the purpose of having a digital version of the curriculum, but for the purposes of what changes in teachers, and what changes in students as this process unfolds.

I love the technology. I used to be the computer professor, that is where I was originally tenured, but my interest is in learning and the digital ink interface is a portal to some very amazing things. That is where my passion lies and why I am in this field.

Jonathan Grudin [3]: Thank you. When, as I said, in 2013, all three of the schools here had already put their own videos up on the web that showed uses of both digital ink and OneNote. And I think as we start looking towards the future, I can see a couple of different audiences—users/perspectives of digital ink—where it would be interesting to hear thoughts from the panelists. In terms of the research perspective, Eric is in the research field and there are a number of people here from the research community. As Rob said, one of the things that we are here for is to find out what you are doing and what might be coming along research-wise. I thought it would be interesting to hear them say, based on their experience, maybe what some of the new ideas are for and where it will be particularly useful to see research directions go. This could include continuing on the directions that some of you are currently pursuing and might also include some areas that we have not heard from.

So, I will open it to panelists: If you could corral these and other researchers and get them working on some new projects or continue in the current directions, what might you find useful?

Mark Payton [4]: Can I address that in general terms? One of the things that I think is really critical is to define what is important in education. Education is not about facts; it is not about memorization; and, it is not about understanding theorems. **Education is all about communication and relationships**. Learning is those other things. Learning is a part of education, but **education is primarily about the teacher and the student, the teacher and the teacher, the student and the student—and the relationships and interactions between them**. At least, that is my experience. And particularly as a preschool teacher, that was my philosophy. The tools have to support that. We have to get out of the way, and the tools have got to support that.

I think one of the biggest hindrances to the acceptance of the Tablet PC has been that it does not get out of the way and the software does not get out of the way. DyKnow is an excellent example of that. DyKnow is phenomenally powerful software. There are some wonderful things that it can do, and the teachers who master at least the part of the interface that they need, are doing great things with it. Where it fails is that for far more teachers than not, it is too big of a cognitive cost for them to use in the classroom.

And, we as geeks—not that I have ever studied computer science, but I have been a professional programmer, kind of self-taught—love to do problem solving, love to work the algorithms, and love to figure out cool tricks to make things happen. But must keep first and foremost in our minds, as I told the kids in South Lafayette: **the interface is the program; the interface and interaction with the software is what it is all about**. Otherwise it is never going to be terribly successful, unless it is so far better than anything else that they are going to use it regardless of the costs.

I really encourage people to think about research, much as Kimberle Koile does. Kimberle [looks at Kimberle in the audience]—slug me in the shoulder if I get it wrong. Kimberle made a comment that she does not want the software grading. She is using that software to facilitate the understanding of the teacher as to what is actually going on for the student, or at least words to that effect. And, that is what needs to be a focus of research. I would certainly encourage every researcher here, and also all the hardware vendors here to think about getting your stuff out of the way and letting it be imminently forgettable—but letting the experience and the improvement of communication and relationship building in the classroom be the thing that drives the acceptance of it and drives the use of it.

Nickolas Jackiw from Audience: Can I respond to that?

Jonathan Grudin [3]: Yea, Sure.

Nickolas Jackiw from Audience: I am struck by your comment that learning is a part of education, but that education is really about collaboration and communication. At first blush, I think instead, that learning is an essential and defining part of education, distinguishing education's primary purpose from any other social or collaborative group's.

If you look at the killer applications of software more generally—email, the web, Facebook—they are all communication and collaboration technologies, so the learning is relevant at some point. I am struck by the history of educational technology broadly, particularly how, with the advent of the microcomputer, there was a tremendous interest from education in learning and domain-specific innovation and education that the programs of the 1970s and 1980s. They were still large scale CAI[2] and Learning Management Systems, but there was this proliferation of ideas about disciplines, subjects, and curricular content. That seems to have really receded in the modern era, and we are back to now talking about deep infrastructure tools for workflow management for generic communication, for generic note taking, but **we have lost any sort of a belief that technology will revolutionize or innovate in students' learning of specific subjects**.

And that has really faded away. I work in Silicon Valley. I am very tied into *edu-tech start-up* culture. There are domain-specific solutions but they are all the Pinterest of mathematics: Can we take 30-year-old math textbooks and digitize them and disseminate them that way? I am interested in your perspective on the relative relation of technology to learning and collaboration. I am also interested in the panel's perspective on this observation, and whether this is a failing: is this a problem? If so, **is it a problem of education or is it a problem of technology**? Who is dropping the ball in saying that there is a relevance of things like digital ink directly into content domains? We have

[2]Computer automated instruction.

seen some examples of that. I do not want to say it does not exist. But as a historical change in the focus of educational computing, it is definitely on the wane.

Eric Hamilton [1]: I will say I don't know if I would use the dichotomy. I think there is an awful lot that is domain-specific in the edu-tech community and in the development community, and there is a lot that is generic that is content-free. I would resist, the same way I resist *do things differently* versus *do different things*. I think we do both. I would resist maybe the characterization that most of what is taking place is *content independent* versus being *content specific*. I think, as we develop and go forward, we are going to be encompassing both of those, because there is a lot that takes place in both categories.

I was also struck by Mark's comment that learning is a part of education. It is an interesting question. But again, I would be hybridized: sometimes education should have a focus that is not completely directed on learning. At other times, someone that calls himself a learning scientist may say that is the main event.

Rob Baker [2]: I will provide my attempt to address that. I would agree with you. In general, in most implementations, and I mentioned it when I was speaking, empathy is a significant issue. I have stood in the front of the class. I am a classroom teacher, and I think I am a good one. And I think what makes me really good at what I do, because my decisions and focus are always through the eyes of a classroom teacher. We know that the role of the teacher should never go away, but it should shift.

Unfortunately, integrating technology usually means putting a textbook on an iPad. So you can see a textbook on an iPad, which is not very relevant or powerful. And there are all kinds of things related to that. I would point out that I am positive that OneNote is the best piece of software Microsoft has ever developed. There is nothing even close in my opinion. I would argue that until two years ago Microsoft did not even think that was the case. Part of our struggle was trying to convince them that you have something here. I would argue that OneNote was definitely not developed with education in mind. I would also argue that I have some impression of how many engineers are now focused specifically on OneNote in education in comparison to what it used to be. I think that education is a focus now. I would argue that, now, education is actually a focus of the whole company, which I think is a great thing. But it did not start that way. OneNote was not developed trying to help educators collaborate, engage, and remove constraints, but it is freaking awesome at that.

So, my answer to domain-specific technologies would be like FluidMath. FluidMath was born here at Brown; we love FluidMath. And that is the kind of technology to take notice of. There is gonna be some residual stuff. Not everything is innovative and awesome, but the technology that matters, the integrated technology that can make a difference, is when you can do something in a way that you can not do otherwise. Just using technology for technology's sake does not make sense.

There are no mandates at my school, and never have been for the last 20 years. The cool stuff is the authentic stuff. And you can not tell teachers what is going to be awesome; you can orient them to it, inspire them a little bit hopefully, but they know what they want to do to enhance it. FluidMath—and I will do a little bit of this at my presentation, if you are here—it transforms how you can teach and engage, as well as how the students learn. I do not want to tell my kids stuff; I want them to discover it. I want to enable my kids to fail miserably, but be able to recover from that failure—the more you iterate easily and recover and learn from that, the better off you are. And when you have a tool like FluidMath where it gets all the junk out of the way, all the constraints of doing "what if" with graphing equations, it can be really powerful. I am not sure what exactly we are talking about, other than I completely agree, and I think the people in this room are aware of that. I think many ink-enabled tools would help teachers do a lot of things.

Cal Armstrong [5]: I might actually disagree with you. Over the last five years, I have been observing that teachers in the classrooms and teachers outside of my school are finding almost standard-specific apps to use on their device to deal with the particular content area. When I walk into several classrooms outside of my school, they would have seven apps rotating through to get at those particular content standards. This is as opposed to, for example, FluidMath and OneNote, where you are just sort of dealing with the more open-ended tools. My argument would be that you would like to have fewer tools just so that there is less of a cognitive demand on teacher and student, but the reverse is happening in a larger market place. There is a ridiculous number of apps that teachers and students have to deal with on a daily basis.

Mark Payton [4]: I am on Cal's side on this particular one. It is interesting to know that OneNote is the thing that is getting such tremendously broad acceptance in schools. It is because it is a really brilliant general-purpose application that can be, just because of the structure, can be customized to the needs of each academic discipline.

A great definition of technology is whatever has been invented since you were born, so we tend to think about technology as these devices. But the reality is that the blackboard behind us is technology. The printing press is technology. The pencil is technology, and, at some point, in the history of education all those things were introduced and were somewhat disruptive. We do not know what else was introduced that did not succeed. There is probably a tremendous amount of stuff, but the fact that it did not succeed is telling. I think we are at a stage right now where the technology that we are seeing contains a plethora of applications coming out that are domain-specific, but that impose that same cognitive cost on the population. If you have to learn 50 different applications because you are studying 50 different things, you can not master any of them. You probably cannot be a jack-of-all-trades with most of them.

Eric Hamilton [1]: ...unless you become a master at mastering.

Mark Payton [4]: Exactly, unless that is your focus. I will go back to my statement, that education has always been about what happens between the teacher and the student. I was a self-driven student, so I could pick up a textbook and learn stuff, but I kind of sit at one end of the bell curve. My son is at the other end of the bell curve; he cannot sit in a classroom and learn things. He is a really smart kid, but he cannot sit and read; he has got to get his hands dirty with things. But in that middle there is a large group, and for them, relationship with the teacher is so critical. Technology has to support that. Any of us can think back to our high school days: do we remember the blackboards on the wall? The projectors? The individual textbooks that we read? Except for maybe a novel or two, I do not; I remember the teachers. I remember the journalism department where I worked in high school. I remembered the relationships I developed, working very closely with my peers to put up a very good school newspaper. All of that had technology that supported it, but that was **the role of the technology, to support these relationships, to support this interaction**.

Rob Baker [2]: That is the good way to put it. I am very skeptical of any article or presentation that has a number in it—the 20 top apps to do X, Y and Z, the 20, the 30, the 100. It is, to a large extent, a search for typically more than one app to try to do what you want to do. What you should be doing in the classroom is collaboration, engagement, and other things that matter.

Mark Payton [4]: It is telling that the pencil, the chalkboard, the projector, the printing press; are all very general-purpose technology. I think the applications have to follow that model as well. We need to be able to adapt it to anything. One of the great things I'd love to see coming out of OneNote is for it to become a vehicle for development of interactive texts by classes and teachers within classrooms. It has the capabilities needed for doing that. It is increasingly getting them. And that, I think, will be a perfect showcase for how you can have one tool that can give you this marvelous language textbook with videos in French and text in French as well as all the things that are going to support that; it would also have organic chemistry, supporting both the hard sciences, like physics, and also the fluid communications side of things. The tool is getting there. OneNote might be certainly the best for doing that.

Ken Forbus from Audience [7]: So one of the exciting things about digitization is that, as Tom Stahovich was talking about, the ability to analyze process as well as product. So—speaking as someone who ate my own dog food last quarter and used the sketch worksheets in my knowledge representation class—what I found was backwards. What we really needed to be using was an assignment where we did not give students feedback. If they got feedback, they actually got close to perfect scores. They learned everything we were doing in the worksheet. And everyone was happy. However, analysis showed that even though we thought we had abstracted things enough, we had not. It was still very tedious to go through the teaching of the abstractions. So now, everyone is using educational software that can provide infinite detail. But that leaves the question: How much time on average are teachers going to spend to look at the student results? This is needed to gauge how much summative information is useful. Obviously, smaller is better if you give them a pie chart that they process best. But, how can you control the level of detail when they want more?

Rob Baker [2]: I can speak to that. Here is what happens, especially in bigger school systems. You need formative assessments. A great classroom teacher, I think, is like a litmus test. You want to know if you are doing a good job. Who knows what, and when do they know it? That is really what it is all about. If you have 120 kids a day and you want to do something like that, you use multiple choice, because that is easy to do; you can click on a button; you can have clickers. And you can see generally what they are doing. But that is almost useless. I mean they are trying to do something. But it is the process that matters. A good example would be working with Don Carney of FluidMath. Rote learning happens. You have got to practice what you do, and it varies greatly based on whether you are in the 2nd grade or a senior in an AP class. What would better help us is if you could **capture all of the process, self correct, using ink alone to reduce the cognitive load**. The teacher could quickly see all the wrong answers by problem or by student. Or, what if you just saw all the ones that are wrong since that is what you care about? You cannot just give multiple choice sheets out; that is more of what you subsidize. Providing details about what was wrong would let the teacher see who is not doing it right, and we are fans of that. You have the students do some stuff; you get great data that can be analyzed by many members of this audience, but just focusing on the subset. I wanna see if 90% of my kids missed number 7. Then I can just quickly scan, because I have details about the process. If you don't have the process, the feedback doesn't help you that much. So, we are greedy in a lot of ways. You guys are working on these problems. I do not know exactly what the format will be, but I think it varies greatly. I do not know if that answers your question but, personally, we want to emphasize that **process is important**. We do not want to see all the good processes. We want to focus on and see who is not doing it right. We want to see patterns and adjust based on that. We think digital ink is the way to do that.

Eric Hamilton [1]: I will add to that.

First of all, Andy, are you waiting to make a comment? Okay, I just wanted to acknowledge, and to make sure that we get to you.

Ken and Rob, I think that in the future, a lot of these questions about processing information will come from richer, more sophisticated representational systems.

An analogy I use: we ran a number of workshops at different conferences on assessment through visualization. I love the example of MLB.com, which is the most sophisticated professional sports site in terms of the kinds of information. What is amazing about MLB.com, a good metaphor for the education community, is that when websites first came up for professional sports, you would get dribbles of information. You would get an equivalent of discrete text pieces of information, you know, ball 3, strike 1, that kind of information. If you look now, for the same amount of cognitive processing, you get a lot more information.

When you go to a traffic intersection at a busy street, we are processing incredible amounts of information, much more than a teacher processes looking at a test score. We know how to process lots of stuff. A quarterback knows how to process all the hundreds and hundreds of interactions going on a field and make snap decisions.

I believe that in the future the representational systems for data will become more sophisticated and enable the kind of selective interpretations—we will know how to look for something in those representations. I am not trying to say it is a rosy future; what I am trying to say though is that the representational systems for collapsing data, using different colors and spatial arrangement for the information that comes to the teacher will be much more sophisticated than they are now, I hope. I think we'll enable some of the things that we can't do now because we don't have the technologies or the spatial sophistication to lay them all out.

Cal Armstrong [5]: I am going to look at it from your statement from my lens of teacher change. I think at the heart of your question or your comment is a fundamental lie of technology and publishers is that anything we introduce to the classroom is going to make teachers' lives easier and allow them to spend less time. Teachers, good teachers—which should be an oxymoron—teachers are automatically good. **Teachers will always spend every minute of their waking life thinking about their students**, and **when we introduce technology we allow the teachers freedom and how they choose to spend their time**. The information that we get from technology makes the teacher's choice better and makes students' learning better as a result. Regardless of the technology you give to us or research you present to us, **teachers will always work a ridiculous amount and make those decisions based on students they have in front of them. What I hope for is that technology will give us enough time that we can spend much more time on the relational aspect with the student and less time on that whole assessment process**.

I was brutally honest with my teachers when we introduced the OneNote notebook. It was strictly a *we are saving you time* with this OneNote notebook. We are taking that motivation back, instead emphasizing that we are making sure you are looking at assessment differently. **It is a zero sum game in our school with any technology, because really that is the way good teachers are**. They dedicate their time during the school year to their students.

Rob Baker [2]: That is brilliant actually. We talk a lot about that too. A lot of this talk about preserving class time, as we say, is an interesting dichotomy. Our school is known for technology, but the reason we are known for technology is because we are not trying to use technology to save a significant amount of time. But rather we use it because it really does simplify processes. It's the whole: "*you get more of what you subsidize*". If I can iterate, and if I can get some feedback and collaborate more efficiently, then maybe we can stop and do something else in all these other classes and get outside and get hands-on learning and active learning. And, your 9th grade boys can stand up and move around the room cause they have to, or they would explode. You know, all that stuff.

Andy van Dam from Audience [8]: I have a number of points to make, so better cut me off when I overstay my welcome. First of all, thanks for the tributes to MathPad, FluidMath, and Fluidity. A comment about that: we never got support from NSF for that project. We got a couple of years of support from Microsoft, and then it was terminated. And it was largely a bootleg effort. Fluidity itself, a little spin-out from my group has been struggling and lives basically from hand to mouth and from day to day on small government handouts. Despite the fact that many teachers love it, it still has not really penetrated in any large numbers and there is not any significant income from it to sustain the few people that continue working on it. It is a cottage industry. ChemPad, which we started multiple years after—

Mark Payton [4]: —We used it. —

Andy van Dam from Audience [8]: —could not get funding from NSF, Microsoft, or American Chemical Society, and so it died. I spent four years on it, and it had a sophisticated simulation engine behind it. I was forced to abandon the effort because we could not get support for it. This goes back to the comment that we made in the research panel yesterday, which was that there is no shortage of great ideas and people to execute on those, but there is not any money in this field for researchers to produce the next generation of tools.

This leads me to the following observation: And for those in the early 1970s Alan Kay published this wonderful article, on Dynabook. And for those of you not familiar with it, there is a little hand-drawn sketch of two little kids sitting in a meadow with grass and flowers and they have what we would recognize instantly today as a Tablet PC on which they are drawing. He itemized some of the things it should be able to do, and one of them was—and that is key to the whole vision—that there should be a lot of simulation-based tools, in other words, tools that have sophisticated models, data representation, process representation, and that can run to simulate a biological process, a chemical process, a physics process, and so on. That vision is almost unrealized today.

You guys are in love with OneNote, I think it is very appropriate; it is a great thing, but it is just a scratch on the surface of what software should be able to do to bring things alive for students and teachers and to allow sandbox experimentation with the basic concepts in physics, chemistry, biology, geology, as well as historical reenactments, and so on and so forth. We are at—if this is an exponential—nowhere near the knee of the curve despite decades of work in various areas that should feed us.

I do not know where I am going with this except to sound like I am whining about the lack of support; others join me in that whine. The lack of support is a serious inhibitor of progress in your missions to try to bring the best of technology to the teaching and learning environment.

Rob Baker [2]: Yeah, you are us; we are the same thing. The two big things that it used to be, and it is coming around. You come to Country Day, and you see what we are doing and what kind of work I am in. There is a lot of obscurity: no one knows that a stylus and active digitizer are different than a pencil/finger on a stick. As far as cost penalty, there used to be a significant cost penalty to having an active digitizer. That is now gone. But again, I go back to obscurity—no one knows the potential. I don't care about Microsoft. If somebody comes out with something that is 50 bucks and it is better (which I hope), then I would recommend that. But right now, the Surface 3, with 30% off, is cheaper than the cheapest iPad. You have to go Chromebook to get a cheaper device if you want to deploy something. But it has not caught on, not as I would have hoped. I am not sure a lot of people know about it, frankly. So, I could not agree more; we say the same things all the time.

Andy van Dam from Audience [8]: Rob, I was really puzzled by your startling numbers, about the number of tablets out there and the number of teachers using them. I must have misheard at the very beginning of your summary, you were talking about hundreds of thousands. What was that figure?

Rob Baker [2]: That was teachers using the OneNote notebooks, class notebooks, and the number of students. That is the number of people using shared OneNote notebooks—almost half a million teachers and about three million students.

Andy van Dam from audience [8]: Really?

Rob Baker [2]: I have got a slide in my presentation in which I will talk more about it. And again, did you know that? Did this group know that? If anyone should know, then this group should have heard that.

Cal Armstrong [5]: To be fair, that may not be a pen-active device.

Rob Baker [2]: Yes, absolutely.

Andy van Dam from audience [8]: Ah, okay, big difference. So, my last comment, or plea. Since you cannot supply the money that we need in this field. Can you at least give us requirements? Can you say, assuming funding would be available and assuming talent is available to work on this set of problems, capabilities, issues, what is next on your wish list? What would you like to get a little further along on the exponential? I am not expecting an answer, not in real time. But, as a homework exercise—like to the student.

Cal Armstrong [5]: My hesitation is that if I would ask anyone of my faculty, out of a hundred faculty, I would get two hundred different answers. And that is going to happen regardless of what school or discipline they are in. Every teacher wants to do something in a particular way different than any other teacher. If OneNote has done anything, it allowed them to be that flexible. I do want to go back to your earlier point. I am a classroom teacher, all my buddies are classroom teachers. When we tried to introduce graphing calculators in the early 1990s, TI spent a bazillion dollars. The Ontario government convinced all the publishers to include graphing calculators in textbooks, so all textbooks had graphing calculators. There was funding for every teacher in Ontario to have graphing calculators. When I walked into the classrooms in 2009, after 20 years of continual pressure and money—and I went into 20 classrooms in the city of Toronto—I did not see a single graphing calculator in use. To hope that you would have progressed just because you really wanted it, it doesn't happen. It really has to come from the teacher in the classroom being able to make that change themselves, because they feel it is important. You can take a horse to water, but you cannot make them drink. The only way to get them to drink is to feed them salty snacks that they want. You have to design those salty snacks. That was what OneNote was for us. We tried something different, it was not a salty snack, and they threw it back at us. So, you need to know what those salty snacks are.

Rob Baker [2]: That is what we have been doing for 13 years. I agree completely. I do know that process is big, and like Cal said, there is currently no way to measure that. That is why we hunger for general tools when possible because no one can know exactly what we need. I cannot know what my teachers want to do, even just in my one little school, ignoring the whole planet. But process is a big part of teaching and learning. And in my example, FluidMath, just showing the wrong answers is great, but you guys are at the point now that you should be assessing that ink for me. Does that make more sense? The only thing I know is that no one would think it is the same thing, no one wants the same thing. So, it is a moving target as it was. So, it always was a slider mentality for us. But the ink, the process is a huge part of this! And there are two different big skews. Again, OneNote is a big cognitive load. If you talk to good educators, they should be aware that **cognitive load really affects the classroom. It affects the weaker student way more than a super star student.** You need, when you can, to let a student really focus on multiple things on the same page at the same time. I have been telling Microsoft to hire Don Carney and buy FluidMath, and I just want it in OneNote. I do not want to leave OneNote to get assessment. I just want to make a new special FluidMath page. And when I do it, it just flies in, because at the end of the day, all the work that I did is somewhere where all the kids can access. Or I can have the kids do that but since it is in OneNote, in this collaborative space, and I can switch through live and share with the class, and show common misconceptions with real examples, instead of the teacher saying that, "a common misconception when graphing a parabola is. . . " If I can just say, "Johnny, Sally, and Cindy, and all did it right away," it is a lot different than having me saying it. I know it sounds like a Microsoft love affair with OneNote. No, it is not. A lot of stuff we do has nothing to do with OneNote, but at the end of the day putting things in there lets us use the other stuff more efficiently, because you have to collaborate, share, and show and all that stuff.

Andy van Dam from Audience [8]: Just a very brief response: The whole vision that we started with after the Tablet PC came out is that you would have some kind of scaffolding like OneNote. And then it is Pages—if you use a page metaphor, based on context, whether you were doing mathematics, physics, biology—the appropriate engines would be activated. The sketch recognition would tell you what the content was and the simulation engine would do the solving, all within that framework, whether you think of it as an unbounded canvas or a constraint page in a notebook. But that integration of different modalities is as far away as the individual components and the individual disciplines are. But, I think it is a very easy vision. Realizing, of course, that it will take huge amounts of efforts and money.

Mark Payton [4]: You are not alone in that. I actually heard Ken talking—was it to Ian from the OneNote team?—about that very thing—was it not, Ken? About being able to put Cogsketch into a OneNote page. I think even Microsoft is kind of thinking along those lines, that there must be a generic platform for the academics area-specific educational tools if they are going to continue with that.

I am going to take a slightly different path at this point and say that I am not sure if there is anything that you as researchers can do at this point, because I think the problem lies elsewhere. You are the victim of other people's mistakes throughout the last 13 years. Microsoft introduced this really fabulous platform, very flawed at that time, you know, XP was better than Windows for pen computing in the 1980s by a long shot, but still flawed. They introduced it; they drove a lot of support with grant money, etc. Then, essentially—my perspective, I don't want to offend anyone from Microsoft here—they did not realize what they got, turned their back on it, and handed their lunch to Apple in the Tablet space. Apple did not take their lunch; Microsoft gave it to them. This was a mixed blessing, because I think the phenomenal success of the iPad kind of woke the whole industry up, and Microsoft has been doing some great things, but Microsoft is not alone in this.

The main reason I brought this device [holds up TC1100] is because the TC1100 in many ways remains the best Tablet PC ever made, not just the Wacom digitizer, and HP's history with this is exemplary of what the whole industry has done. Compaq designed this, and that really shows. Compaq was a brilliant design company. They did so many things right with it. If I could get batteries and Windows 10 running on this with any kind of performance, I would probably still be using it. 13 year old batteries do not do much for you. HP bought the device, re-branded it as HP, and made no real changes to it. Compaq addressed the shortcomings in TC1000 prior to the sale to HP. And then HP proceeded to release the whole series of machines. Each of which, as a Tablet PC, was absolutely worse than the one before. The TC4200 series, and then the 7000 whatever series, 27XX series, just went downward, not even in a spiral, just in a downward plunge, in terms of what that did for you. Conversations I had with Motion Computing—brilliant tablet device—I could not get them to understand that as good as their tablet was, they needed a keyboard to go with it for the function in education.

Jonathan Grudin [3]: This has a detachable keyboard. You did not see it. I saw.

Mark Payton [4]: I mean, it is everything. Fujitsu has built some wonderful Tablet PCs over the years. But even the designers there—and I've talked to some of the engineers there and argued that basically they needed to go back to this as a starting point, both in terms of how the keyboard works, as well as how the slate device itself works and feels in your hands. And, honestly, for Lenovo, the same thing. There was a phenomenal cluster of bad decisions in terms of design for so many things.

Eric Hamilton [1]: Mark, I guess I will depart a little from your line. I do not know that these companies gave it to Apple. I do think Apple took it. I think Apple has some intrinsic corporate and individual genius, they awakened the world to the power of design. And if design entails giving the user something that gives the user feedback, and changes the way the user looks and works with the technology, Apple won hands down. It was a hard competition. They also came up with the fact that—with the realization or the insight—you don't need a heavy-duty processor to excite the masses. And so they came up with bicycles without needing to have a hot rod. And people bought it. They came out with good products. They came out with products that were phenomenally successful. They drew resources. They drew government grants. They had a cost factor that schools could afford, that was much more favorable. Even if you did not think they were as educationally powerful, it was much more favorable to what schools could afford and what people wanted. I said this last night and I got some flak for it, but I think that much of the future for education with tablet computing comes down to what companies like and Wacom, because those are the main entrees into digital ink—what they can provide. The Surface Pro, for my money and for the money of a lot of people, is the first time that the tablet model actually works, and excels, and is better than the corresponding Mac product. I know there were some great products before, and I have to think that some of the HP 2700s were not as bad as going downhill, but the fact is that what academics wanted and what we might value, is not what was working in the market place.

Mark Payton [4]: That's actually, the point that I was getting to.

Eric Hamilton [1]: Okay.

Mark Payton [4]: You said it probably better than I could have. And so, I did not mean to disparage Apple. Apple does a brilliant job at design. Microsoft is catching up on that count, finally, and that is all to the good. Where I was going with it and what Eric actually said quite well was that Apple was able to capture the mind-share through brilliant product design, brilliant marketing, and admittedly a large media bias towards Microsoft, a large education bias towards Apple, and a large education sphere bias towards Apple. They were able to establish a large enough base that a lot of developers jumped on board, and therefore, there was a lot of money available. I do not know whether it was research money, venture capital money, or whatever, but they were able to do that. We need that same ecosystem with regards to digital ink. Maybe, Wacom will do that. Certainly that gives us a cross platform.

Eric Hamilton [1]: Somebody needs to do that. It will be a big part of the ecosystem.

Rob Baker [4]: We need the whole ecosystem. And it is not just on you. It is a cart and horse situation.

Jonathan Grudin [3]: I would like to. I do not want to distract from all these nice things being said about Apple or anything, but I do want to make a comment and offer hopeful encouragement for those of you doing research.

I spent a lot of time at schools, and I spent time in Chromebook schools, and iPad schools, in Kindle schools even, in netbook schools, as well as tablet PC schools. And, clearly, if you have a good teacher and all the kids have a device, it can just be very impressive, whatever the device is. If all the kids have it, and the teacher knows how to use it, and knows how to organize the pedagogy around it. It can be quite electrifying for any of these devices. And I think, the grounds for optimism, which ties into what Bill Buxton was saying that we really should be thinking in terms of prospecting rather than inventing.

The key thing is that every kid has this device, that they carry it around, in early grades they use it in the classroom every day, or they take it home with them in middle school and high school. That really changes the playing field. It makes the pen much more useful than if you are just using it in one class per day or one class per week. It makes OneNote much more useful, you can use it in all of your classes. We are moving in that direction very quickly out there in public schools, where a lot more are going one-to-one. I think that the math and chemistry modeling, had pretty limited use if kids didn't have devices that could be used at home, and for every class, if they can just get a laptop cart now and then.

We are moving; we have our famous Moore's Law; we have the cost curve that is coming down and the capabilities are going up. I think we want to be thinking out ahead to what is really going to be a relatively near future, in which all the kids can have devices. We are finally coming to the Alan Kay and Nicholas Negroponte vision of one laptop, where you finally do have a device for every kid that has capability. And when we finally get there, people will be back, hopefully, rediscovering some of this work that was tried and set aside. Because, I do not think that researchers have to wait completely for the technology companies or others to come around to it before they realize that. I mean, Apple has come up with the pencil, and Microsoft is working incredibly hard now to develop for the education market, so I think that the infrastructure for the research is going to be there. And it is going to be there probably faster than we might sometimes fear.

Ken Forbus from Audience [7]: I would just like to make one comment on that. There is one thing I would disagree with slightly. We can do a lot better now than simulations. It is sort of interesting when people think about education and computers. They think simulation or animation. Alright, we know from a lot of cognitive psychology that, actually, animation is really bad for learning. It is actually much better to have discrete snapshots, where the qualitatively distinct states of the phenomena are visible and make it very easy to compare across snapshots. Cognitive science on this is really clear. Animation sucks. You cannot tell what to look at. Unless you know what you are doing, you miss things. If you can control it, you get a little bit better learning. But there are a lot of ways to deal with knowledge that do not require simulation. In fact, they are better for students. If you look at the content of middle school science, and my colleague Bruce Sharon has done this, it is all qualitative knowledge at that point. They are still trying to get straight the idea of an intensive vs extensive quantity. Being able to capture qualitative causal models is what learning is all about at that level. Research in qualitative reasoning has actually developed ways of doing that, so you can have human-like explanations.

Even in engineering, when looking at engineering design classes, we said, "Would you like a simulation under the hood."

And the instructor said, "No. We want students to be thinking causally about the designs. To be putting their explanations of what they think is happening, and a simulation will be misleading because there are always extra parameters, there are always assumptions, and so you end up having them trust the simulation more than developing their intuitions."

And so, I think we really need to broaden ourselves away from the *and finally we will be able to simulate.* There was a nice vision back then. But in the meantime the cognitive science has come up with the whole bunch of results and a whole bunch of new ideas on what you put in these systems. That is another exciting thing.

Cal Armstrong [5]: But, do not worry, the marketers will say that is engaging the students. That will be the selling point, right?

Andy van Dam from Audience [8]: May I respond to that? We are talking about having a bunch of tools in our tool kit, Ken. I certainly did not mean to imply that simulation should be the basis of everything we teach or have kids experiment with. Notice, I said *simulation.* I did not say *smooth animation.* That is one way of representing the output of a simulation: having stop motion, having discrete frames, things that help you see changes over time. Those are all ways of rendering what the model is producing as output. We are talking about just having a bunch of different techniques, all of which get the students to do the cause and effect, causal relationship model-building in their heads. It is obviously not the model in the computers that matters. Ultimately, it is the mental model that the students form over time. My point was simply that those of us who worked on this arena understand the expressive power of having a model that you can interact with, experiment with, and that augments what you do with physical labs. These are the virtual labs that have been discussed over so many decades and my point was that kind of toolset is largely still unavailable even after 50 years of talking about it.

Eric Hamilton [1]: I would say, I would resist the *either-or* in terms of animation and simulation. You made that point very well, Andy. We take a portfolio and for any given physical phenomena or complex phenomena, different strategies, different tools might have different use or meaning. Sometimes the simulation does represent discrete steps when it's rendered; other times it might be more dynamic, but I would not rule out any a priori, because virtually any tool works in some cases, and no tool works all the time.

Jonathan Grudin [3]: Okay, we have 10 min left here. So, I would like to let the panelists, if they have something they were burning to say, to get it in now. But, one thought, one possibility here is, because we have a lot of people here from the post-secondary world, *How should they prepare for the students who have gone through your programs and have used these tools and have been able to adopt different pedagogies? How might post secondary schools prepare for this new generation of students?* One reason I ask is because I have talked with one or two students who came through these kinds of programs and got to a community college or university, and then were a little shocked at what they found. Possibly, there is also some learning that might flow in that direction.

Okay, any other comments that our panelists might have to make. I'll start with Cal.

Cal Armstrong [5]: That sort of builds on a question that folks had earlier about what research should be done. I do not think this speaks to any of you in the room, because you guys are all going where we will be in five years. But what I have not seen is the research being done on the kids experiencing the digital environment right now, specifically in OneNote. Rob's school has been around doing the OneNote environment for eight years. We graduate our first high school class that has been existing in OneNote for the first time this year. And I have not seen the research on what actual effect has occurred to our students. We have plenty of anecdotal stuff. But that and two bucks will buy you a cup of coffee. But I have not seen research on what the changes have actually been to our students, other than what we think we know, which may or may not be true. And, building on that, when we did our OneNote project in the summer of 2012 we created the OneNotes for all of our faculty, all of our classes and just sent it out there not expecting a whole bunch. And within two weeks, eight-five percent of our faculty were using it without too much push from us, even without much support from us. We realized that we hit something. We were thrilled. We were elated. We were walking on the clouds. And I realized that I had a responsibility to push this out further. And so, as any good academician would, I started to do a lit review for the article I was going to write, and *I realized when I read it in the 2010 WIPTTE proceedings that I was Gottfried Liebnitz to Rob Baker's Isaac Newton.*

Rob Baker [2]: What a great analogy.

Cal Armstrong [5]: It also said that the universe is trying to tell me something. That independently, two schools recognized the need and saw a solution in OneNote. It is not the only solution, but it was a solution that worked independently in two different spaces for two different distinct cultures. And that says something to me on how important this step has been. That OneNote is providing universally something different to teachers regardless of grade, regardless of subject, and I think in large part it was missed for eight years. Fortunately, numbers are increasing. But I do not think that the research community—again, that's not you—it is the more reflective educational research community that has responded to the changes that are occurring right under their noses. I encourage you to walk over to your education department and start knocking on the doors and saying, "Hey, have you heard what is actually happening in education?" Because, Lord knows, education faculty have little clue most of the time.

Rob Baker [2]: Yeah, I can speak to it personally. My oldest son just graduated from the Cincinnati Country Day School. He had a Tablet PC his entire career and OneNote for the vast majority of the classes, not all of them. He is now at Alabama with a Surface Pro 3, and in general, no one else has a Tablet PC. And he is struck in awe of all the crazy, ridiculous hoops the professors and other students will do to get handwriting in the process, digitized, and in there. And it is painful and absurd to him; it is really interesting that he knows nothing else.

I am sure at some level he had some concept, but when he gets there, he is like, “Dad, the professor is writing; he’s got a camera; and then he is taking a picture of it; and putting it in a PDF; and he is putting it on Blackboard; and then we have to go there; and I go there and I save it and I print it on OneNote, so that it is with everything else.”

So, the environment matters. So, he is doing less; he is absolutely not using his tablet as well as he could, because if no one else is digital, at some point it is too painful to do it the better way, right? If the strategy does not take advantage of the great stuff, then it does not happen. And no one wants Apple to come out to respond to the Surface pro 3 more than us. I am tired of arguing with fanboys who love the thing and, God bless America, it is a beautiful, awesome product. It is the BMW; it is a status purchase at a lot of levels, I believe. But, I just wanted to have an active digitizer, touch, and a detachable keyboard, so that *they can do what we know is a huge deal*. I do not want it to matter.

So, we get kids. I just had one student at Yale. You know, all the time we get emails: “Oh, Mr. Baker, can I still get my AP statistics on the OneNote notebook? The tech department here reimaged my Surface and I lost the copy.” And I am like: “Oh, dude, I don’t know; let me check.” And luckily that was in our old system. It is on our file server; so we zipped it right out and sent it to him.

It is changing. Our kids want to keep it. They want to refer back to it; they want to have it; they take it to college, just because it is awesome. It is organized, and sorted; that does not happen with paper.

Cal Armstrong [5]: Grade 10 students who say, “Where is my grade 9 math notebook?” I mean, that used to be burned at the end of June. And now they are showing up in September when they get their new laptop, making sure that their notebook is there for access. Those are the anecdotes that we see. What effect is it actually having on them?

Jonathan Grudin [3]: I would like to see the quantitative data that Cal asked for. If you have anecdotes, I will buy you a cup of coffee.

Cal Armstrong [5]: The LMS[3]—we invested a whole bunch in Blackboard, and then we invested a whole bunch in developing something in Sharepoint. OneNote destroyed that. We found the LMS obfuscates the digital content of the course. It hides it; it structures it in a way that is foreign to both the teacher and the student, and it is likely the worst thing that can happen to K12 education. I cannot speak to distance, I cannot speak to higher ed. But in K12 education it is likely the worst thing you can do to bring the LMS into school, because it imposes an artificial structure and unnecessary steps to the entire learning process.

Rob Baker [2]: Yes, files are the enemy of education. That is why LMSs do not take when you get a big class and spend money on LMS. But when it is authentic, when you go to a reading and it is just there and it’s an active note-

[3]Learning management system.

taking exercise for the teacher. When you go to a piece of paper to do what you were told to do, and the teacher can just see without anybody doing anything. That is transformational. And, really, the workflow... You know, it was really cool, because we were so well informed, because we had laptops for six years, and we recognized what the ink would do.

That is why I said the workflow thing, which Cal just tweeted out. I saw a tweet. I do not know how he is doing that up here; he is really good at that.

But, the best example I have, and I will give it to you, I will mention it in the keynote too, is when we got Tablet PCs, we thought we were awesome because my English and History teachers could offer the form of assessment that they wanted to. Email me your rough draft in Word; open it up in Word; and mark it up the way you want to. You just got a big cup of Starbucks; you are in your pajamas at home. The teachers are living a dream except that they were files, so now you save it as a different name, maybe with the word "with ink" at the end. You save it somewhere and then you email it back. And you still have 79 more students to do. That is why the constraint thing—that is why I think we are workflow specialists—because now everyone just says, all right, just print your rough draft to OneNote, and a teacher literally just goes in on a piece of paper and does what they want to do and then they go to the next piece of paper and do what they want to do, and they go to the next piece of paper. And then, when they are done, everybody has what they want to have and you get more of what you subsidize. I cannot emphasize that enough.

Mark Payton [4]: I think an underpinning subject to what these guys both just said is that any educational environment is going to be able to function only at the level of the least common denominator of the platform you are on. And so, all schools these days, public and independent, secondary and higher ed as well, seem to recognize that it is important for students to have a device. But, in a BYOD world you do not know what that device is, and it may be somebody's—like my really cheap Windows 8 phone here—and they cannot do anything across the board that I cannot do on this device. This device cannot do ink. This device is a pain to type on because of the screen size. And even schools—independent schools, primary schools—that go with a one-to-one program with a standard device can only do what that device is capable of doing.

One of the beauties of the Tablet PC is that it has got all the capabilities of any of the devices out there, plus this phenomenal digital ink. I do not know what the answer for higher ed is, but the privilege of working in independent schools, is that we can say we are going to be a one-to-one program, and we are going to get these relatively expensive machines, and we are going to invest in professional development, and we are going to have a classroom specialist go around to help analyze what is it that you are trying to achieve educationally and help you figure out the best tech, even if it is no tech, to achieve those ends. And that is a luxury that has let us be at the forefront of what we hope is a massive wave of implementation. But the obstacles that higher ed faces... I mean, what higher ed schools will be able to dictate to their students: you are going to buy this machine. Like it or not. Most public schools cannot do that and they cannot afford it. I do not know the answer, but I can tell you that is what needs to happen: for the power of devices like this to take what we are able to do and translate it to the next level of education, or to translate it broadly into public education.

Jonathan Grudin [3]: We have Andy and Eric...

Eric Hamilton [1]: I defer to Andy.

Jonathan Grudin [3]: Okay, you've got closing remarks.

Andy van Dam from Audience [8]: Sorry for monopolizing. I just want to emphasize that our crowd allows us to compare an Apple product that does not support digital ink and pen properly with a Tablet PC, whether it is Microsoft or Fujitsu, or HP. This comparison is irrelevant; as if they were the same thing and you guys know better. Do not fall into that trap. **It is all about digital ink and stylus and touch, all working together.** And that ecosystem, bears no resemblance to the iPad ecosystem, which is about media consumption. In education we are not about media consumption. And that distinction needs to be just made over and over again. One of the ways in which Microsoft lost out was that it did not make that distinction. It did not understand what it had in digital ink. And one of the great things about this conference for me this week was to learn about the recent announcements that Microsoft is finally, after ten years or more of abandoning ink as a first class citizen, coming to recognize that it is a central value proposition and simultaneously, it is finally fixing the simultaneous pen and touch problem. Those are great steps in the right direction. But that is what we all as a community need to be pushing. The device wars do not matter. The capabilities of those devices do.

Jonathan Grudin [3]: And it is great that Apple has come up with a pencil; it also validates this area.

Andy van Dam from Audience [8]: Jobs had to die for that.

Jonathan Grudin [3]: Thank you all very much for a stimulating morning.

19.3 Panel Member Bios

Dr. Jonathan Grudin Jonathan Grudin was the moderator of the Education Panel and the co-chair of the WIPTTE 2015 conference in Seattle. Grudin is a Principal Researcher at Microsoft Research in the fields of human-computer interaction (HCI) and computer-supported cooperative work (CSCW). Grudin is a pioneer of the field of CSCW and one of its most prolific contributors, and was awarded the inaugural CSCW Lasting Impact Award in 2014 on the basis of his work. Grudin is currently also an affiliate professor at the Information School in the University of Washington. Previously, he was a Professor of Information and Computer Science at University of California – Irvine. Grudin received his Ph.D. under Don Norman in Cognitive Psychology from the University of California San Diego.

Calvin Armstrong Cal Armstrong is a mathematics educator & the Director of Teaching and Technology Innovation at Appleby College and Teacher Leadership Program coach at the Institute for Advanced Study's summer Mathematics program. While developing the progenitor of the OneNote ClassNotebook is likely his more notable accomplishment (in a situation with Rob Baker that parallels Newton & Leibnitz and the Calculus), he has been integrating technology, from the earliest graphing calculators to the latest in pen-based tablets and cloud services for more than 20 years. He is a devout pragmatist and critically eyes technology with the lens of researched-based pedagogies. Outside of the classroom, he is an Auxiliary Constable with the Halton Regional Police.

Robert Baker Rob Baker is responsible for creating some of the most powerful teaching and learning environments in the world. He has been working on developing technology programs both in the United States and internationally focused on embedding a range of new technologies into the learning process. Rob is a Director of Technology at Cincinnati Country Day School. He is an educator which has allowed him to look at initiatives through the eyes of a classroom teacher focused on pedagogy, rather than technology. His work at Country Day shows what is possible when school deploys the kind of hardware, that removes all constraints and allows teachers to use the modality that fits the task. Some of Robert's most important work has been with Microsoft on replicating Country Day's shared OneNote environment, which he developed and implemented in 2007. The fruit of that labor – the Class Notebook Creator App for Office 365 has made this a powerful framework for teaching and learning available to every teacher and student in the world.

Dr. Eric Hamilton Eric Hamilton is a professor of education at Pepperdine University in Malibu, California. He holds a courtesy appointment in mathematics. He was originally tenured in computer science and mathematics at Loyola University Chicago, where he founded and led an award-winning citywide consortium to promote the participation of underrepresented minorities in science, technology, and engineering professions. He has served in the US federal government's senior executive service as a division director for research, evaluation and communication in the National Science Foundation's education and human resource directorate. Following an appointment at the US Air Force Academy where he ran the Center for Research on Teaching and Learning, he came to Pepperdine University, where he is currently based. He specializes in research around the creation and use of interactive digital media, and recently led a major initiative of the US and Finnish governments in areas related to education innovation and technology. He has won US and overseas patents on pen-based computing in education. He currently is a Fulbright Research Fellow funded by the US State Department serving in collaboration with the Namibian Ministry of Education, Arts and Culture and with the University of Namibia, where he serves as a Visiting Professor. Current work also includes leading a four-country digital-media making collaborative funded by NSF. Dr. Hamilton studied at Northwestern University and the University of Chicago.

Mark Payton Mark Payton was the Director of Technology and Library Services for Whitfield School, an independent 6–12 school in St. Louis, MO, where he developed their version of the shared OneNote techniques that helped prompt the development of the OneNote Class Notebook. His time at Whitfield completed his seventeenth year as an IT Director in independent schools, having previously been at schools in Vermont and Madaba, Jordan, and he now works in healthcare IT. Prior to education he was in business as IT Director for Burton Snowboards and as an independent software developer. With a degree in early childhood education, he has taught subjects as varied as basic reading & arithmetic, Introductory and AP Programming and Christian Theology to students of every grade between Pre-K and the undergraduate university level and leveraged his varied classroom experience to inform his IT work. He has been interested in pen-based computing since the days of the GRiDpad and Windows for Pen Computing and has been a member of the WIPTTE steering committee since the conference's inception. Mark Payton was the co-chair of the WIPTTE 2015 conference at Seattle and the WIPTTE 2014 conference at Texas A&M, and has been on the WIPTTE/CPTTE conference committee since the beginning.

Acknowledgements The speakers especially thank Dr. Anna Stepanova, Dr. Tracy Hammond, Dr. Jan Hammond, and Mark Payton for their help in this transcription as well as Lisa Manokofsky for her help in providing the videos for translation.

Chapter 20
CPTTE Research Panel: Frontiers in Pen & Touch Research

Andy van Dam, Bill Buxton, Tracy Hammond, Kimberle Koile, and Thomas Stahovich

Abstract At the 2016 CPTTE Conference on Pen and Touch Technology on Education, five expert researchers and educators spoke at a panel in front of the audience of CPTTE attendees where they discussed past influences and future directions. The members of the panel were, as seated from left to right, were Bill Buxton, Tracy Hammond, Andy van Dam, Kimberle Koile, and Thomas (Tom) Stahovich. The discussion also included comments and questions from Ken Forbus (2017 CPTTE conference chair), Jonathan Grudin (2015 CPTTE conference chair), and Aaron Adler (2014–2017 Paper Chair). Topics included sketch forensics applied to education, representation fidelity, ubiquity, domain-specific versus domain-independent, accessibility concerns, funding concerns, and future directions. Special attention was also spent discussing ink accessibility and how it might improve the lives of those facing disability.

A. van Dam
Department of Computer Science, Brown University, Box 1910, Providence, RI, 02912, USA
e-mail: Andries_van_Dam@brown.edu

B. Buxton
Microsoft Research, Microsoft, One Microsoft Way, Redmond, WA, 98052, USA
e-mail: bibuxton@microsoft.com

T. Hammond (✉)
Sketch Recognition Lab, Department of Computer Science & Engineering, Texas A&M University, College Station, TX, 77843, USA
e-mail: hammond@tamu.edu

K. Koile
Computer Science and Artificial Intelligence Laboratory, MIT, 77 Massachusetts Avenue, Cambridge, MA, 02139, USA
e-mail: kkoile@mit.edu

T. Stahovich
Bourns College of Engineering, University of California, Riverside, CA, 92521, USA
e-mail: stahov@engr.ucr.edu

T. Hammond et al. (eds.), *Frontiers in Pen and Touch*, Human–Computer Interaction Series, DOI 10.1007/978-3-319-64239-0_20

20.1 Introduction

At the 2016 CPTTE Conference on Pen and Touch Technology on Education, five expert researchers and educators spoke at a panel in front of the audience of CPTTE attendees where they discussed past influences and future directions. The members of the panel were, as seated from left to right, were Bill Buxton, Tracy Hammond, Andy van Dam, Kimberle Koile, and Thomas (Tom) Stahovich. The discussion also included comments and questions from Ken Forbus (2017s CPTTE conference chair), Jonathan Grudin (2016 CPTTE conference chair), and Aaron Adler (2013–2017 Paper Chair).

Topics included sketch forensics applied to education, representation fidelity, ubiquity, domain-specific versus domain-independent, accessibility concerns, funding concerns, and future directions. Special attention was also spent discussing ink accessibility and how it might improve the lives of those facing disability.

20.2 Panel Transcript

Andy van Dam [3]: Hello again, folks. It's a small, intimate group, so it's good for interaction. As in the minimalist instructions I gave you, I will give you five minutes for self-positioning. Then we will let the panelist query and argue with each other, and as soon as possible we will open the floor for questions. So, in no particular order, we will start with Tom. Go left to right, my left (Fig. 20.1).

Tom Stahovich [5]: I am Tom Stahovich from UC Riverside. I really wanted to prepare for this last night. So, I did my homework; I asked Bill what to say. So, my thoughts about where the future of this field is, are: I think that the best illustration of this is, if you are familiar with the *unreasonable effectiveness of data*? I think that's really, what I see is the future of this field. You know, we are now starting to get very large datasets. Tracy is collecting datasets from children; Kimberle is as well. I have about 15 million pen strokes from university students. The more data you have, the simpler the models can be

Fig. 20.1 Members of the education panel, from *left* to *right* are [1] Bill Buxton, [2] Tracy Hammond, [3] Andy van Dam, [4] Kimberle Koile, and [5] Thomas Stahovich. Their bios are found at the conclusion of this chapter

and the better they work. I really think that—now as everybody carries a touch device in a pocked and pens are ubiquitous—we now have so much data available to us that I really think that advances are going to come quicker as we get access to these massive datasets. And then, Bill mentioned that if we want to see the future of what this field is going to be, we just have to look in the past. I realized that I spent a lot of time creating technical drawings. When I create an assignment or an exam for my classes, I spend a lot of time actually drawing figures, and if you look at Ivan Sutherland's SketchPad from the 1960s, we haven't gotten past that. I would like to see sketch-based tools that will allow me to create quality technical drawings, not artistic drawings, but technical drawing with a pen. So, those are my thoughts.

Andy van Dam [3]: Thank you. Kimberle?

Kimberle Koile [4]: I spent a lot of my research time thinking about how to build tools to support teachers and learners. In particular, I have looked at how to take advantage of domain knowledge in order to improve a machine's *understanding* of what the students have created. Specifically, our domain has been upper elementary math, focusing on multiplication and division, and, the representations themselves matter more than the final answers.

I got some good questions at the last session. They pointed to the use of these kinds of intelligent systems for evaluating students. When you are evaluating hand-written exercises, representations, and answers, there is a danger that you either over-correct and you bias your interpretation based on expected answer, and you give a student more credit than they should have; or you make a mistake.

My position is that I am not interested in evaluating students. Rather, I am interested in supporting the teaching and learning in the classroom. I am interested in getting the feedback to the teacher, so she could identify certain students, or she can make pedagogically interesting choices in the classroom, with data-driven decision-making in the classroom. We are at the point where we can do that. It doesn't just have to be Big Data analytics. Rather, it can be really at the level of understanding the visual representations that students have created, and use that to enhance both the teaching and learning. In fact, I am always amazed that so many teachers, sadly, don't understand the math very well themselves and so, they view this particular system as a way to get a sort of *in situ* professional development. The system can almost work as a virtual math coach, making suggestions such as: "These four examples are interesting as a subject of a conversation about multiple problem-solving strategies, and here are two problems that look similar, but they're actually not." The system can use the information to pair students, e.g., think-pair-share. We can do those kinds of things in the classroom.

And, I think we need to be careful about how we are using this information; privacy issues come up: "Do I want my machine to know exactly how to identify my love letter to someone? I don't know." It came up earlier this morning. It's interesting. It's fascinating. And, I do think, as Tracy mentioned, that **this technology can and should do more than paper and pencil**. But also, I think,

it needs to be done in a thoughtful way. It needs to be grounded in good, pedagogical strategies, in what we know about how people learn, and in continuing practice and design groups that include teachers, practitioners themselves. Because, sadly, they are often the disempowered ones. The computer shows up in their offices, and they are just expected to use it. They don't have any choice what to get; they don't usually have a choice about what curriculum goes on it. Then, when the districts lose money, the people who are laid off are those who bridge the gap between the technology and the teachers. The teachers don't have a lot of help in how to use technology effectively in the classroom.

I think the onus can be on us to make it easier by building software that has good pedagogical strategy built in—has good curriculum built in. Sometimes the curriculum is awful—we don't need to have that conversation right now—but, curriculum that is based simply on having kids fill out worksheets, ad infinitum, and getting them to learn long division by third grade, before when they have a deeper understanding of what they are actually doing, I think does a disservice to them, and in the long run, they may fail in algebra. There are all sorts of studies indicating that deep early understanding of mathematical operations and arithmetic leads to success. I am conflicted about the trade off between intelligent tutoring systems and supporting the teachers, because I am a classroom teacher, so I love supporting them, especially with technology. But there is so much serendipity that happens when learning in a social context that I think being in a classroom is important.

But that is not to say that a classroom model is necessarily scalable. When you don't have teachers, when you don't have access, then I think that there is something that can be said for building these intelligent tutors. But, **the system should act as a teacher, not simply as something that would continue to give hints**, nor something that has a structured decision tree that is based solely on a particular model that may or may not **fit that particular learner**.

Andy van Dam [3]: Tracy?

Tracy Hammond [2]: I thought you were up.

Andy van Dam [3]: No, I am just a moderator.

Tracy Hammond [2]: Okay, So as I said this morning, I feel that **computers are meant to facilitate human-human communication**. That is their main goal. HCI[1] was initially termed CHI,[2] until people decided that the human was more important than the computer and should be put first. But I argue that the computer should not be part of the interaction at all. The important interaction/communication is between two people—**that communication may be between your future and past self**. Quite often, though, communication is between a teacher and student. That's a nice model of communication that surrounds many-many hours and years of our lives. A lot of us have sixteen

[1]Human-computer interaction.

[2]Computer human interaction.

years or more of education, and a number of hours more as we continue to learn throughout our lives.

Education is pervasive, and a huge place where computers can improve our lives through improved communication between people. So, that's crucial. The other thing is that intelligent sketching systems such as those that use sketch recognition have the opportunity to make a significant impact on education, greatly enhance human-human communication. Even more, I argue that all sketching applications need to have some intelligence in them for them to be successful, because just mimicking the paper is not going to change people's mental models enough to make them want to use the new device.

But then here we are mixing two very difficult and disparate ideas. So, I have degrees in Math and Applied Physics—and a degree in Anthropology, teaching me to look at things from sides. Usually people are one side of the spectrum and will classify themselves and put themselves in a box as either a *people person* or *math person*. If I have to choose, I would say that I am a math person. That Anthropology degree was not easy for me. It really pushed me to think in way that I was not comfortable with. Anthropology was a hard degree for me. Writing that thesis was hard. I cheated a little bit; I did my thesis on EthnoMathematics.

Kimberle Koile [4]: Well, all of the world is math.

Tracy Hammond [2]: But, as we are developing, we have to remember to really push that mixture, really thinking about the human element, really thinking about how we can make smarts in the underlining devices and bridge those things.

Then, there is this third thing, which is feedback, which Eric Hamilton asked me that question at the end of my talk. He said: "You didn't mention *feedback*." My response should have been: "**None of these is worth anything without feedback**." The whole point of having intelligent computers is for them to do something with your input. Having these little boxes say, "I am really smart. Ha-ha. I won't tell you how I am really smart, but I know how I can help you." The whole point is that this system can then facilitate a dialog with you, which is feedback. The smarter we can make that computer understand what's going on, learn more about you, learn more about what you are doing, what you are trying to do, where are you going, where are you having difficulties, that is how we can make a computer to facilitate the learning process. And that's all about feedback and how that feedback should be given.

So here we are again. We are straddling the ideas of AI, of smartness, of knowing what it is that I am trying to say, as well as how I should be saying that, which is the human part to be able to make that heard. People say things in many ways, even the same thing in five different ways. People are really good at only hearing what interests them. They are even better at tuning out that which does not interest them or is not accessible to them. **New information needs to be presented in a digestible way, and in a way that appeals to their sense of reason**. It's a lot of straddling these different notions.

Okay, Bill's up. Bill is going to tell us that for the most part we have developed nothing new in a long time, which mostly is true, but...

Bill Buxton [1]: No, I am actually, well, what's the definition of new?

It's, as Kimberle said, about representation. I want to point out that representation isn't just a notation and methodology. It's actually in the body language in the action and the pragmatics of interaction.

So, pretend this is a PowerPoint slide that I want you to create. (See Fig. 20.2.) So, this is what it takes to do it if you are using PowerPoint. (See Fig. 20.3.) Everyone of these boxes is explicitly specified with explicit paths. The tools you have to access, so on and so forth.... However, if you do it just by writing with a pen, that's what it takes. (See Fig. 20.4.) And it's a factor of ten. So, the first thing is that if you take some query to write an analysis, you can actually start to quantify and predict and then test. So before you decide, you can make *a priori* predictions and test. And that's data that is useful. And, so, if you just compare the two, that gives you a sense.

The second thing I want to talk about is—we talked about recognition and all these things machines tell you and so on and so forth. Maybe we should spend some time thinking about where we are putting it based on what I just said.

So, back in 1993, there was this thing called Aha![3] (see Fig. 20.5), which was shipping—I still have it; it still works. But instead of trying words—trying to understand the semantics of what you are doing—it actually recognizes the morphology of what you are doing (See Fig. 20.6).

Therefore, it understands what a word is, what a paragraph is, what a bullet list is (See Fig. 20.7). So, for example, we can build a note processor so that you can delete a sentence or insert a word and it will actually do word flow,

Fig. 20.2 Bill: "So, pretend this is a PowerPoint slide that I want you to create"

[3]Microsoft in an agreement to acquire Aha! Software, April 9, 1996, http://www.nytimes.com/1996/04/09/business/microsoft-in-an-agreement-to-acquire-aha-software.html.

Fig. 20.3 Bill: "This is what it takes to do it if you are using PowerPoint"

Fig. 20.4 Bill: "However, if you do it just by writing with a pen, that's what it takes"

despite the fact that it has no idea what you wrote. It doesn't even understand the characters. It can't read it. It knows nothing about it. It just understands the morphology.

This is actually not hard to do. It's really important. So you can come back and, the same thing here if you are doing bullet lists. You can insert a list or take it away. So the operational aspect of getting concept of getting things down is simple. If you had this in OneNote, the world would be very different.

By the way, the irony is that Microsoft bought this company in 1996. And Aha! is not alive nor acknowledged in any of our products.

And I want to finish up by just saying that this comes back to what we saw in Michelle's talk yesterday about "God's in detail". So we all think we know about writing and note-taking, a lot of features in OneNote and so on. But let's watch this.

Fig. 20.5 Bill: "Back in 1993, there was this thing called Aha"

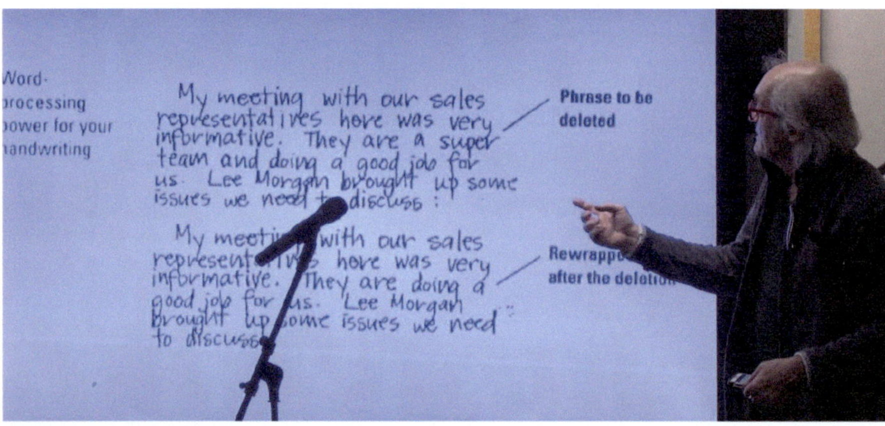

Fig. 20.6 Bill "Instead of trying words—trying to understand the semantics of what you are doing—it actually recognizes the morphology of what you are doing"

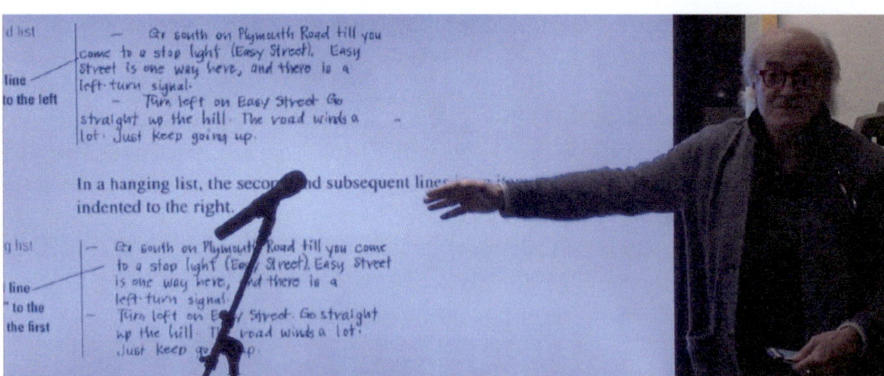

Fig. 20.7 Bill: "Therefore, it understands what a word is, what a paragraph is, what a bullet list is"

Fig. 20.8 Bill: "So pretend I am writing. Or pretend I have just finished writing my notes"

Bill Buxton [1]: So pretend I am writing. (See Fig. 20.8.) Or pretend I have just finished writing my notes or a thing for Craig Mundie. And I am finished. And this is what I have. A 9 by 11 sheet all nicely lined out. Terrible handwriting. but no one can understand what I wrote. But this isn't what I wrote. I have a friend in DPR,[4] and he did this study. This is a replication of what he did. Carbon paper on the bottom of the writing surface with a sheet of paper here. So that is actually what the carbon paper said. (See Fig. 20.9.) And what you will notice when we compare the two is that the carbon is way smaller. It is narrower. And it is rotated. Note this clearly shows that while I was writing, I was manipulating the paper, rotating it, and sliding it left and right as well as up and down, so that the zone of comfort when writing is much much smaller than the surface I am writing on, even though I have this wide larger desk cover that was perfectly available to me. And this works all the time. What this says is the "God's in detail". Understanding that writing is a bimanual action, that involves this coordinated skill, involving two hands, using different functions: touch and stylus, if you will, gives us a whole different insight on how we write with a pen.

Bill Buxton [1]: The whole point of representation is—you cannot decouple motor sensory skills of the action from the cognitive skills that you are trying to develop. And if you are trying to make a note on a slate, and you don't take into account how narrow it is, then you are now prejudicing against everything you are trying to accomplish no matter how neat you write, no matter how good your recognizer is, not matter how fast your computer.

Tracy Hammond [2]: That was great.

[4]Microsoft digital pen research.

Fig. 20.9 Bill: "That is actually what the carbon paper said I wrote"

Andy van Dam [3]: Well, my compliments. You were all very efficient. You can now discuss amongst yourselves, or we can throw it open to the floor. What would you like? Statements, please.

Tom Stahovich [5]: Well, I think what Bill just pointed out there about the mechanics of writing is what I am sort of seeing in my work too. When I look at students' written answers to engineering problems, I can with fairly high accuracy determine if their work is correct without understanding the content, but by looking at the mechanics of putting that content down on the page. How do they move around on that page? Do they start at the top and work to the bottom, or do they bounce left to right, up and down? I look at the timing of the work. So it's the mechanics of getting the ink on the page—from that we get about half of the correctness of answer just by looking at the mechanics.

Andy van Dam [3]: So have I.

Bill Buxton [1]: I have a statement about this. It was said earlier that this stuff should do better than paper and pencil. I am willing to accept that, as long as you at the same time take the contrary position that this will always be less than paper and pencil. And as an academic I can always justify that by sighted pens. And the question is: What does it depend on? And even then if I am doing math using pen and paper with a system with perfect mechanics, if I am teaching long division and using Roman numerals, the complexity is going to be equivalent to third year calculus as opposed to using decimals. I'm not—so again, **representation is independent of the tools**—but actually at the representation that is involved. These things all change together, and have a domino effect. And that is having a holistic view. And generally none of us, and I am appreciating that some like Tracy, have these weird backgrounds. But you can't assume that. So **you need a Renaissance team**. Without Renaissance it is not possible. You need the collection of skills.

Tracy Hammond [2]: Yes, even with an Anthropology degree, I am more one side than another.

Bill Buxton [1]: We all are.

Tracy Hammond [2]: We all need complementary expertise.

Andy van Dam [3]: So I had an observation which is the most startling thing I learned at this conference. It came out of the talks by Tom and Tracy this morning, which is that, you can make deductions and inferences and predictions out of observing things that you might not have thought were first order in the actual recognition algorithms themselves. That's freaky. They showed statistical significance of some of these predictors. And to me it opened up the idea that we as computer scientists are very focused on the algorithmic side of the equation, being time and space efficient for example, coming up with more efficient representations and so on. But there seems to be this whole new branch of research that says: "Here are the incidental things you can get, almost for free, by analyzing certain aspects of the process." And that to me was a revolution.

Kimberle Koile [4]: That is actually where we are heading at the moment, too. **Rather than just looking at the final representation, we capture the process**. I think early on, the first thing that we—Steve, our lead programmer, and Lily, one of the researchers at TERC—noticed was that kids who struggled had a very long history of interaction. They deleted a lot of things. They started over. They scribbled out.

Andy van Dam [3]: False starts.

Kimberle Koile [4]: False starts. Right. Then the question is **how do we identify a false start**. Because in some cases we had a student who did three different things, but the first thing was to get the answer, the last two were to check the work. So we have to identify something more than just that they did several different things. There have to be characteristics of what is means to be a false start, and that is where getting a human looking at it to identify possibly what those are is an important part of the process.

Tracy Hammond [2]: Aaron Adler came up to me after the talk just this morning. And he said, "I was going to ask this question, but **do you remember we used to spend forever trying to remove those hooks that are now so valuable**." Initially when we were trying to recognize what was drawn, the easiest way to do that was to attempt to get the strokes as close as possible to the perfect version of the shape, which meant getting rid of the scribbles, getting rid of the false starts, getting rid of the extra note-taking doodling marks—which happens to mean that you are thinking and trying to solve the problem—getting rid of hooks, getting rid of all of the other stuff besides those points that represent the ideal of the shape. But **all of that extra information really tells us how do we perceive our task, what are we working on, what are we doing, what is**

our mind doing while we are solving our task—when we are going from here to here. And it is really exciting to look at these—no longer trying to get this weird thing that someone drew as close to perfect as possible to what it *should be*, but **instead, bask in those imperfections, and use the imperfect parts of the sketch to understand what is going on in the user's mind**.

Kimberle Koile [4]: We had conversations of that same ilk. We decided that **we don't prettify strokes**. Those jumps on the number line are what the students put there. They own it; it is theirs. **There is a sense of ownership** every single time one of the sketches goes up anonymously in a classroom. **It doesn't need to look pretty. If it looks pretty, it makes kids perseverate** because they think theirs isn't pretty enough. So they erase it. And they will put it back. And again they erase it and then put it back. And the eraser is both a good and a bad thing. It is easy to erase, but sometimes it is also too easy to get fixated on it. So we made a decision early on that we don't prettify the ink. It is what it is. We try to get them to move on—to use it as representation and not as something that has to be perfect.

Andy mentioned something earlier today, that in some cases people might have some fine motor skill issues with pens. We saw this with the number lines. It looks as if it is easy to draw, but if someone has fine motor issues, it is difficult to get from one tick all the way to another that they have targeted. There were two tricks: One was to put the dot down first where you want to go, so you can see it. That helped some of them. But for the kids who really had fine motor skill issues, we redesigned the number line. There is a second one that you will see this afternoon. It is called an auto-number line. Instead of them making an arc, they make a tick where they wanted the arc to start; they make a tick where they wanted the arc to end; and the machine draws an arc between them. And so it's making ticks along and they have to understand that they start at zero and that they have to be consecutive ticks.

There was an interesting issue that we are still debating with people who use this that—so what do the kids erase then? To us, they erase the marks that they make, so they don't erase the jumps because the computer wrote those; they erase their own tick marks. But then the flip side is, well, since they can erase the arcs when they drew them, maybe you should be consistent and they should be able to erase all of the arcs, including those drawn by the computer. Again, those are usability issues that change expectations, and we have to ask the question: "What's the right model to use?" And I don't really know; we just need to get it out there and see how it's use?

Tracy Hammond [2]: And what you are talking about there, beautification, is something that we have battled with over and over again with the Mechanix software. There is so much evidence that says—leave the original strokes. But those honors kids—every time we ask for feedback and suggestions, they say, "If you could just clean up my drawing and straighten my lines, it would be the perfect software. Just let me draw a perfect straight line." Which, then of course, you lose all of that other magic. And trying to explain to them, that, "No, no, we are not going to do that because it is much better pedagogically for you to work with your own strokes."

Kimberle Koile [4]: "Turn your OCD down just a bit today."

Tracy Hammond [2]: Exactly. Those honors kids are really obsessed with having perfect drawings to turn in. It is interesting; they are trying to create something that aligns with their imagination of what that perfect thing looks like. It is that same thing; it is like us at the beginning of that recognition process.

Kimberle Koile [4]: So, I would like to see a show of hands. If you drew an arrow and it is kind of ugly, how many of you would really—now fess up—would really like the computer to just clean it up and make it really look nice for you. [Some hands go up.] Maybe a third? Okay.

Tracy Hammond [2]: I confess, I might. I don't know. Context would certainly play a part.

Andy van Dam [3]: Let's do the opposite. How many people would not like it. You want it to stay as you drew it. [The vast majority of the audience's hands go up.]

Ken Forbus from Audience: And it differs by culture. Military commanders never straighten their ink. Subordinates, it turns out, straighten their ink.

Tracy Hammond [2]: Really? Interesting. How about the perfectionist?

Bill Buxton [1]: I believe it is really important, that as long as we teach, is to not want that, because I believe there are reasons for it. When I am teaching—at the moment I am currently teaching design, but I would say this applies for technical things as well—in my class, I take marks off for good work. [Laughter in the audience.] **I take marks off for good work because I want to teach you how to explore and how to check, to take multiple stabs at everything you do. I would rather that, instead of doing one thing perfectly, I want to see three things that approach the problem in different ways that show diversity of mind and exploration in creativity**, whether I am solving a specific technical problem, whether you are designing a circuit, or designing a car. If you can't work quickly and fluently, I think you are in trouble.

The second thing is that I really worry about data collection and measurement because we basically evaluate what we can measure. So many of the things that are important are hard to measure. And so you have to really look out for that. I am old enough that I learned how to use a slide rule. I was talking to Andy about this the other day. Some of us have learned to use a slide rule. I have found that no matter what you have got in calculations, that you learn different skills. It's really hard. Your answers are only good to three significant digits. But the fact is that if you have to know how to estimate within an order of magnitude your results, and you never make mistakes, then you will make the calculator that produces that perfect answer. But we have lost something along the way. I am not saying to go back to slide rules. But I am saying is that notion of quick estimates, and the idea of being able to do three different problems and get the same order of result. I would rather get close and have that confidence interval, otherwise you get into some very serious challenges. I have seen that once at my company where we did something last week that was probably worth a hundred million dollars of just screw up in terms of funds because somebody bought something that they didn't check and they didn't evaluate. That tells you what it does to your brand. It is absolutely outrageous and incompetent. It is inexcusable, and it should never have happened. But it happens all over the place.

Kimberle Koile [4]: This whole debate about sketching versus computer systems has been going on in the architectural design community for decades—the power of drawing and the power of sketching. My PhD thesis is actually on computer-aided architectural design. It was about the design component and what it meant to design and the importance of sketching, and the serendipitous things you can find when you look at a sketch that hasn't been made perfect. There is a famous Frank Lloyd Wright story. He claimed that he designed something on purpose, when it looks like it was just kind of an ink blot on the page. Those kinds of things you can't get to unless you sketch. **There is something about a computer, a CAD[5] drawing, that makes it look finished even though it is not.** To your point, it looks finished; but it isn't finished even though it looks finished.

Bill Buxton [1]: So the point is this. **You need a full palette of skills. You do need the drafting skills.** That's the paint process. But the rule of thumb in my opinion should be this: **You should never represent or render an idea at a fidelity that is greater than the fidelity of the idea that you are thinking about it at the time**. The drawing should never—I always try to do hand sketches if I am doing a talk or something where I have fuzzy thinking—and **the minute you start seeing it perfect on a spreadsheet or with perfect lines, it brings an authority to it**. We all explored that in high school or grade school when we had our big sister or brother type our essays for us, because we knew we would get ten percent higher marks just because it was typewritten rather than handwritten, especially for those that were pencil challenged like Andy and me. But we know that, that is gaming the system, and that it is wrong. That is just exploitation of weaknesses. But when it actually comes down to it, **a half-baked idea should have a half-baked representation**. And that is why at my own industry, when I see envisionment videos that look like they are real, I am outraged. They should be done in a cartoon shader, or something like that to say this is not real, this is not product, this is fake. I am trying to give you a concept, not trying to show you something. And if you don't have the imagination to dare to cross, that is too bad. But that error is far less bad than the error of having you believe that this is where we are right now, and sets expectations that we can't meet.

Ken Forbus from Audience: So I want to argue that actually in the long run this is going to become a very heavy AI area.

Bill Buxton [1]: Why?

Ken Forbus from audience: Because if you look at the whole product versus process argument, we have done NSF research showing that the copy order

[5]Computer aided design.

of a thing in biology or geoscience, includes this expertise. **Ask students just to copy a diagram, and you can tell if they understand the concepts on the diagram.** Very simple, robust, easy to do. But more subtle things involve looking at timing and how often they ask for feedback. We have done this with CogSketch. If you look at teachers' lives, they have got a lot to do already. If you then say, "Now you are not just getting the finished product, you are getting the whole process," that is a lot to ask them to look through. So **when we are thinking about feedback, it is not just feedback to the student immediately, but also feedback to the instructor for assessment purposes**.

Tracy Hammond [2]: Absolutely.

Ken Forbus from Audience: Many of us are already going down that path, and I think that is a very important path. The other reason why we need stronger AI is that we keep building domain specific systems most of the time. And we can't do that any longer. **We can't have a different system for every domain, every grade, every field of education. We have to start building more general systems.** And that involves having much more context in the machine's head. The software needs to have the conceptual models that we want students to have, with the same breadth of knowledge that we expect from students. So instead of having a separate sketch understanding system for each domain, we customize systems for domains by adding to their knowledge. We have to focus more on that, instead of people building one-offs one at a time. That can't be the end state. It just can't.

Kimberle Koile [4]: That's my opinion about rule-based versus model-based thinking too: With a model of some sort, where the model is the infrastructure of the software that you are building, you could imagine that it would be able to operate in both math and anthropology or math and literature, rather than having them completely separate.

Tracy Hammond [2]: And, Ken and I do a lot of similar work, and he knows my work well. And he knows that this generalization was the inspiration for my own work on LADDER—trying to figure out what can we gain from just looking at the way people sketch, learning even from a single example. Can we learn how to recognize new domains just by having people look at how people draw in the domain? And how far can we get doing something like that, which was the goal for LADDER.

Tom Stahovich [5]: In the beginning, I made a comment about the increasing availability of large datasets. I actually think that this will contribute to our ability to build systems that are not narrowly focused on one single domain. The systems that I have that can predict the correctness of a students' problem in a statics course—what grade the instructor will assign them—the only thing that my software knows about statics is that there are diagrams and equations. It doesn't know anything else. On some of the exam problems, my software is as accurate at assigning grades as my teaching assistants. And it is just because I literally have millions of pen strokes to train my system from. So I think that moving forward as we now have pen devices everywhere, I really think that we are going to get to a point where we can use very simple models that apply to lots of different areas and then we simply customize them with data.

Jonathan Grudin from Audience: Back to the cleaning up your sketching issue, there was a discussion at dinner the other night about tremors, and people who—Andy was volunteering—but there were others. Do we feel that for people with disabilities of different sorts that there shouldn't be any cleanup? Is there a possibility that there is kind of a middle ground like the work that Michel or the South Park style animation? Is it possible that you could clean things up without making them look perfect? Might there be a middle ground there?

Tracy Hammond [2]: Or smoothing. Yea. I like that idea. There has been some work done replacing sketches with sketchy-looking letters or objects. It does bring on some questions as per sketch-ownership, but in certain instances that has great potential.

Kimberle Koile [4]: Let them decide. Let them decide how clean they want it to be. And give them a clean up button where they push it until they are happy with it.

Jonathan Grudin from Audience: Or a slider.

Tracy Hammond [2]: A slider. Yea, I like that idea. So you can leave your own individuality, but possibly applying a smoothness filter on it. I agree with Kimberle that in these situations, "Ask the user" is the answer.... Except for high school honor students. [Laughter from the audience.]

Kimberle Koile [4]: They get this "absolutely perfect" button at the top.

Actually, we evaluate the "fuzziness" of the solution at the time and that calibrates the button as to how cleaned up it can get.

Bill Buxton [1]: There are examples where—and again we don't want to swap something that works with something else. It is a question of representation and of understanding the place. There is a theory that I would like to introduce, and feel free to look it up in the Oxford English Dictionary. Many of us know of the term *ubiquitous*. And in fact I want to point out that in '88–'89, when Mark Weiser was starting to talk about ubiquitous computing, all of us who worked with him at the time had to look up the word in the dictionary. Many of us did not know what that word meant. And by the way many of us at the same time also did not know what a red circle with a white line through it meant—*Do Not Enter*—because that was a recent thing that we all take for granted now in street signs. But the point is this: If you use the Tablet PC and you use the stylus instead of the keyboard, you notice that there is this intermediate thing when you do your thing in really crappy handwriting, amazingly it echos back in cursive script if you used cursive script, but it makes it legible, and yet it maintains the class of representation. It's not your handwriting, but you could imagine that it could capture the essence of your writing, but make it legible. I mean I can't read my own handwriting, because I am taking notes really quickly on this thing [holds up 6 inch smartphone] on the fly. So there are places like that. There are no hard rules. The rules start when you think about the sense of place. So *ubiety*—ubiquitous means everything everywhere—*ubiety means basically* *the right thing at the right place*. This essence of place—what is appropriate for the place? If you extend the word of place to social place and cultural place as well as geographical place, it becomes a really meaningful term, far more meaningful than something like natural. **My sense is that ubiety is perhaps the most important thing. And that drives how we make the decisions about what the representation should be.** When I should do a rough sketch, and when I really do need a more formal representation.

Tracy Hammond [2]: What I like about this is handwriting that is representative of the state of the process is that it evens out the playing field for people with different handwritings. So, the keyboard in a sense was the ultimate evener, even though it took the initial extra effort of translating into the keyboard, everyone's handwriting looks the same. This way, people aren't penalized for their ideas based on their handwriting, because people do make judgements. Because of the fact that our handwriting progresses as we get older, and in some sense gets *better* [air-quotes], although that is very arguable. At a young age, we are trained, at least for kids—

Kimberle Koile [4]: Not any more.

Andy van Dam [3]: Yea, I was going to say that handwriting is no longer taught as a skill. Typing has replaced it.

Tracy Hammond [2]: Yes, and I have alternative issues with that. But there is still the sense that adults at our age are sensitive to differences in handwriting, that better handwriting seems to imply better ideas, getting back to the same idea of the older sister typing out an essay to try to game the system. I like the idea of using representative cursive to even out the playing field.

Tom Stahovich [5]: This gave me a new idea for a feature for MS Word. MS Word has the ability to measure the writing quality. We could detune the fonts based on the writing quality.

Emily Tran from Audience: If you have other ideas for Office please send them to me because I work with Office. My other question—my actual question—is related to accessibility [points to Jonathan Grudin who introduced the topic]. I was wondering if you think **ink is inherently inaccessible** or have you ever seen signals of a demand for ink to be accessible? So for example you may have a student, if one student in your class has some sort of disability, and everyone else is on Surfaces and they have ink, what do we do for them, because now they are not following along with the ability of the ink. Maybe there is a solution that scales from when you get old and you get carpal tunnel or when you have low vision. What are your thoughts on ink plus accessibility beyond just smoothing out your handwriting and making it a little bit prettier?

Tracy Hammond [2]: In terms of providing ink accessibility, you could do a translation of their personalized input modality to sketched inputs. A lot of what sketch recognition is translating between different representations of knowledge, often translating between computer-understood text and human-understood image, such as in the canonical *car-on-the-hill* example. If you can translate between different types of representations of the same knowledge, then you can do things like have one user use a keyboard, an eyetracker, or other modality. Then, that input can be translated to produce the image that represents their ideas in the representation formality that is appropriate—changing the handwriting to be the appropriate level of messiness as Bill had mentioned. This allows their representations to be on par with others in the classroom and enable effective sharing of their ideas, so they can be interacting through voice, through the keyboard, through BCI, through whatever modality is most appropriate for them to most effectively express their ideas. As long as you can translate to various representations, I think that you can provide accessible means to allow people to interact using the same representations or display their information similarly.

Kimberle Koile [4]: So Aaron, you did work on gesture and voice and sketching.

Aaron Adler from Audience: Yea, there is some—I mean it depends on the context. Handwriting—you could obviously do speech recognition for. But with the sketches, there are some things which are hard to express. **There are some things that are a lot easier to draw than they are to talk about.** There are some things you want to label with your voice because that is easier than drawing. And there are some things you want to draw because it is hard to use your voice. If I am trying to draw a box, trying to draw that with your voice would be near impossible. You can draw a box very quickly. There is a balance there. Certainly doing the text would be a lot easier to tackle for those having trouble with the pen.

Tracy Hammond [2]: Absolutely, but I do think one could figure out a way to **translate from any modality**. The issue is that each disability is going to be completely different in how you tackle that and what modes of interaction are possible. Eventually, perhaps it might be easier to think the box with your mind and just traverse the path with your head, which would be better than trying to describe a verbal representation of the path you are trying to take. We are a long way from that, but eyetracking might give us a faster path, using some combination of button clicks and eye paths. Each disability is completely different, but I think at the core of it, we will be able to translate between modalities. This will be difficult, and as Aaron said, there is a reason why we sketch. Certain things are much easier to draw than they are to type or enter in another modality.

Kimberle Koile [4]: There is the storyboard idea that we heard this afternoon, too. If there is a core set of predefined skills or boxes, we are back to domain. I appreciate the notion that there could be this domain independent system. But, on the other hand, domains are so rich in their own ways that we need to try and find a way to be able to take advantage of domain knowledge when we can, without getting to the point where it means that we can't scale anything. I could imagine boxes, hills, or in Ken's world of geology, there is a vocabulary there that we might be able to take advantage of.

Bill Buxton [1]: I am on the advisory board for Microsoft for accessibility. We do a lot in that space. I am glad you asked that question because we should be asking that question on every decision that we make. How does this expand, understanding what—but also not set up goals that we can't achieve, and to take another approach where they can be achieved. You can decompose it. I claim that nobody in this room knows what a keyboard is. Within 5 min I can deconstruct a keyboard such that nobody knows what it is. For example, if I have a graphical QWERTY keyboard for persons with ALS[6] who you can only use their eyes to type, is that typing? Well, I would probably change the design of the keyboard—the same thing for other types of keyboards. If I have a tremor and if I put my hand down and just type using Morse code, would that be a keyboard? What about a chord keyboard?

There are all types of variations. Part of making successful accessible systems is knowing what representation to use and how it maps onto the space of the abilities that might be relevant. Then, make sure that the underlying structure can accommodate those tools, as opposed to making any one tool serve all purposes. That will never work.

It is a really important thing to do and to pay attention to. The same concerns exist even when I am using the pen to write text; your answer is going to be very different between a keyboard and a pen if I happen to have to write with Chinese characters versus the Roman characters. It's a very different thing. It's going to be different whether I am using Graffiti type of single stroke character shorthand—

I have a wristwatch—a King '84—with a capacitive touchscreen where I can just write with my finger right on the screen, character on top of character, and it will do recognition and capture all of that but I don't have to look. I used to be doing notetaking on my wrist in 1984; you can do that today on your iPhone. I am just pointing this out, because—oh by the way, it costs $99 and the battery life lasts over a year, and still works today. When we are working with students that are working on a computer, we are teaching them about the importance of drawing and sketching with a pen, and other things. At the same time, we are reminding them that this focus is one tile on a much larger mosaic. How does the mosaic fit together so we have **appropriate tools in our palette** by the time we graduate and go into the world as professionals?

You know the old story that if all you have is a hammer, everything looks like a nail. It is an old cliche, but it is really rings true for a lot of people. It is really challenging to add semantics. Whether you are in university or grade school, there are already huge demands on you, so you have got to do this collectively. This is far more about the cultural environment of the learning system than a specific department or even a university or your company or your grade school. We have to ensure that we have got the right curriculum to get balanced students out into the world. It is a really interesting problem. It is all about exposure and experience, so you can viscerally experience what the differences are and why you have to make that decision.

Tracy Hammond [2]: I just want to echo about how different each type of disability is. But disabilities are not the only reason why we need these new

[6] Amyotrophic lateral sclerosis.

devices. One of my students who is working on developing a new eyetracking device for people who have functioning eyes and feet, but no hands. For the longest time I thought that was too small part of the population to make an impact—thinking about how paralysis usually happens from bottom up. In turned out that I was totally wrong about that once I found the enormous number of Vietnam Vets and Diabetes and others who are affected in this way.
But now, having just had a child, I realize the need is even broader, as I find myself constantly holding my infant, wishing I had his system so I could get some work done with my eyes and feet, while my arms are occupied.
As I am holding or feeding the baby, I try to read email or possibly get any sort of work done whatever, and I can't, since standard ways of interacting with a computer assume full continuous use of **both hands**. **There are so many types of situations that unless you are in them at the time, you don't realize how you are affected.**

Bill Buxton [1]: The rule of thumb is that what you really want to do if you are a designer building out these system is to realize that **the edge conditions are really interesting**. So when you go into certain disabilities, you have to focus on one thing, but in almost all cases, everyone is going to benefit—there is some place where we will need it, it is just a matter of degree. We are all blind when we are sitting there operating our phone while we are driving a car. Right, no hands, no eyes, or else you lose your driver's license. When you do build a solution, you have got to think about economies of scale. You can make the solution can fit the people with specific extreme disabilities and needs, but you can make it so that it doesn't cost them ten thousand dollars for a keyboard because you've got the economies of scale. That's the tactical part of it.

Jonathan Grudin from Audience: There are only about 5 min left. I would be interested if people have any thoughts, having come to this area more recently than anyone on the panel, and being very impressed with how significant this research is, do you have any thoughts on how to get more research funding? Or what might be done? I see people shaking their heads. Maybe there is nothing to be said. But I would be interested in your thoughts because I have been giving this some thought, both within my place of employment where it is not too bad; you know, it is kind of on an upswing. But in general, what we might do to fix it?

Tom Stahovich [5]: So I can comment about my research. Getting funding—you know, I have gotten funding—but it has been work. The approach that I am taking, this big data approach to education, is not something that educational psychologists appreciate. It is just a completely different way. Traditional research is hypothesis driven. My hypothesis is that if you get data you will figure out what is going on. That just doesn't work for an educational hypothesis. I am hopeful that as results from this area start to become known, that funding will become easier, but that is just a hope.

Kimberle Koile [4]: From where? The government?

Tom Stahovich [5]: Yea, so yes. I live off of NSF funding.

Kimberle Koile [4]: I think an awful lot of us do, but it is getting harder to get.

Tracy Hammond [2]: I agree, and personalized learning is an NSF grand challenge which our work should fall under. It is considered to be something that NSF is focusing on. And they claim that it is something that is very important to them. And they are funding me too. I also get NSF funding to support my lab as Kimberle and Tom do. I think and hope that we are going to have a big change now that people are getting past the "I can do everything with my finger" mentality, which really hurt us for a long time. We got over the Newton failure, sort of. Now even Apple is coming back and saying, "I admit that we might need a pen again after we swore off the pen for a good 20 years." So I think that we are in a really great place of change. Last year you called it the Perfect Storm, and I think we are getting on that wave of change and that everyone is starting to feel it. I see lots of people jumping on the pen bandwagon.

Kimberle Koile [4]: But there's no money.

Tracy Hammond [2]: No, but **my hope is that as more people start seeing this as an important problem, the money will start to flow**. Because that is where the world is going.

Kimberle Koile [4]: But I think that's the question for us and for the corporate world. For corporate folks out there, I would love for you to weigh in. What is it that we need to do in order to get the corporate world interested, invested, passionate, really wanting to support us, so that we can support them doing something? We as academics are not as well connected with that world and what their needs are. It's great getting NSF money because you can basically do what you want, but there is a responsibility I think now for us. If we do really want to scale this and get it out there, I think, as Jonathan points out, that we need to find different ways to do that; partnerships don't mean that we all need to start our own company and get out there and try to be businessmen, because some of us would be really terrible at it. It is the partnership kind of thing that I would love to be able to think more about, but I don't know anything about it.

Ken Forbus from Audience: I think NSF can work. But it is equally problematic in a couple of ways. So we are just coming off a ten year NSF center, which is a mixed bag, mostly good, and you can accomplish things that you can't do anywhere else. But there is no second round of Science of Learning Center funding. And every plan they have talked about so far has been stone soup. So while they talk about the importance of personalized learning, it hasn't been followed by money. And part of the big problem is that there is some politics there. And I don't know how you get funded for something that you do within your own university because I have been told if I put something in like that, "Ah, you are just padding your own vest; you have to go work in four different schools." This can be done, but it is a trade off. I think there is a bigger funding issue, and corporations can help some, especially—more with engineering than with anything else. But funding is an issue. It is very hard to get money to do things even if they are popular.

Bill Buxton [1]: I think the key thing is—I think there are three things that I would say. One is: we have to learn to separate our motivations from from what lies behind them. We have to go meta. If we don't explain—we have to become much better story tellers in terms of what is the higher thing and how does what I have done support the interjection and help lead us down the path towards what our deeper missions are? And what I would say within the user interface community and interaction world is that to researchers I say **instead publish two less peer-reviewed publications a year and write for the local newspaper**. The best thing I ever did was starting writing a column for Business Week. It got way more impact than all of my academic things put together, at least in terms of consciousness raising. That is probably an exaggeration, but I'll say it just for impact. The other part is: recognize that ultimately it is not about any technique, it is not about the pen. And that is where Steve, you know, you can't allow people from marketing taking history or psychology. It is ultimately about people. It is how we think, how we learn, how we problem solve, how we communicate, and how we function as a society. It's not about the pen or the computer. These things are just prosthetics. And ultimately it has to come back to that. That's why I am outraged a little bit about most of what has been passing through machine learning right now, because it is putting computers against people. I can beat the world champion at Go. I can drive better than a human. I can play chess better than a human. I can beat you in Jeopardy. But what we know is that I would like to think that we are smarter together than we are individually. That isn't recognized that two smart machines could probably be smarter with specialized integration that is specially architected to work collectively. But when we superimpose the society of machines with the society of people, you create that intelligence—we call that integrated intelligence—you have the chance to do something much better. That is very different than what is the current norm. Educators, who are the people who most know about this stuff, aren't speaking out. And not being blown over by these people who go "neural networks" and "deep learning" and so on and so forth, and say "Excuse me, you know nothing about intelligence." Your definition of intelligence is pathetic. Your speech bots can't even understand an indirect pronoun, a reference; we can't go and say, "what is this?" when I point to something on the very screen that bot is executing on. It is all limited. That is not a complaint. It is an observation. It's the fact. And somebody has to speak the case for the larger principles. When you start to do that, you get the attention of school boards, administration is educated, and you have a better chance of at least getting your arguments going. But you need to articulate it better, and don't expect it to happen over night.

Andy van Dam [3]: I am going to call it to a close. Thanks to the panelists and the audience. You are more than welcome to gang up on any or all of them as we exit the door over to lunch.

And just about tonight, I should have mentioned it this morning, we are going to have a little bit of entertainment which involves you providing entertainment with some assistance. You'll see what I mean. It should be a very interesting dinner.

20.3 Panel Member Bios

Dr. Andy van Dam Andries van Dam, is the Thomas J. Watson Jr. University Professor of Technology and Education and Professor of Computer Science at Brown University. He has been a member of Brown's faculty since 1965, was a co-founder of Brown's Computer Science Department and its first Chairman from 1979 to 1985, and was also the Brown's first Vice President for Research from 2002–2006. His research includes work on computer graphics, hypermedia systems, post-WIMP and natural user interfaces (NUI), including pen- and touch- computing, and educational software. He has been working for over four decades on systems for creating and reading electronic books with interactive illustrations for use in teaching and research. In 1967 Prof. van Dam co-founded ACM SICGRAPH (the precursor of SIGGRAPH) and from 1985 through 1987 was Chairman of the Computing Research Association. He is a Fellow of ACM, IEEE, and AAAS, a member of the National Academy of Engineering, and the American Academy of Arts & Sciences. He has received the ACM Karl V. Karlstrom Outstanding Educator Award, the SIGGRAPH Steven A. Coons Award for Outstanding Creative Contributions to Computer Graphics, and the IEEE Centennial Medal, and holds four honorary doctorates from from Darmstadt Technical University in Germany, Swarthmore College, the University of Waterloo in Canada, and ETH Zurich. He has authored or co-authored over 100 papers and nine books, including "Fundamentals of Interactive Computer Graphics" and three editions of "Computer Graphics: Principles and Practice".

Dr. Bill Buxton Cal Armstrong is a mathematics educator & the Director of Teaching and Technology Innovation at Appleby College and Teacher Leadership Program coach at the Institute for Advanced Study's summer Mathematics program. While developing the progenitor of the OneNote ClassNotebook is likely his more notable accomplishment (in a situation with Rob Baker that parallels Newton & Leibnitz and the Calculus), he has been integrating technology, from the earliest graphing calculators to the latest in pen-based tablets and cloud services for more 20 years. He is a devout pragmatist and critically eyes technology with the lens of researched-based pedagogies. Outside of the classroom, he is an Auxiliary Constable with the Halton Regional Police.

Dr. Tracy Hammond Director of the Sketch Recognition Lab and Professor in the Department of Computer Science and Engineering at Texas A&M University, Dr. Hammond is an international leader in sketch recognition and human-computer interaction research. Dr. Hammond's publications on the subjects are widely cited and have well over fifteen hundred citations, with Dr. Hammond having an h-index of 19, an h10-index of 35, and four papers with over 100 citations each. Her sketch recognition research has been funded by NSF, DARPA, Google, Microsoft, and many others, totaling over 3.6 million dollars in peer reviewed funding. She holds a Ph.D. in Computer Science and FTO (Finance Technology Option) from M.I.T., and four degrees from Columbia University: an M.S. in Anthropology, an M.S. in Computer Science, a B.A. in Mathematics, and a B.S. in Applied

Mathematics. Prior to joining the TAMU CSE faculty Dr. Hammond taught for 5 years at Columbia University and was a telecom analyst for 4 years at Goldman Sachs. Dr. Hammond is the 2011–2012 recipient of the Charles H. Barclay, Jr. '45 Faculty Fellow Award. The Barclay Award is given to professors and associate professors who have been nominated for their overall contributions to the Engineering Program through classroom instruction, scholarly activities, and professional service.

Dr. Kimberle Koile Dr. Kimberle Koile is a Research Scientist at MIT's Computer Science and Artificial Intelligence Laboratory and a lecturer in MIT's Department of Electrical Engineering and Computer Science. Her research, in both industrial and academic settings, has focused over the past 30 years on building intelligent computational tools for complex human tasks. Her research interests include pen-based computing, machine understanding, educational technology, ubiquitous computing, knowledge-based systems, human-computer interaction, and computer-aided design. She has been involved in a wide range of projects related to topics that include teaching and learning, "intelligent"Ì rooms, architectural design, rational drug design, and molecular modeling. She currently leads an educational technology research group, focusing on using tablet computers to increase classroom interaction and learning. In her NSF-funded project, *INK-12: Teaching and Learning Using Interactive Ink Inscriptions* (http://ink-12.mit.edu), she and co-PI Andee Rubin of TERC have developed a pen-based wireless classroom interaction system and investigated how such technology can support students and teachers in creating, viewing, and sharing visual representations that facilitate learning multiplication and division, and what role machine analysis of those representations can play in making students' mathematical thinking visible. Dr. Koile has Ph.D. and S.M. degrees in Computer Science from MIT, with specialization in Artificial Intelligence, and a B.A. in Chemistry from UT Austin, with a minor in Education.

Dr. Thomas Stahovich Dr. Stahovich received a B.S. in Mechanical Engineering from UC Berkeley in 1988. He received an S.M. and Ph.D. in Mechanical Engineering from MIT in 1990 and 1995 respectively. He conducted his doctoral research on computational sketch understanding techniques at the MIT Artificial Intelligence Lab. After serving as an Assistant and Associate Professor of Mechanical Engineering at Carnegie Mellon University in Pittsburgh, PA, Dr. Stahovich joined the Mechanical Engineering Department at UC Riverside in 2003. He also holds cooperative appointments in the Computer Science and Engineering Department and the Electrical Engineering Department at UC Riverside. Dr. Stahovich conducts research in computational design tools and natural user interfaces. Examples of the former include: LearnIT, a system for capturing and preserving design procedures; RedesignIT, a system for managing design changes in large-scale engineered systems; and ExplainIT, a system for automatically documenting designs. Work in natural user interfaces focuses extensively on the creation of computational techniques and user interface design principles to enable natural, pen-based software for engineering and education. Recent work has also begun to explore the creation of multimodal interfaces in which interaction relies on sketching, speaking, and gesturing.

Acknowledgements The speakers especially thank Dr. Anna Stepanova, Dr. Tracy Hammond, Dr. Jan Hammond, and Mark Payton for their help in this transcription as well as Lisa Manokofsky for her help in providing the videos for translation.

Chapter 21
Learning Through the Lens of Sketch

Tracy Hammond

Abstract Dr. Tracy Hammond gave a keynote on morning of the second day of the conference. She spoke about how both her research and the field of Sketch Recognition evolved over the last decade. One motivation of her career was to develop algorithms that provide insights into human brain activity and also develop applications that improve human-human communication. Her initial work focused on domain-independent recognition methods, while her current work focuses on developing systems to improve education. She also shows how sketch recognition methods can advance both sketch forensics and activity recognition, providing inspiration as per how this can allow for surprisingly intelligent personalized feedback. This chapter provides a lightly edited transcription of that keynote.

21.1 Introduction

Andy van Dam: Good morning, folks. Welcome to our 2nd action packed day. As I am sure you know, one of the key technologies in making our field work, is gesture and sketch recognition. And this morning's keynoter is a real expert in Sketch Recognition as well as many aligned areas. She runs a big lab at Texas A&M University specializing in Sketch Recognition with many students working there with her. She has a very eclectic background, as I was very interested to discover when I read her bio again this morning. She has multiple degrees from Columbia, one of them in Anthropology, another one of them in Math. She is practically unique among computer scientists in that she spent a number of years at Goldman Sachs. She has her Ph.D. from MIT in addition to these undergraduate and Master's degrees. I think we are going to learn a lot from listening to her. Welcome back, Tracy. She was here in 2007 presenting her Ph.D. work... (Fig. 21.1)

Tracy Hammond: That's right.

Tracy Hammond (✉)
Sketch Recognition Lab, Department of Computer Science & Engineering, Texas A&M University, College Station, TX, 77843, USA
e-mail: hammond@tamu.edu

T. Hammond et al. (eds.), *Frontiers in Pen and Touch*, Human–Computer Interaction Series, DOI 10.1007/978-3-319-64239-0_21

Fig. 21.1 Andy introducing Tracy

Andy van Dam: ...on sketch recognition. And she stuck with it. She will tell us, I am sure, lots of things we need to know.

Tracy Hammond: Okay, so, let's start this the Texas A&M way. Howdy!

Audience: Howdy!

Tracy Hammond: Nicely done. Okay so, I am Tracy Hammond, and yes, my lab is large. Over the years there have been a many people who have contributed to things I am going to speak about.

Andy already did a great job talking about how my path has really impacted why I am in the field that I am in, nudging me in this direction over the years (Figs. 21.2).

In college, I entered computer science kicking and screaming. I originally planned to study neuroscience or psychiatry because I was really interested in how we think. But, I was *really* good at math. I loved math, and it was a passion of mine that I couldn't stop, finishing most of the major by my second year.

During that second year of college, a close friend and suite-mate of mine emphasized, "You need to try computer science. I know you. You will be great at it, and you will love it. Trust me."

To which I replied, "Absolutely not. I want nothing to do with that black box." I preferred math where all of the details were out in the open.

But I took a class anyway.

After my first class, my friend, smirking, asked, "How was it?"

I responded with irrational exuberance, "Oh my God. I love it. It is so exciting. **Computer Science is essentially mathematics while providing instant gratification**, because you immediately get feedback as per whether or not your algorithms work."

From that day on I was hooked, and it really changed my life. I abandoned going into medicine. From my new perspective, I saw that with Computer

Fig. 21.2 Hammond's educational path

Science I could **implement algorithms that mimic the way I think the brain is working**. Then I can find out instantaneously whether that algorithm would in fact work, using those results to get insights into how the brain works. My skills were in mathematics, and with computer science, we can use mathematics to develop and test algorithms about human perception.

21.2 Sketch Recognition at the Crossroads of AI and HCI

Tracy Hammond: So a little bit about Sketch Recognition.... Sketch Recognition is essentially at the crossroads of two fields, Artificial Intelligence and Human Computer Interaction (Fig. 21.3). People define AI in many ways, but the way I like to define it is: **"The field of AI is trying to solve a problem that is inherently really simple for a human but for some reason incredibly difficult for a computer."** This includes problems in vision or other classical AI problems. These problems are usually those things that humans can do without even seeming to think about. Humans can do these intense computations instinctually or almost magically. When we start to look at how the brain works, one can't help but wonder how are we able to do these magical things—these things that seem so difficult to compute, yet trivial for a human.

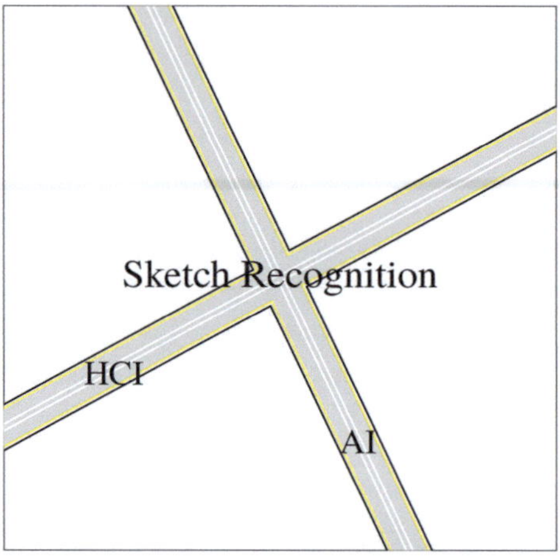

Fig. 21.3 Sketch recognition is at the crossroads of artificial intelligence and human computer interaction

Fig. 21.4 Hammond's AI Dogma

Hammond's AI Dogma

- Artificial Intelligence focuses on problems that are trivial for a human to solve but difficult for a computer to solve.
- Developing computer science algorithms to mimic human activities can give us insight into how the brain works.

And that is one of the main motivations for why I am in this field. I ask myself, "**Can I develop computer algorithms that can mimic human activities? And can this give us insights in how the brain works**?" (Fig. 21.4)

The other mover of the field of Sketch Recognition is Human Computer Interaction, or HCI. HCI was originally called Computer Human Interaction, or CHI, which does roll off a little easier—although I still get new students from time to time who pronounce CHI as "chee" instead of "kai."

As CHI stands for Computer Human Interaction, someone along the way noticed that the computer was listed first and getting the emphasis. That seemed backwards since computers are supposed to benefit humans, not the other way around. And, so, it got flipped around, and we now call it HCI, which takes more syllables off the tongue, but is more relevant.

I argue that this title is also incorrect. Because, I don't know about you, but I don't care how I get along with my computer. I may be sad when I lose my computer, but that is more about the data and access than the computer itself. And when I abandon my computer for a spanky new one, I am downright happy.

Thus, I feel strongly that the field should really be called **improving human-human interaction or communication with technology**. It's about how computers can facilitate communication. All of those "killer apps"are

Fig. 21.5 Hammond's HCI Dogma

Hammond's HCI Dogma

- I don't care about how well I get along with my computer.
- The focus is on "Improving Human-Human Interaction through Technology."
- Focus on the WHY, not the WHAT in terms of pattern recognition.
- The goal is not to replace humans (we have plenty), but to complement them.
- We can do more.

really those apps that have somehow allowed us to communicate with each other better. And by, "with each other," I include communication with ourselves. **Many of the best apps allow us to communicate with future or past selves**, providing intuitive ways for us to record our current ideas so we can tell ourselves these ideas a little bit later. This includes things like Microsoft Word, Gmail, your calendar, or many of other things that are helping us work and communicate with our future and past selves (Fig. 21.5).

Right now "Big Data" is all the rage, and many people are stepping into the field of machine learning to do effective predictions. In my research, I am less focused on the "what," and more on the "why". It's great to have something that works really well, but I always ask, "Why is that working? What is it about these features that happen to be chosen? Why do they work better than another set of features?" Even our first version is built with the neural network, we will attempt to dive down into the features, trying to reduce them, reapply them, until we can figure out what it is about this selection of features that is performing the magic of recognition. And then we try to see if this gives us any insight into how our own mind works. Again, the idea is never to replace humans. We, actually, have a lot of humans on this planet. From my perspective, the goal of AI is not to replicate a human, but to understand them and complement them. Computers can do many things much better than we can, and when we can figure out how to use them effectively to do that, harnessing their complimentary skills, it's quite exciting.

21.3 All Forms of Communication Are Not Equal, AKA Why Do We Sketch?

Tracy Hammond: All forms of communication are not equal. There is a reason why we sketch. Look at the text in Fig. 21.6. What is it? Some people in the audience will recognize this text—I see you, Aaron. But for the rest of you, it is hard to tell.

All forms of communication are not equal

Body: Polygon, center (2.28, -1.90), points (2,0, -1.12) (2.88, -1.73) (2.48, -2.05) (3.22, -2.19) (2.78, -2.73) (1.38, -1.91)
Body: Polygon, center (2.73, -3.99), points (1,-2.41) (4.88,-4.97) (4.72, -5.38) (.81, -5.57) (0.68, -2.67)
Body: Circle, center (2.48, -2.44), radius .35
Body: Circle, center (1.77, -1.98), radius .3
Pinjoint: center (1.82, -1.94)
Pinjoint: center (2.53, -2.38)
Anchor: center (1.61,-4.37)
Gravity on

Fig. 21.6 All forms of communication are not equal

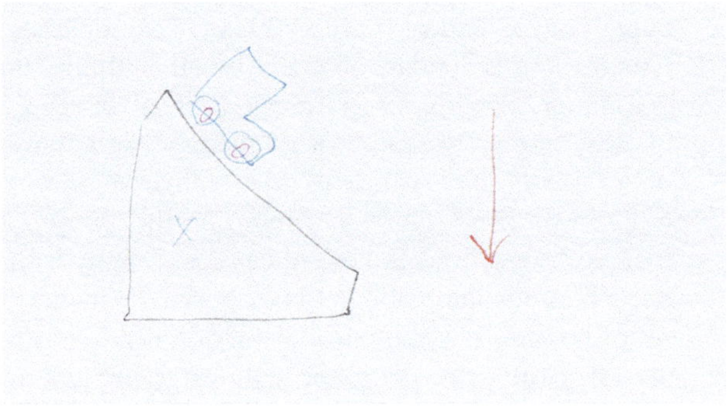

Fig. 21.7 A sketch is more intuitive

What about this image in Fig. 21.7? Which one is easier for you to process and understand the meaning of?

In essence, **Sketch Recognition is trying to translate how we think to how computer thinks**, so the computer can better help us. It is easiest for the computer to process the text in Fig. 21.6, while it is easiest for us to process the sketch of Fig. 21.7. They represent the same thing, but they are two different manners of looking at that same item.

21.4 Sketch Recognition Is the Automated Recognition of Hand-Drawn Diagrams

Tracy Hammond: Sketch Recognition is the automated recognition of hand-drawn diagrams. Sketch Recognition attempts to process messy hand-drawn images. It tries to break down corners. It tries to recognize things the way that we recognize them. It combines shapes into objects, into meanings. In Fig. 21.7, a downward arrow means gravity; polygons means bodies; and circles are also bodies. Circles can mean multiple things; they can also imply a pin joints. When we see the shapes on the page, they have meaning to us. If we can teach a computer to be able to understand it or translate it to a form it can easily process, it provides the computer with substantially more power to aid us.

That canonical, car on the hill example, that I was able to implement using my domain-independent LADDER system in under 2-days, was the subject of Christine Alvarado's thesis [4], as well as a test domain of many other graduate student's theses and dissertations under my Ph.D. advisor, Randall Davis, at my alma mater, MIT, including Aaron Adler, who is also in the audience [1].

My dissertation work, LADDER [35, 37, 42–46], focused on trying to generalize many of the sketch recognition algorithms to attempt domain independent recognition [49, 57], and provide a language to quickly create domain specific systems[36, 38, 39, 48]. With LADDER, I was able to create a system that recognized simple hand-drawn mechanical engineering diagrams and animate them according to the laws of physics in under a day.

21.5 Do We Really Need the Pen? Yes

Tracy Hammond: In the field of Sketch Recognition, we tend to focus—although I'll argue this later in this talk—we tend to focus on the pen. I do have a little bit of dogma about the pen, which results because of the controversy over the pen over the years. Especially when the iPhone first came out and also when the iTouch and iPad came out, people argued, "Why not just use the finger; isn't the finger much easier?" The field had a little bit of struggle during that time; it almost went under, I would say, when this conference first started and a little bit before. There was a lot of struggle in the early 2000s, with much of it stemming from the Apple Newton's failure and discontinuation in 1998. People were arguing that the finger is more intuitive than the pen. I do believe that we had to go through that, because you sort of need to perfect the simpler interfaces before you can get people to use harder more sophisticated ones. Also, recognition wasn't ready yet to be able to use those systems well.

But the pen is not going anywhere. It was invented thousands of years ago. Somebody decided to pick up a sharp bone from the ground—the first pen and drawing—and found that drawing using a sharp tool on the surface of a moist clay tablet was easier and more effective than drawing with your finger. Writing utensils have existed ever since.

The pen gives you more dexterity; you can do a lot more with the pen than with you can with a finger. People have argued that pens are going away, but that's just plain silly. We will always be using pens.

Sketching is pervasive in education, and crucial to it [90]. Last year at WIPTTE there were several nice talks that emphasized that students learn more when they are using pens to take notes on paper than when they take notes with a keyboard. Another argument that we have heard over the years is that the keyboard is going to replace the need for the pen. That is also silly. There are lots of reasons why the pen is sticking around.

Hammond's Pen versus Touch Dogma

- Sketching is not going to disappear.
- Not all forms of communication are equal. Some things are better expressed through sketching.
- Sketching improves creativity [40].
- Text is not going to replace a sketch.
- The finger is not going to replace the pen.
- Students learn more by taking hand-written notes.
- Students perform better on problems when doodling.

21.6 Sketch Recognition Techniques

Tracy Hammond: Sketch Recognition has a variety of algorithms that are broken down essentially into three different types.

First, we have appearance-based algorithms. This class of algorithms represents traditional solutions used by researchers in computer vision. These techniques often come from computer vision researchers dabbling in sketch recognition, thinking it is the same problem. The appearance methods look at these sketches based on what ink is on the piece of paper. This makes a lot of sense; you can get a lot of information from what ink is on the piece of paper. These methods can make significant headway on certain types of drawings, such as zip codes, and other cases where the items in the sketch are very visually similar, such that with minor deviations, you can generally superimpose to drawings of the same item. In Fig. 21.8, you can see a simplified example of what a bitmap image of an arrow that might be used as a template for searching for arrows. Note that it would have trouble looking for arrows of other shapes, with curvy tails, or what not.

What is on the page is only one part of sketching. There is also a lot of information that can come from watching the act of sketching itself. Tablet PCs provide us with timing information that can be used to analyze the path that the sketcher took. Gesture-based methods analyze the path of the pen to perform

Fig. 21.8 Arrow bitmap template

```
0000000000000000000000000000000
0000000000000000000001100000000
0000000000000000000000011000000
0000000000000000000000000110000
0000000000000000000000000011000
0111111111111111111111111111100
0000000000000000000000000011000
0000000000000000000000001100000
0000000000000000000000110000000
0000000000000000000011000000000
0000000000000000000000000000000
```

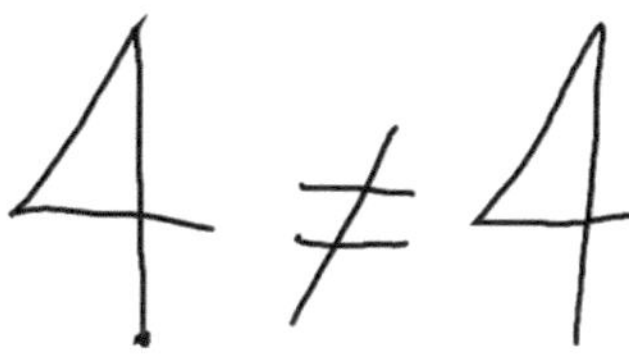

Fig. 21.9 A gesture-based recognizer would not consider these two fours to be the same shape

Fig. 21.10 The letters "*A*" through "*G*" in the Graffiti language. The *dot* specifies the starting point

recognition [15, 16]. This works remarkably in many instances because people tend to draw the same shape in the same way most of the time. There are some cases when this does not work. For instance, two "fours" that look identical but drawn in opposite directions would not be considered the same shape. (See Fig. 21.9.) In 1997, many of us owned and used a Palm Pilot. There was not a keyboard in the Palm Pilot. Rather, users sketched the letters using the Graffiti handwriting language [75] in which each letter was reduced to a simpler form that had a distinct path from all other letters. Figure 21.10 shows how the letters "A" through "G" are drawn in the Graffiti language. This meant that users had to be trained before they could successfully "type" using sketched letters. This was a big push forward in the field, and people still use similar methods in many recognition systems today. What excites me about these methods is that they provided insight into how things were drawn, and I'll show you interesting ways to use this later.

A third method—which my Ph.D. dissertation focused on—was geometric-based recognition, looking at how to describe things geometrically [41, 85]. This sparked a lot of insight about how people perceive shapes and the world around us. Geometric-based recognition uses the geometrical algorithmic definition of particular shapes to recognize the shape. For instance, the definition of a circle is such that all points are equidistant from the center, defining the

Fig. 21.11 A pendulum could be defined geometrically as a *circle* connected to a *line*

center either as the center of the points or using other more mathematical methods. Then primitive shapes are combined into higher-level shapes with broader meaning using constraints. For instance we can define a pendulum as a circle connected to a line as in Fig. 21.11.

Each of these methods has drawbacks and advantages, and the most effective systems will use a combination [18]. Each method is more flexible in one way, but less flexible in another. Gesture-based methods can be messier, but have to be drawn using the same general path. Vision-based methods allow the user to draw using any path, but the shape has to look quite similar. Geometric based methods allow shapes that look different to have the same meaning, but the primitives need to be drawn such that they are effectively recognized.

Benefits/Drawbacks of Recognition Methods

Appearance-based:

- Good for shapes drawn in different paths
- Doesn't work for varying shapes (like arrows)

Gesture-based:

- Most useful for forensic methods
- Useful for trained editing commands [25, 72, 132]
- Usually requires user-specific training

Geometric-based:

- Provides greatest drawing flexibility
- Often requires neater sketches
- Can provide insight into perception

Gesture methods tend to be most useful for forensic uses, such as trying to find out who is drawing, since everybody has their unique style. Geometric methods tend to be most useful for diagrammatic drawings, where each drawing produced is very different, but follows the same underlying set of principles, and the diagrams are built from a set of components.

21.7 Constraints

Tracy Hammond: My work on geometric methods for low and high level recognition brought about paleosketch for low-level recognition [83, 84], and LADDER for high-level recognition.

Defining the constraints for high-level recognition had surprising HCI implications. A major goal of LADDER was to build a language that was intuitive for people to use, but also that matched how they naturally perceived shapes [61].

One interesting thing that came from this was the constraints for angle specification, which is something that naturally has to happen for most shapes. Coming from a mathematical background, when I first thought about building a recognizer for a triangle, I thought to myself about things such as: "Triangles sum to 180. An isosceles triangle has three 60 degree angles. So, I probably need a way to specify a 60 degree angle. 30, 45, and 60 degrees seemed to be the perceptually important angles that need to be easily specified." And so, in the first version, those specific angles could be specified. As I had people start writing descriptions and recognizers, I started to get the impression that there might be some problem with my angle assumption.

This concern prompted me to try a user study where I showed people random angles on a screen and asked people to tell me what the angle was. People from different backgrounds were in the study, mathematicians, computer scientists, artists. When looking across the participants, ANOVA showed the participants not to be statistically significantly different from each other. The average error was 7 degrees, with a variance of 25 degrees. But angles close to the horizontal or vertical, something called a singularity, had statistically significantly less error than those further away. We divided the angles tested into two groups, those within 10 degrees of vertical or horizontal, and those further away. The angles within 10 degrees had an average error of less than 3 degrees, and a variance of less than 5 degrees. The other angles had an average error of almost 8 degrees and a variance of almost 26 degrees. Note that there were 3.5 times as many in the far category as the close category due to random selection of the angles. Figure 21.12 shows how the variance increases as we get further from a singularity.

So we are very good perceptually at recognizing horizontal and vertical. That makes sense because the way our eye muscles work primarily constrains eye movement to a combination of horizontal and vertical eye motions, making it natural to be predisposed to identify horizontal and vertical lines. So, you are

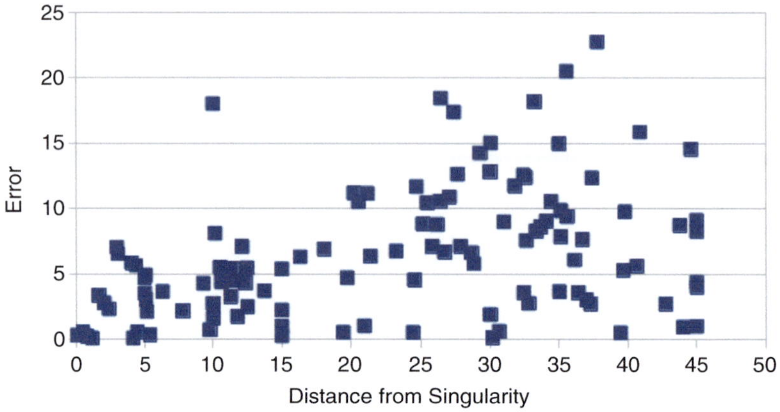

Fig. 21.12 Users get worse at identifying angles as they strayed from a singularity

predisposed to be able to recognize these types of angles. But, any other angle, we are really-really bad at. It is quite common for people to be over 20 degrees off. And this was when they were trying their best to be precise and accurate.

This study was done over 10 years ago, but I still see this in common everyday life. As I work with people, I see people estimate things, and they are quite prone to this particular error. I may ask them to draw a 30° angle, and I see them draw one that is clearly closer to 60°. Those instances remind me of this study.

Coming back to LADDER, one of the things we removed from the language was the ability for people to precisely specify an angle, since those differences are not perceptually important to the user. In LADDER, you cannot specify a 60° angle. Rather, you can only specify those things which are perceptually important: "horizontal," "vertical," "positive slope," "negative slope," "acute," "obtuse," "parallel," and "perpendicular." Those are the things that we can distinguish in our mind, and so we should have the language map that.

It turned out that there are a bunch of other constraints that were also problematic from an HCI perspective. When we are thinking about "above" and "below," and what is considered "inside" or "outside," what is considered "bigger than" or "smaller than" are all perceptually important and surprisingly contextually dependent. It depends on the shape/form of the shapes as well as their relative sizes. Relative size issues are why—for those of you who knows Long versus Rubine set of gestures—why Long chose to use logarithmic features to help provide a scale in some of his features, including size as well as angle from a singularity. Our perception of whether or not a constraint is true, changes greatly on the relative sizes and forms of the shapes.

21.8 Mechanix

Tracy Hammond: Using this constraint-based system to describe shapes, we built tons of systems. Most of the systems were focused around education. One of them you heard about yesterday from Randy Brooks talking about Mechanix [13], which allows students to get immediate personalized feedback on hand-drawn homework problems in mechanical engineering [32, 64, 77, 87, 118, 119, 121, 122]. Thus far, as Randy mentions, we have had some success. In controlled studies, we have seen that people who use Mechanix tend to have a greater pre-post test improvement. It also seems to increase homework completion/motivation in at-risk students, which is also quite exciting [5–9, 34].

This is the first place that we have started to look at sketch forensics. Last year Seth Polsley presented here at WIPTTE how Mechanix data can be used to predict various things [89]. One of the things that Seth showed is that there is this sweet spot between 1 and 2 min, or even more specifically between 1 min and a minute and a half where you get the most benefit from submitting your homework for feedback. If students submit too often, or if they don't submit often enough, the statistics show that they are not going to be able to complete the problem set. So we can predict whether or not they are going to be able to complete their homework set from factors independent of the content of their actual homework. There is some logic behind this. Those students who are submitting their homework for feedback every ten seconds are probably trying to game the system; they are not taking enough time to actually think about the problem. Those who take too much time between submissions are not taking advantage of the personalized feedback system, and when they finally submit the problem and find that their submission has many errors, they may get frustrated and give up.

21.9 Generating Description from a Drawn Example

Tracy Hammond: We built more systems using the constraint systems. One of these systems could recognize hundreds of shapes [51, 52], which shows the power of this type of system. When building a system with many shapes, writing a shape description for each shape could be quite tedious. It was much easier to draw the shape than to describe it. This was work that was done both with and inspired by Olya Veselova [124], a colleague and friend of mine from MIT, now at Microsoft, and WIPTTE 2013's keynote speaker for her work on OneNote [56].

So of course it is easier to draw the shape than it is to describe it. And a human can see a single example of a shape and understand the essence of that shape and the allowable variations, but that is very difficult for a computer. If you think about it, almost any single shape has many possible variations within

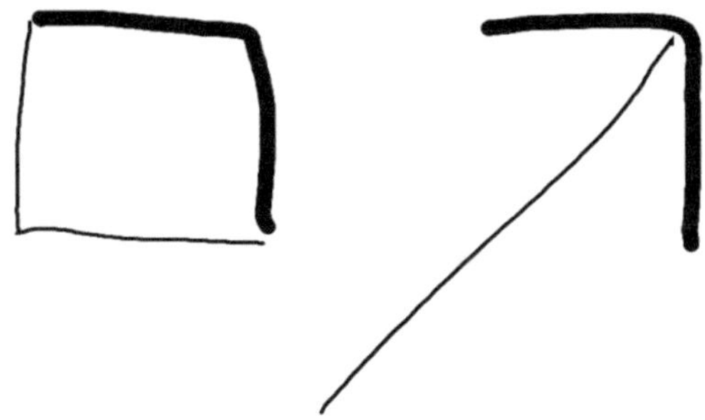

Fig. 21.13 The *bold lines* in the *rectangle* and the *arrow* are both perpendicular

the constrains of the shape, even a simple arrow or triangle. And any instantiated instance of a shape has hundreds of constraints that can be used to describe it, even for a simple three-lined shape. When someone draws a canonical example, how do we know which constraints are important? Can we rank them? As in Olya's work, we can look at Gestalt principles, and use them to determine which features are more perceptually important. Humans pay attention to things like similarities, grouping, whether or not things are touching, singularities—which are things I mentioned before—if you have horizontal or vertical lines, those tend to be very important. If you draw a shape where any of the lines are vertical or horizontal, it is rarely on accident; it usually implies that every instantiation of that shape will also have horizontal or vertical lines.

There are still many difficulties, though. The same constraint could be very important in one shape, but not very important in the other. Looking at the bold lines in Fig. 21.13, the same constraint holds for both shapes—we have two perpendicular lines. The perpendicular line constraint is important in the square, but not important in the arrow. What matters in the arrow, is instead that the two smaller angles are approximately equal, but the point of the arrow can be wide (obtuse) or narrow (acute). Again, "equal" has different perceptual threshold in different shapes.

It's impossible to draw different examples for all the different types of arrows that exist. But we can look at the canonical example; often people will draw what is most important in that first example. That first example provides so much more information than feature examples.

I have to admit, that I tried it the other way first. I built a system where the user could draw all possible variations so that the system could learn the correct constraints that describe a shape. And, while that provided value up to a point, **fixation proved to be a very big problem**. Users thought they were drawing different arrow examples as in Fig. 21.14), but from the perspective of the constraint system, they were drawing the same shape in the same way. The users were confusing constraint variations and handwriting variations. For example, one user drew a right pointed arrow with two strokes over and over. When we asked them why they draw the same shape over and over, they replied, "Oh, no. They are different. Look, there is this little space right here. And this one is longer than this one."

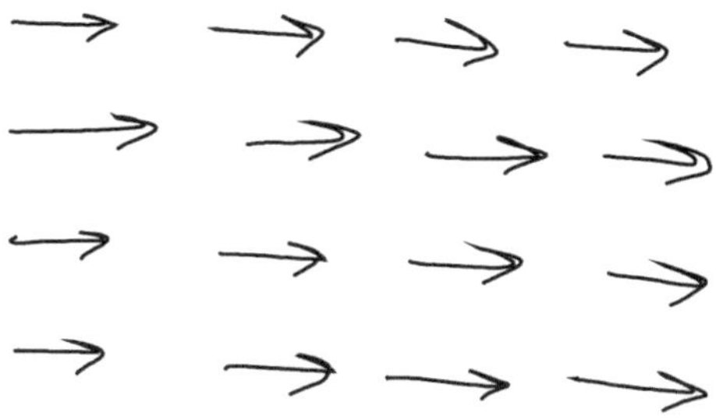

Fig. 21.14 Examples of "various" *arrows* drawn by a user with fixation

But from the system's perspective, the set of constraints were all the same. Puzzled, we asked them, "So the arrow always has to be facing the right?"

And they responded, "Oh, yes, it can point any direction."

But they didn't draw any in another direction. Other variations, such as those that might occur in the shaft of an arrow prompted similar responses. So, overcoming fixation is a barrier. And we learned that perhaps a dialogue with the system might be a better way to get variations. Another piece of work from my Ph.D. was to use the canonical example to generate near-miss shapes [47]. Using that first example combined with a ranked list of constraints from Gestalt principles, the system was able to generate shapes similar to the first drawn example and attempt to pre-categorize the shapes as either in the shape class or not in the shape class. Then confirmation by the user helped to define the shape, and was quicker and more effective than having the user generate lots of variation (although the user could continue to do that). That taught us a lot about how people perceive the world around them.

21.10 Teaching Drawing

Tracy Hammond: Our other work has also provided insights into human perception. One thing that has come up over and over again, is that: **We don't draw what we see; we draw what we perceive.** And **what we draw provides insights into what we perceive.** This became quite valuable in our iCanDraw program that taught students how to draw faces—Manoj Prasad, over there, was a co-author on this work [28, 29, 55, 105].

When people draw human faces, classic problems that show up—as many of you may know—are: People tend to draw the eyes too high, and they draw the eyes too close together.

There is a reason for this.

You have probably all heard that the eyes actually lie at the halfway point in the middle of the head. When you draw them you are supposed to divide the head in half, and draw the eyes there. Mathematically, it is pretty close to perfect. Your eyes are right in the middle of the head. However, no one ever draws it this way unless they have been trained to draw it this way.

The reason for this is that it is not aligned with our mental model of the face. We look at people all the time, every day; it's not like the face is new to us. But what do we focus on? We focus on the parts that are communicating with us, which are the eyes and the mouth, and maybe the nose a little bit. For the most part, it is the the eyes and mouth that are important. These parts of the face are perceptually important, so we center them. We center the two pieces that are communicating with us and they come out to the forefront.

How we perceive an object greatly affects how we draw the object on paper, unless we are a trained artists. A lot of artist training is about training yourself to view the world differently.

So, we were able to use what we know about how people perceive the face, what mistakes they commonly make, and mathematical rules to help frame their mental models differently and teach them how to draw. And more specifically, teach people how to draw what they are seeing by giving them personalized feedback. And faces are filled with lots of mathematical principles. For instance, **the space between your eyes is the width of your eye**. So there are lots of neat principles to help reframe your mental model.

One interesting tidbit to think about that may change your mental model about where eyes lie on your head is this: **If you were going to put two cameras on a balloon or ball, where would you put the cameras to best see the things directly in front of it, at a similar height? You would put them in the middle of the balloon/ball, because that is where the cameras would be most effective/efficient. Nature is remarkably efficient.**

21.11 Teaching Perspective

Tracy Hammond: After this system, we built several other systems teaching people how to draw [24], and even systems recognizing people drawings in 3D [88, 110–112, 114].

One system that combines the two, that you will hopefully get to see tomorrow at some point is Persketchtivity (or the alternative spelling PerSketchTivity explains the origin of the name: teaching **Per**spective **Sketch**ing to increase Crea**Tivity**). Swarna Keshavabhotla—in the audience—is a co-author on that work. Persketchtivity teaches people how to draw in 3D perspective, starting from the basics, such as lines and circles [59, 126]. The idea behind Persketchtivity, the act of drawing many solutions quickly can aid in creativity. Sketching is a very important part of creativity, and the inherent imperfectness of sketching can actually provoke new ideas [27, 40, 74].

When teaching people to draw in 3D, one really important thing is teaching people to focus on those perspective points, and to draw lines that point to them correctly. As they progress through the lessons, they get feedback on their technique, not only accuracy and speed, but also style, such as line order and direction. Watching them draw and create the right feedback is exciting. We can see what people are perceiving, how they look at the diagram, and how that perception and focus changes over time.

In essence, we are trying to get novice sketchers to sketch more like experts. But that is not an obvious path. Expert sketchers aren't just more accurate. What makes an expert? What differentiates a novice person and an expert? You might think that accuracy would be the number one thing that differentiates experts from the novices. But it's not. What would it be? Speed—I heard a couple of people in the audience say it—is one big differentiator, and probably the biggest one. So is fluidity, which correlates highly with speed.

We computed p-values from a t-test comparing experts versus novices. To show significance, you are usually looking for values such that $p < 0.05$. With a p-value of 0.5, the data has a 5% chance of coming from that same class through randomization. You can see that average speed and minimum speed are big differentiators. The bold ones are features that are statistically significantly different between novices versus experts. Only maximum speed is not very different, which implies that consistency of speed is more important than the maximum speed. That leads us to our next set of features.

Factors related to speed fluidity are also big differentiators. Speed fluidity is essentially a measure of how constant or consistent is your speed.

Stylistic features that relate to planning are also quite important for rectangles. In terms of stroke order, what is most important is that experts tend to draw parallel lines together, whereas novices tend to draw rectangles in one continuous line. Experts use multiple lines to draw their rectangles, so they can draw parallel lines together and focus on the perspective point that they eventually meet at. Novices focus on those supposed 90° angles. But while we perceive 90° angles, they don't really exist in our three-dimensional perspective vision. So again, novices are drawing what they perceive, not what they see. Experts instead focus on which lines are parallel and extend to the same vanishing point. A square is made up of two sets of parallel lines, each set going off to its own vanishing point.

We don't get the same statistical significance in cuboids, but that is because the experts may start at different parallel lines, but the drawing differences still exist.

That is why from the rectangle, we can see how the process of how it is drawn gives evidence of what is going on in your mind, how you plan, how you think about those shapes, and what you focus on.

Accuracy—looking at maximum deviation, average deviation, and Hausdorff similarity—is a lot less of a differentiator, which we can see by the lack of bold. The accuracy of lines is important, but I think that is more related to fluidity. None of the other shapes show accuracy to be a big differentiator, except for rectangle, which is a simple combination of lines. We see once the number of lines increases, such as for a cuboid, that distinction goes away.

Accuracy is still important, but experts care less about whether they drew the perfect circle than if the lines extend off to the correct vanishing points.

Looking at fluidity of accuracy, here we get statistical significance again. Consistency is much more important for novices than is accuracy itself.

21.12 Corners

Tracy Hammond: As part of recognizing geometrical shapes, one thing that we often have to do is break strokes down into multiple shapes. We have developed a couple of algorithms [127, 129–131], but one piece of information, which actually came from Herot in 1976 [58] and again from Sezgin, Stahovich, and Davis in 2001 [102], is this insight about corners and speed.

I have been using this insight over and over again to really think about how people perceive corners. So, corners. We see a square in Fig. 21.15. Figure 21.15 also shows the direction graph of that square. Let's look at the direction graph. The numbers on the left are degrees. The square traverses a total of 270°, which makes sense if we start and end at a corner. If we take our pen and start traveling right, such as by following the path of the top line of the square, we would be traveling in the direction of 0°.

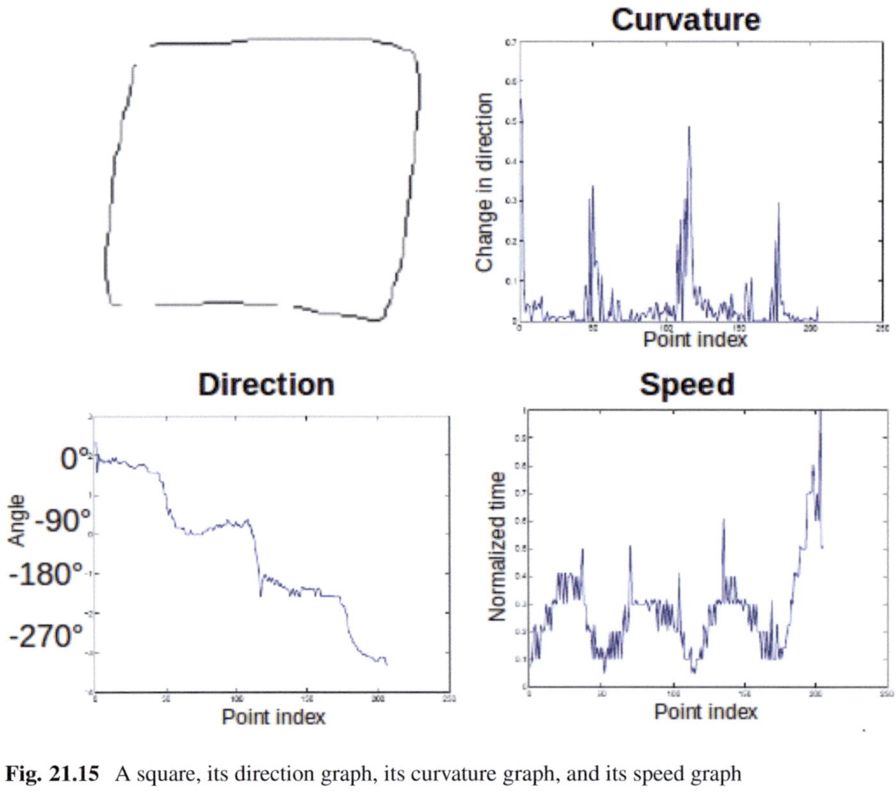

Fig. 21.15 A square, its direction graph, its curvature graph, and its speed graph

At the end of that top line, you suddenly make a sharp 90° turn down. Following the line on the right of the square down, we are now traveling in the direction of 270° or −90°. Note that 270° and −90° are the same direction. Our direction graph plotter picks whichever one makes most sense given where the user was drawing before so we don't have huge jumps in the graph that are not reflective of how the user drew.

Notice how the direction graph has a sudden drop of about 90 degrees at around point index 50. Around point index 110, we see another quick drop again of about 90 degrees. Now we are traveling to the left on the bottom line, traveling at 180° or −180°. Around point index 180, we make another quick turn to go up, which is 90° or −270°. Note that the last line was drawn the quickest, which is why that part of the direction graph is so short.

That's the direction graph. We can look at the first derivative of the direction graph and get the curvature graph shown in Fig. 21.15. The curvature graph easily visualizes the changes in the direction graph. Every time we get a large change in the direction graph, we get a spike in the curvature graph, because it is the first derivative, a graph of the change of the direction graph. Notice that there are three big spikes representing changes of direction at point indices 50, 110, and 180. These spikes represent the three corners of the square. There are no four corners, because we started and ended at a corner.

Figure 21.15 shows the speed graph of that same sketch. What Herot, Sezgin, Stahovich, and Davis noticed was that, when we look at the speed of the pen, there is this magic that the pen slows down when we take a corner. Notice that we have dips in the pen speed again at point indices 50, 110, and 180 again. Also notice the high speed at the end of the stroke, which was something we also noticed from the direction graph. We also see a pretty smooth speed increase and decrease for each of the lines, showing speed fluidity.

The direction, curvature, and speed graphs provide wonderful visualizations into forensics about how the user drew, and insights into their planning process.

21.13 Sound

Tracy Hammond: One thing that we were able to do with the insight of that the pen slows down when you are taking a corner is to build algorithms that recognize what you are drawing from the sound of your pen.

How does that work? It works because the sound of the pen is very much dependent on how fast the pen is going. So, you can actually hear the corners. When you sketch an "A", it makes a very different sound than when you sketch a "C."

We were able to get pretty high accuracy using this principle along with other sketch recognition techniques to process these corners. We were able to find the corners, and also distinguish between lines and curves, since they sound different as well, because their speed profiles are different.

Using these techniques, we got an accuracy of over 80% for the 26 letters of the alphabet [71]. This is quite exciting because certain shapes, such as a D or a P, really sound identical, as would X and Y, or C and U. These results are from using a single microphone. We originally tried it using four microphones to try to hear from each of the microphones through triangulation, but that required constant calibration. From an HCI perspective we felt that if you need to place four microphones and calibrate between them, it offers very little advantage of just using a tablet to collect the sketch data. Any case when we would want to recognize the sketch by sound would probably only be in the case of one microphone.

21.14 Oversteering

Tracy Hammond: Going back to these corners, and this idea of you slowing down. We noticed, as part of our PaleoSketch algorithm for recognizing low level primitives, that not only do you slow down, but when you zoom into the motion it shows that you over-turn for all corners except for those in rectangles.

The analogy is very similar to when you are driving in a car. When you approach a corner, you over-steer at a corner that you do not do at a curve. On a curve, your hand is more steady. But for a corner, you over-steer and then you correct to eventually settle on the right amount of turn. The same thing happens with your pen. And it is only viewable on a microscopic level. We were able to use it to recognize arcs versus polylines. It ended up boosting our recognition rates up tremendously.

Look at Fig. 21.16. You can see the peaks in the direction graph that over extend the actual angle to be covered.

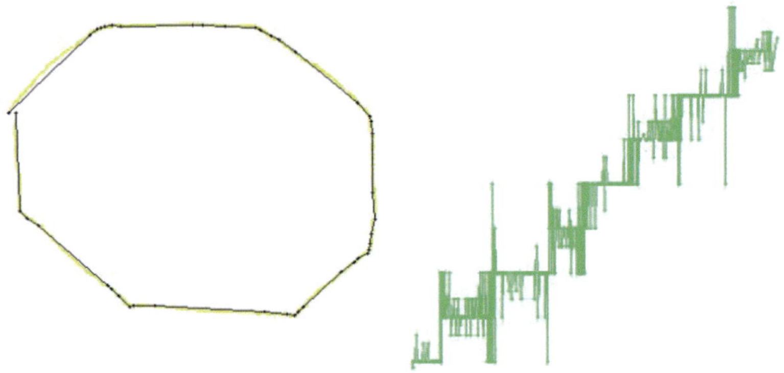

(a) (b)

Fig. 21.16 Corners in shapes other than *rectangles* show evidence of over-steering. (**a**) Drawn figure. (**b**) Direction graph (*left axis*: angle, *bottom axis*: point index)

If you remember this example of the rectangle in Fig. 21.15. There are not so many peaks. Rectangles are different. Which polygon did you draw more than any other polygon? Rectangles. We draw a lot of boxes. Our world is filled with boxes. “Little boxes on the hillside, little boxes made of ticky tacky, little boxes on the hillside, Little boxes all the same…” that song from Weeds by Malvina Reynolds. Our world is filled with boxes, and we are constantly drawing boxes. We are perceiving them, and it is two singularities combined together: horizontal and vertical. We have a lot of practice perceiving them and drawing them. So, this is the only shape where oversteering is not seen every time. It still shows up from time to time, but not as much. When drawing squares, we don’t have to mentally prepare ourselves for drawing corners as much as we do for other shapes. But every other shape, we have these extreme preparations. It seems like we are really mentally trying to decide how much we should alter the turn.

One of the things we have learned from this is that we need to have a rectangle primitive because it is so perceptually important. A square/rectangle is not just the composition of four line. It is much more. It is perceptually its own shape to us.

As far as recognizing the over-steering that exists in corners, we developed two features, NDDE and DCR. Brandon Paulson was the Ph.D. student at the time who identified these [80].

NDDE, normalized distance between direction extremes, looks at the total stroke length and then compares it to the distance between the direction extremes. So we are looking at the tallest point versus the shortest point in a stroke or substroke. Figure 21.17 gives an example. We can see that NDDE values are lower for polylines than they are for curves. DCR is the direction change ratio. Looking at the direction graph, what is the largest jump that we make, normalized? DCR is the ratio between the largest change in direction and the average change in direction. (See Fig. 21.17.)

So, it is really looking for these spikes. How long are the spikes compared to the average spike/change.

When we compare corners and arcs, shapes with corners had these really large spikes, and that is because when we draw, we are hesitating about and really thinking about what should that angle be. We over-correct, because we are over-thinking. We over-steer and get it to the right angle.

These features were really valuable in recognizing primitives [54].

21.15 Children

Tracy Hammond: These features were also valuable in predicting children’s ages. We have been working for awhile with children—Kimberly, I don’t know if she is here, but she is an expert in children recognition. We worked with her when I was at MIT. Since then we developed our own child database, with elementary

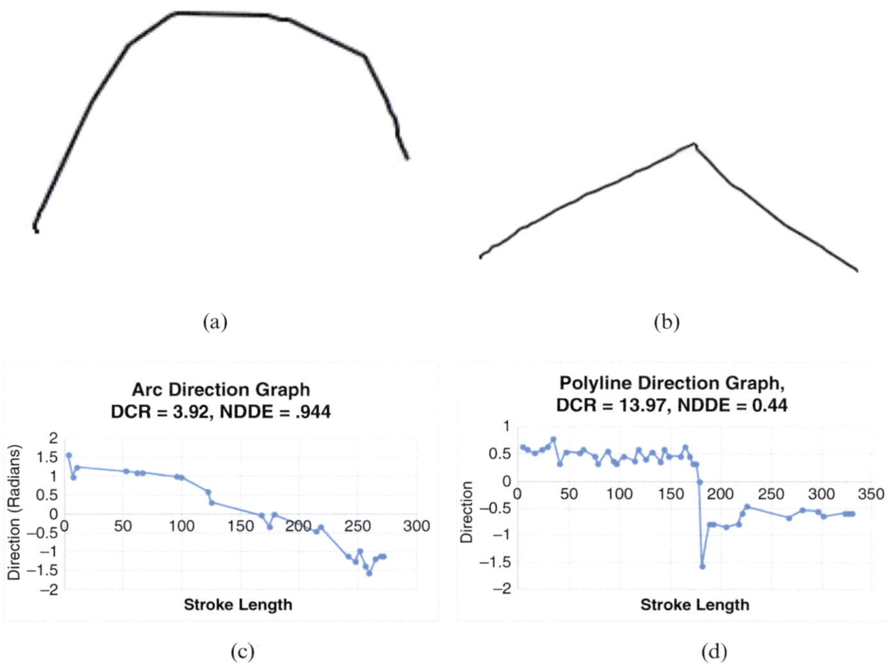

Fig. 21.17 NDDE (normalized distance between direction extremes) = stroke length between direction extremes/stroke length. DCR (direction change ratio) = largest change in direction/average change in direction. (**a**) Drawn arc. (**b**) Drawn polyline. (**c**) Direction Graph for Arc. Direction extremes are at stroke length's 0 and 256. Largest change in direction is around stroke length 264. (**d**) Direction extremes are at about stroke length's 34 and 180. Largest change in direction is at about 180

school children. We were really interested in developmental ages. We focused on getting sketches from people from ages 2 to 8. We already had a lot of sketch data from adults, but we wanted to see how children drew.

There is very little data of kids' pen sketches. Lisa Anthony has some kids' sketches using fingers, but fingers and pens are very different. You need a lot more dexterity and fine motor skills to draw with a pen. Other datasets exist for older children or adults [62, 86], but right now we have the only large digital pen sketch dataset for young children.

We currently have over 2000 sketches from 109 different children aged two through eight. For 70 of those children we also have 210 paper sketches linked by ID so that we know which are drawn by the same person. For 89 of the children we also have gender information. We started looking at what can we learn about how people perceive related to age. Table 21.1 shows the current dataset total.

Children draw differently as they age [65–69, 125]. Obviously, people draw better when they are older. It is no surprise their accuracy is better. They also draw faster, with higher pen speed as they get older. But the total time they spend

Table 21.1 Children's sketches dataset

Data set	Number of volunteers	Number of sketches
3–4 years old	62	1260
5–6 years old	22	467
7–8 years old	25	531
Adult	4	320
Total	113	2,578

drawing a given shape, letter, or number is about the same. But as the children get older, they draw the object much larger. So the older children take up more space, have a bigger stroke length. When young children copy a shape, they draw the shape very small. They use lots of small strokes. And they draw the shape very slowly. They use a lot more ink. They are focusing really hard. They are trying really hard. They are struggling. And that struggle shows up in these and other features in their writing.

What is also exciting is that the two features we talked about before, NDDE and DCR, tend to show up at a particular time in developmental processes. This is really cool, because it gives more credence to this idea that kids learn to plan what they are drawing. And that, as they start to perceive that they are turning a corner, it shows up in their drawing.

Randy Davis, my Ph.D. advisor, was looking at hooks and how they start to disappear in elderly persons with dementia [17, 70]. Randy gave a keynote talk at WIPTTE 2013 about these hooks. Hooks tell you where you have been and where you are going [56].

We have been looking at hooks and when they appear for young kids. Hooks are evidence of low-level planning. If you look at the end of the first line of an "A," you can see a hook showing where you are going to start the second line. Those hooks provide evidence of planning. Very young children and elderly persons with dementia do not have hooks in their drawing. Rather, after drawing the first line, the child will lift the pen, and then try to figure out what to draw next. Randy found that you lose hooks after the onset of dementia. We have found that children start to have hooks in their drawings somewhere between ages five to six. At age three, children don't have any hooks in their strokes. By age seven, essentially all of their strokes have hooks.

Accuracy of a drawing correlated very highly with the presence of hooks. All of these related to the general development of fine motor skills.

So when we look at all of these features together—NDDE, DCR, hooks, speed, accuracy, size—we can predict how old a child is by looking at their sketches. We also think this could be used to help identify kids who might not be on target with their fine motor skills.

When we provide age to the classifier, we are also able to predict gender. Girls tend to develop fine motor skills faster than boys do at young ages, with stroke density being the biggest distinguisher, along with DCR. There are various explanations for this.

We don't yet have enough data to confirm another hypothesis, but we had several Korean and Chinese children in our dataset, and they seemed to have higher motor skills than their counterpoints. I am currently calling this the *chopsticks hypothesis*, as I think that children who grow up using chopsticks have better fine motor skills at a young age. All of these differences flatten out as they age. By the time kids are 7 or 8 years of age, their writing is pretty indistinguishable from adults.

Before this, we did some work trying to identify adults by their handwriting. For that work, the strongest identifier was pressure and tilt [30, 31]. For each stroke drawn, we can tell whose stroke it is with 97.2% accuracy out of a set of two people, and 83.5% accuracy out of a group of 10 people. When you look at multiple strokes together, assuming all the strokes from a document came from the same person, that accuracy approaches 100%.

21.16 Shape Versus Text

Tracy Hammond: Another goal of ours was trying to distinguish shape versus text without trying to understand the actual text and without having any type of domain information. This is a common problem in diagrams, and can be quite difficult when the text labels are just one or two letters long and/or in blocked caps. But it seems possible, because when we look at diagrams even in a different language, we can distinguish the text from the shapes. The standard way people solve this problem is by running all of the strokes through a handwriting recognizer, such as the Microsoft handwriting recognizer. And the Microsoft recognizer is great at recognizing text, a very good job. But as soon as you put a rectangle around it, it freaks out. So we tried to figure out other methods to allow us to get a basic idea if something is text or not, so we can make more effective use of the existing handwriting recognizers.

Looking at text and shapes, we noticed that the shapes have less density than text. There seemed to be more information per letter than per shape. Shannon entropy seemed like an appropriate measure of what we were seeing. Shannon entropy is the expected value of the information contained in a message or flow of information.[1]

This hypothesis was true, and there seemed to be more entropy for text than for shapes, implying that there is more information in text than there is in general iconic shapes. General iconic shapes tend to be simple and repetitive, even resistors, which seem quite busy, have a lot less entropy than text, because they are quite repetitive.

So we developed a method to map strokes to entropy [12], reducing each stroke to a string of letters representing the angle between the points. We trained

[1] https://en.wikipedia.org/wiki/Entropy_(information_theory).

it on 162 diagrams consisting of military course of action diagrams and free-body diagrams containing 756 strokes—225 shape strokes and 531 text strokes. Strokes were grouped by time and space. Using 10-fold cross validation, we had over 92% accuracy using a simple threshold. But the nice thing about this method is that each stroke has a score, so we can use that to gain a confidence measure of if something is text or shape. Arctan approximates the function quite nicely.

Another explanation of Shannon entropy is how much something can be compressed. So we thought we would compare this to GZip, putting each stroke group into its own file and compress the file, and the curves were remarkably similar.

21.17 Activity

Tracy Hammond: Sketch recognition, when we focus on the path of the pen, is essentially activity recognition. So we have applied these same features looking at other types of human activity paths. Looking at our paths of 3D motion, just as with the 2D pen motion, we can recognize what we are doing, what we intend to do, how we perceive the world, and who we are.

We used these methods to provide insight to other spaces. Folami Alamudun, one of my Ph.D. students who just graduated, tried using these techniques on eye tracking data. After looking at eye tracking of users looking at magazines, gender differences, sexual preferences, and other preferences become quite apparent.

But we were curious what we could determine about a situation without the context of what they were actually looking at, rather looking at the path of the eyes on their own, using entropy (or specifically fractal analysis), shapelets (inspired by the Paleosketch primitive shape recognizer, and other sketch features). Specifically, we had 10 experts at various levels look at 100 mammogram x-rays and diagnose to the best of their ability. The fractal analysis showed predictive power to (1) determine the density of the breast they were looking at, (2) determine the pathology of the breast they were looking at, (3) determine the experience level of the viewer, and (4) determine what their diagnosis was going to be. Which when you combine 2 and 4, that implies we had predictive power of whether or not the person was going to be wrong [3]. **So without context of what someone is looking at, through the path of the eyes along, you can gain information about the person themselves, what someone is looking at, and how they perceive that object.**

Shapelets, looking at small repeated path types in people's eyes provided similar insight, but also showed some ability to differentiate individuals [2]. That made us wonder what other sketch features could be used to identify the person in order to authenticate them. To do this, we looked at saccades.

Saccades are interesting; they are the path your eye takes to get from one dwell spot to the next. Your eye does not take a straight path. That is because the muscles in your eyes travel horizontally and vertically only, as I mentioned earlier. These muscles are strengthened and impacted by all of the things you have looked at over the course of your life. So everyone's muscles are different, stronger in some places and weaker in others. The state of those muscles affect that path you take when traveling from one dwell spot to the next. Looking at those saccades and using standard sketch recognition features from Rubine, we were able to identify who looked at each mammogram x-ray with 89% accuracy, where 10% would be the accuracy for a random or a majority classifier.

Bringing this back to education, we can use this to evaluate expertise and predict if people are going to answer correctly or not—all without looking at their answer or the content. Imagine what we can do from an academic standpoint when you do include the content.

When we extend this back to sketching, what can we learn about how people look at a diagram when they are solving it? Can we teach people to look at diagrams more effectively?

We have tried some eye tracking with Mechanix [63]. We looked at trying to predict how hard the problem is. That way we could provide help if they are struggling. Both saccades and fixations allowed us to predict the problem difficulty. Pupil diameter also is helpful, but only in contexts with constant light, which makes it less useful in many scenarios, since controlling external light won't work well in standard classroom scenarios that would use a camera from the monitor.

We also found that the saccades and fixations can be used to predict what you are looking at, whether the sketch, the image, or the text of the question. Our eyes do very different things when we are looking at these different things.

As far as uses of these technologies. As cameras that support eye tracking are built into more and more screens and laptops, the saccades could be a method to ensure continued authentication. MOOCs often require users to take a picture and type a sentence at the start of a test, but the user could switch with another person right after that. Saccades might provide a way to verify the user continually through the exam; similarly pressure and tilt of the pen could be used to do the same thing. Continued facial recognition could also be used, but it feels more invasive for them to keep a video of me than just a tracking of my eye motions.

And finally, I will mention that with Daniel Goldberg, we have been extending these same features and algorithms to recognizing human motions using the accelerometer [33]. We started with musical instruments [76], office activities [78, 82], followed by exercise [11]—recognizing jumping jacks, squats, push ups, sit-ups, and others, using traditional sketch recognition methods to look at the path of your hands and your body doing these various exercises. More recently, with my MS student Josh Cherian, we have also been looking at teeth brushing, various cleansing activities for elderly [14]. We have been looking to see what interventions, perhaps using haptics [20–22, 91–94, 100] or other techniques [60, 98, 99], even completely changing the UI [19, 23, 26, 123], would be most helpful. And next, we aim to see how recognizing activities can be used to improve educational pursuits.

To finish, I want to reiterate, that all of the work I have talked about could not be done with the many students and collaborators I have worked with at Texas A&M. (See Fig. 21.18.)

Fig. 21.18 SRL, members, alumni, and collaborators

21.18 Questions

Andy van Dam: Lots of information. What can you say is the biggest implication of your research?

Tracy Hammond: We are learning that we can learn a lot more from people's sketches than just what they are sketching. This could have far reaching benefits in education—in how we can really start to predict what students' educational needs are and react to them. And this would be across any domains, even physics, games [79], music [10, 104], biology [117], searching [81, 128], logic [101], software [50, 53], language learning [103, 106–109, 113, 115, 116], digital citizenship [120], etc.

A lot of the work we see is really focused on replacing the pen and paper with something that is identical to the pen and paper—trying to get as close as possible to pen and paper. I think we need to be better than paper, not identical to paper, else we should just be using offline recognition methods [95–97]. People are resistant to change. If you want them to do something differently, we need them to have a reason for it.

This talk, I think, shows many ways in which digital ink can be more powerful than static ink.

Sketch recognition along with the underlying sketch forensics can make sketching on a computer so much more powerful than sketching on a piece of paper. We can go far beyond than just the traditional animations—making your sketch come alive—that we usually show with sketch recognition.

Computers come with hiccups. Sometimes there is data loss, although people also lose pieces of paper all the time. But for people to switch, it needs to be valuable in a different way than paper is. This talk attempted to show the unexpected values of digital pen input.

Digital pen input can help us learn more about the people interacting with our devices. It can teach us how people perceive the world around us.

But beyond the academic value of learning more about people, it can be used to provide systems that are interactive in a whole new way. Sketch recognition can be used to create *game changing* applications, so that the Tablet PC becomes the *game changer*, much more so than what it has in the past.

I argue that many of the applications thus far are trying to mimic paper and pen, but we can do so much more.

Andy van Dam: Please, use the microphone, Eric. Jonathan goes first.

Jonathan Grudin: Very interesting talk. I am interested in that circling back to AI thing that you started with. I was wondering how much it is involved. For example, when you discovered the over-correcting on the rectangles in corners, was that a hypothesis that you generated and then tested, or was that something that emerged from data that showed that.

Tracy Hammond: For that particular item, it was a little bit of both. We were focused on trying to distinguish arcs versus corners, since that continued to be a tricky thing, since standard corner finding algorithms would have trouble distinguishing the two. So, my Ph.D. student, Brandon Paulson—now doctor, and I stared at hundreds of direction graphs, and we noticed this pattern. Then we made an algorithm—the two equations we showed above—to test that pattern. We tested that pattern on hundreds of shapes, and found it to be significant, and a really effective way to identify corners versus shapes. So that part was hypothesis driven.

But then, these two features became standard features in our toolkit. I had a suspicion that it might be significant, so I asked my graduate student, Hong-Hoe Kim—or Ayden—also now a doctor, to include it in the features he was looking at in regards to children sketching. I did not expect it to have the effect that it did have. And it was only after seeing how it was something that evolved through the life did I make the connection that it was related to planning.

Jonathan Grudin: Was it you who noticed it? Or who did that? Did the AI sort of say: "Here is the pattern that you may want to look at?"

Tracy Hammond: Brandon Paulson noticed its effect in distinguishing corners. It was me who thought that it might play a role in child development. But it was the computer that actually pulled it out. Initially, we were just trying to predict age and gender from a set of features, and we included those two features in our feature set. We wanted to know which features were having the greatest effect. So we applied subset selection, and that is when the computer highlighted these features. It was then that we went back and looked at these features independently to see how this feature evolved over the span of child development.

It was only after it started showing up in multiple places did I start to try to figure out *why* it was occurring. And it was then that I came up with the analogy of over steering, and about preparation—or low-level planning—of turning a corner.

We are now looking to see how this feature may play a part in other domains.

Eric Hamilton: Thank you, Tracy. It was a great talk. You covered a lot of ground. You made a point early on in the talk, and I should have written down the question, but it grabbed me it was the first time you used the word "feedback" in your talk. Much of what you are doing is not particularly related to feedback to a user, but a whole lot is. If you remember what that reference was to feedback, could you clarify it a bit and if you don't, can you give us a few comments about your theory of feedback as it relates to sketch recognition, and how that might apply to change in learning, especially in the context of your comments earlier about what we do and that to change that we need something better. Feedback is a critical element of that, and can you comment on that briefly?

Tracy Hammond: I may not have used the word "feedback" very often in this talk. I may have used words like "the computer reacts" or "interaction." All of that is feedback. The purpose of this talk was to show you ways in which we can be more creative in our feedback, going beyond simple animations of a sketch.

Feedback is crazy important. A lot of what has motivated this work is our work with sketch systems in the classroom. With Mechanix we see how crucial the right feedback is, as I am sure that Randy Brooks, who has used the software in his high school classroom can attest to. Mechanix uses scaffolding in its feedback process, which means it has the ability to do more and less hand holding of students as they start to excel in the topic.

Seth Polsley's work that he presented at WIPTTE last year really focuses on how the feedback affects what the user is doing and how to straddle the gap between too much feedback—which encourages kids to game the system, and too little feedback— which causes kids to get frustrated and stuck. We have spent many hours discussing the right feedback for Mechanix, and many of our other systems. We still have far to go.

The word "feedback" may not have appeared often in this talk, but all of the research I have showed you today can provide information about students that I hope can allow educational developers provide the right feedback.

When we were trying to answer the question, "How can we determine how much trouble someone is having with a problem," it was not just to know the difficulty of a problem—although ETS (Educational Testing Service) may be interested in that—rather, it was to know that we could respond to the student in a way that is most effective for their learning. Their number of submissions, their eye motions, their sketch marks, all betray information about how they are doing at solving the problem. Our feedback needs to be aligned with the state of the student. We can use this data to make sure our feedback is at the right spot on a trajectory.

A huge part of my work, which I did not have time to talk about today, is about interventions, which is feedback that attempts to change the trajectory you are currently on. But to do that effectively, you have to first figure out what trajectory a person is on, which was the focus of this talk. Feedback needs to be personalized, context-dependant, and appropriate in order for it to be effective. Feedback needs to be user-dependant, and at the right time and place. The more information we have, the better—How much are they struggling? How are they trying to solve the problem? What experience do they have? All of these things can help us provide better feedback. Feedback is hugely important.

Nickolas Jackiw from Audience: I was interested in your talk. I was not tracking the dates of the projects you presented, but the efforts discussed in the first half of your talk involved recognizing sketches. Then, there was a shift, to projects about recognizing sketchers—recognizing their gender, guessing their race, estimating their age, ultimately revealing their individual identity. "What is drawn?" became "Who is drawing?" That seems to be a very different problem than the first, and one that is easy to be perturbed by in a sense of privacy concerns. There goes not only the criminally poisonous poison-pen letter, but also the unsolicited letter from a secret admirer. "Right, that was you!" *[pointing to imaginary person]*. Confidential voting records? "Sorry, our algorithms say—these are *yours*!" So I wondered in your work if there was any way the second effort circles back and informs the first. If we know who you are, do we become better at predicting your sketches and recognizing them with that personal information being productive to recognizing sketches, rather than to general surveillance?

Tracy Hammond: You are right, I once gave a keynote to a forensic association! And to your question, yes, absolutely.

I mentioned three types of recognition techniques early in the talk, and of the three, gesture recognition is the most user dependent. One of the difficulties of gesture recognition early on was that we either had to train the user or train the system, since the path has to be the same every time. And if we swap out users, we have to retrain, or alternatively save different models for the different users. By knowing something about the sketcher, we can use that to use what we have learned about how that person sketches to better recognize their sketchers.

But this can go beyond just recognizing a particular person to improve recognition. We can also learn some things about people to create clusters of user groups. For instance, novices in many cases sketch differently than experts, so by learning which group they are in, we can not only use that information to affect the feedback given to them, but also to better recognize what they are intending to sketch, improving and speeding up recognition.

When we are able to use these path-specific methods, we have found that with gesture recognition, we can even tell you what you are going to draw before you are completely done with your drawing [73]. By watching your path, we can predict where you are going to end up. And some of these can still be done on a user independent basis, but if we have profiles of how a particular person or group of people sketch something, we can improve recognition.

For instance, just looking at drawing simple straight lines, experts will have a smoother speed curve—with an increase than a decrease, so for those users it is easier for us to figure out where they are going to complete their line, so we can start recognition earlier and with higher confidence for those users. This is somewhat related to Fitts' Law, which predicts the time it takes to move to a target area based on the distance to the target and the width of the target.

So, yes, these forensic features that help us know more about the user can help improve recognition immensely.

And I believe this is also interwoven with feedback. The more you understand about the user, the better you understand the sketches, the better your feedback will be, and the response to the feedback by the user will in turn help us better understand the user, and so on, we ideally continue the cycle.

And of course, there are privacy issues that come with all of this.

Andy van Dam: That was a terrific talk, Tracy. Thank you very much.

21.19 Keynote Bio

Dr. Tracy Hammond Director of the Sketch Recognition Lab and Professor in the Department of Computer Science and Engineering at Texas A&M University, Dr. Hammond is an international leader in sketch recognition and human-computer interaction research. Dr. Hammond's publications on the subjects are widely cited and have well over fifteen hundred citations, with Dr. Hammond having an h-index of 19, an h10-index of 35, and four papers with over 100 citations each. Her sketch recognition research has been funded by NSF, DARPA, Google, Microsoft, and many others, totaling over 3.6 million dollars in peer reviewed funding. She holds a Ph.D. in Computer Science and FTO (Finance Technology Option) from M.I.T., and four degrees from Columbia University: an M.S. in Anthropology, an M.S. in Computer Science, a B.A. in Mathematics, and a B.S. in Applied Mathematics. Prior to joining the TAMU CSE faculty Dr. Hammond taught for 5 years at Columbia University and was a telecom analyst for 4 years at Goldman Sachs. Dr. Hammond is the 2011–2012 recipient of the Charles H. Barclay, Jr. '45 Faculty Fellow Award. The Barclay Award is given to professors and associate professors who have been nominated for their overall contributions to the Engineering Program through classroom instruction, scholarly activities, and professional service.

Acknowledgements None of this research is possible without the support of funding agencies that have supported this work, including NSF, DARPA, NIH, TDADS, ONR, Google, Microsoft, Rockwell Collins, TEES, and TAMU as well as the fantastic work of the student members of the Sketch Recognition Lab. Special thanks to Dr. Anna Stepanova, and Dr. Jan Hammond for their help in this transcription as well as Lisa Manokofsky for her help in providing the videos for translation.

References

1. Adler, A.D.: Midos: multimodal interactive dialogue system. Ph.D. thesis, Massachusetts Institute of Technology (2009)
2. Alamudun, F., Yoon, H.J., Hammond, T., Hudson, K., Morin-Ducote, G., Tourassi, G.: Shapelet analysis of pupil dilation for modeling visuo-cognitive behavior in screening mammography. In: Proceedings of the SPIE. SPIE, vol. 9787, pp. 97870M01–97870M13 (2016). doi:10.1117/12.2217670. http://dx.doi.org/10.1117/12.2217670
3. Alamudun, F., Yoon, H.J., Hudson, K.B., Morin-Ducote, G., Hammond, T., Tourassi, G.D.: Fractal analysis of visual search activity for mass detection during mammographic screening. Med. Phys. **44**, 832–846 (2017)
4. Alvarado, C.J.: A natural sketching environment: bringing the computer into early stages of mechanical design. Ph.D. thesis, Massachusetts Institute of Technology (2000)
5. Atilola, O., Field, M., McTigue, E., Hammond, T., Linsey, J.: Evaluation of a natural sketch interface for truss FBDs and analysis. In: Proceedings of the 2011 Frontiers in Education Conference, FIE'11, pp. S2E–1–1–S2E–6. IEEE Computer Society, Washington, DC (2011). doi:10.1109/FIE.2011.6142959. http://dx.doi.org/10.1109/FIE.2011.6142959
6. Atilola, O., Field, M., McTigue, E., Hammond, T., Linsey, J.: Mechanix: a sketch recognition truss tutoring system. In: American Society of Mechanical Engineers (ASME) 2011

International Design Engineering Technical Conferences and Computers and Information in Engineering Conference, Volume 7: 5th International Conference on Micro- and Nanosystems; 8th International Conference on Design and Design Education; 21st Reliability, Stress Analysis, and Failure Prevention Conference, vol. 7, pp. 645–654. ASME, Washington, DC (2011)
7. Atilola, O., McTigue, E.M., Hammond, T., Linsey, J.: Mechanix: evaluating the effectiveness of a sketch recognition truss tutoring program against other truss programs. In: 120th American Society for Engineering Education Annual Conference & Exposition (ASEE), June 23–26. ASEE, Atlanta (2013)
8. Atilola, O., Valentine, S., Kim, H.H., Turner, D., McTigue, E., Hammond, T., Linsey, J.: Mechanix: a natural sketch interface tool for teaching truss analysis and free-body diagrams. Artif. Intell. Eng. Des. Anal. Manuf. **28**, 169–192 (2014). doi:10.1017/S0890060414000079. http://journals.cambridge.org/article_S0890060414000079
9. Atilola, O., Vides, F., Mctigue, E.M., Linsey, J.S., Hammond, T.A.: Automatic identification of student misconceptions and errors for truss analysis. In: 119th American Society for Engineering Education Annual Conference & Exposition (ASEE), 10–13 June 2012. ASEE, San Antonio (2012)
10. Barreto, L., Taele, P., Hammond, T.: A stylus-driven intelligent tutoring system for music education instruction. In: Hammond, T., Valentine, S., Adler, A. (eds.) Revolutionizing Education with Digital Ink: The Impact of Pen and Touch Technology on Education, pp. 141–161. Springer, Cham (2016). doi:10.1007/978-3-319-31193-7_10. http://dx.doi.org/10.1007/978-3-319-31193-7_10
11. Bartley, J., Forsyth, J., Pendse, P., Xin, D., Brown, G., Hagseth, P., Agrawal, A., Goldberg, D.W., Hammond, T.: World of workout: a contextual mobile RPG to encourage long term fitness. In: Proceedings of the Second ACM SIGSPATIAL International Workshop on the Use of GIS in Public Health, HealthGIS'13, pp. 60–67. ACM, New York (2013). doi:10.1145/2535708.2535718. http://doi.acm.org/10.1145/2535708.2535718
12. Bhat, A., Hammond, T.: Using entropy to distinguish shape versus text in hand-drawn diagrams. In: Proceedings of the 21st International Jont Conference on Artifical Intelligence, IJCAI'09, pp. 1395–1400. Morgan Kaufmann Publishers Inc., San Francisco (2009). http://dl.acm.org/citation.cfm?id=1661445.1661669
13. Brooks, R., Hammond, T., Koh, J.I.: Score improvement distribution when using sketch recognition software (mechanix) as a tutor: assessment of high school classroom pilot. In: 10th Conference on Pen and Touch Technology in Education. CPTTE 2016. CPTTE (2016)
14. Cherian, J., Rajanna, V., Goldberg, D., Hammond, T.: Did you remember to brush? A noninvasive wearable approach to recognizing brushing teeth for elderly care. In: 11th EAI International Conference on Pervasive Computing Technologies for Healthcare. ICDC (2017)
15. Choi, H., Hammond, T.: Sketch recognition based on manifold learning. In: Proceedings of the 23rd National Conference on Artificial Intelligence, AAAI'08, vol. 3, pp. 1786–1787. AAAI Press (2008). http://dl.acm.org/citation.cfm?id=1620270.1620353
16. Choi, H., Paulson, B., Hammond, T.: Gesture recognition based on manifold learning. In: Proceedings of the 2008 Joint IAPR International Workshop on Structural, Syntactic, and Statistical Pattern Recognition, SSPR & SPR'08, pp. 247–256 (2014). Springer, Berlin/Heidelberg (2008). doi:10.1007/978-3-540-89689-0_29. http://dx.doi.org/10.1007/978-3-540-89689-0_29
17. Cohen, J., Penney, D., Davis, R., Libon, D., Swenson, R., Ajilore, O., ... Lamar, M.: Digital clock drawing: Differentiating "Thinking" versus "Doing" in younger and older adults with depression. J. Int. Neuropsychol. Soc. **20**(9), 920–928 (2014). doi:10.1017/S1355617714000757
18. Corey, P., Hammond, T.: Gladder: combining gesture and geometric sketch recognition. In: Proceedings of the 23rd National Conference on Artificial Intelligence, AAAI'08, vol. 3, pp. 1788–1789. AAAI Press (2008). http://dl.acm.org/citation.cfm?id=1620270.1620354

19. Cummings, D., Fymat, S., Hammond, T.: Sketch-based interface for interaction with unmanned air vehicles. In: CHI'12 Extended Abstracts on Human Factors in Computing Systems, CHI EA'12, pp. 1511–1516. ACM, New York (2012). doi:10.1145/2212776.2223664. http://doi.acm.org/10.1145/2212776.2223664
20. Cummings, D., Lucchese, G., Prasad, M., Aikens, C., Ho, J., Hammond, T.: Geotrooper: a mobile location-aware system for team coordination. In: Proceedings of the 13th International Conference of the NZ Chapter of the ACM's Special Interest Group on Human-Computer Interaction, CHINZ'12, pp. 102–102. ACM, New York (2012). doi:10.1145/2379256.2379286. http://doi.acm.org/10.1145/2379256.2379286
21. Cummings, D., Lucchese, G., Prasad, M., Aikens, C., Ho, J., Hammond, T.: Haptic and ar interface for paratrooper coordination. In: Proceedings of the 13th International Conference of the NZ Chapter of the ACM's Special Interest Group on Human-Computer Interaction, CHINZ'12, pp. 52–55. ACM, New York (2012). doi:10.1145/2379256.2379265. http://doi.acm.org/10.1145/2379256.2379265
22. Cummings, D., Prasad, M., Lucchese, G., Aikens, C., Hammond, T.A.: Multi-modal location-aware system for paratrooper team coordination. In: CHI'13 Extended Abstracts on Human Factors in Computing Systems, CHI EA'13, pp. 2385–2388. ACM, New York (2013). doi:10.1145/2468356.2468779. http://doi.acm.org/10.1145/2468356.2468779
23. Cummmings, D., Fymat, S., Hammond, T.: Reddog: A smart sketch interface for autonomous aerial systems. In: Proceedings of the International Symposium on Sketch-Based Interfaces and Modeling, SBIM'12, pp. 21–28. Eurographics Association, Aire-la-Ville (2012). http://dl.acm.org/citation.cfm?id=2331067.2331071
24. Cummmings, D., Vides, F., Hammond, T.: I don't believe my eyes! Geometric sketch recognition for a computer art tutorial. In: Proceedings of the International Symposium on Sketch-Based Interfaces and Modeling, SBIM'12, pp. 97–106. Eurographics Association, Aire-la-Ville (2012). http://dl.acm.org/citation.cfm?id=2331067.2331082
25. Dahmen, K., Hammond, T.: Distinguishing between sketched scribble look alikes. In: Proceedings of the 23rd National Conference on Artificial Intelligence, AAAI'08, vol. 3, pp. 1790–1791. AAAI Press (2008). http://dl.acm.org/citation.cfm?id=1620270.1620355
26. Damaraju, S., Seo, J.H., Hammond, T., Kerne, A.: Multi-tap sliders: advancing touch interaction for parameter adjustment. In: Proceedings of the 2013 International Conference on Intelligent User Interfaces, IUI'13, pp. 445–452. ACM, New York (2013). doi:10.1145/2449396.2449453. http://doi.acm.org/10.1145/2449396.2449453
27. David, J., Eoff, B., Hammond, T.: Coske-an exploration in collaborative sketching. In: Computer Supported Cooperative Work Posters (CSCW), pp. 471–472. Savannah (2010)
28. Dixon, D., Prasad, M., Hammond, T.: iCanDraw?: a methodology for using assistive sketch recognition to improve a user's drawing ability (2009)
29. Dixon, D., Prasad, M., Hammond, T.: icandraw: using sketch recognition and corrective feedback to assist a user in drawing human faces. In: Proceedings of the SIGCHI Conference on Human Factors in Computing Systems, CHI'10, pp. 897–906. ACM, New York (2010). doi:10.1145/1753326.1753459. http://doi.acm.org/10.1145/1753326.1753459
30. Eoff, B.D., Hammond, T.: User identification by means of sketched stroke features. In: Proceedings of the 23rd National Conference on Artificial Intelligence, AAAI'08, vol. 3, pp. 1794–1795. AAAI Press (2008). http://dl.acm.org/citation.cfm?id=1620270.1620357
31. Eoff, B.D., Hammond, T.: Who dotted that 'i'? Context free user differentiation through pressure and tilt pen data. In: Proceedings of Graphics Interface 2009, GI'09, pp. 149–156. Canadian Information Processing Society, Toronto (2009). http://dl.acm.org/citation.cfm?id=1555880.1555916
32. Field, M., Valentine, S., Linsey, J., Hammond, T.: Sketch recognition algorithms for comparing complex and unpredictable shapes. In: Proceedings of the Twenty-Second International Joint Conference on Artificial Intelligence, IJCAI'11, vol. 3, pp. 2436–2441. AAAI Press (2011). doi:10.5591/978-1-57735-516-8/IJCAI11-406. http://dx.doi.org/10.5591/978-1-57735-516-8/IJCAI11-406

33. Goldberg, D.W., Cockburn, M.G., Hammond, T.A., Jacquez, G.M., Janies, D., Knoblock, C., Kuhn, W., Pultar, E., Raubal, M.: Envisioning a future for a spatial-health cybergis marketplace. In: Proceedings of the Second ACM SIGSPATIAL International Workshop on the Use of GIS in Public Health, HealthGIS'13, pp. 27–30. ACM, New York (2013). doi:10.1145/2535708.2535716. http://doi.acm.org/10.1145/2535708.2535716
34. Green, M.G., Caldwell, B.W., Helms, M., Linsey, J.S., Hammond, T.A.: Using natural sketch recognition software to provide instant feedback on statics homework (truss free body diagrams): assessment of a classroom pilot. In: 2015 ASEE Annual Conference and Exposition. Seattle (2015). doi:10.18260/p.25007. https://peer.asee.org/25007
35. Hammond, T.: Automatically generating sketch interfaces from shape descriptions. In: Proceedings of the 4th Annual MIT Student Oxygen Workshop. MIT (2004)
36. Hammond, T.: Enabling instructors to develop sketch recognition applications for the classroom. In: 37th Annual Frontiers In Education Conference – Global Engineering: Knowledge Without Borders, Opportunities Without Passports, FIE'07, pp. S3J–11–S3J–16 (2007). doi:10.1109/FIE.2007.4417930
37. Hammond, T.: Ladder: a perceptually-based language to simplify sketch recognition user interface development. PhD Doctoral Dissertation, Massachusetts Institute of Technology (MIT), Cambridge (2007). Advisor: Randall Davis, 495 pp.
38. Hammond, T.: Simplifying sketch recognition UI development. In: Grace Hopper Celebration of Women in Computing. GHC, Orlando (2007)
39. Hammond, T.: Workshop – integrating sketch recognition technologies into your classroom. In: 38th Annual Frontiers in Education Conference, FIE 2008, pp. W2C–1 (2008). doi:10.1109/FIE.2008.4720505
40. Hammond, T.: Dialectical creativity: sketch-negate-create. In: Studying Visual and Spatial Reasoning for Design Creativity, pp. 91–108. Springer Netherlands, Dordrecht (2015). doi:10.1007/978-94-017-9297-4_6. http://dx.doi.org/10.1007/978-94-017-9297-4_6
41. Hammond, T., Davis, R.: Tahuti: a geometrical sketch recognition system for UML class diagrams. In: Technical Report SS-02-08: Papers from the 2002 Association for the Advancement of Artificial Intelligence (AAAI) Spring Symposium on Sketch Understanding. AAAI, Menlo Park (2002)
42. Hammond, T., Davis, R.: Ladder: a language to describe drawing, display, and editing in sketch recognition. In: Proceedings of the 18th International Joint Conference on Artificial Intelligence, IJCAI'03, pp. 461–467. Morgan Kaufmann Publishers Inc., San Francisco (2003). http://dl.acm.org/citation.cfm?id=1630659.1630728
43. Hammond, T., Davis, R.: Automatically transforming symbolic shape descriptions for use in sketch recognition. In: Proceedings of the 19th National Conference on Artificial Intelligence, AAAI'04, pp. 450–456. AAAI Press (2004). http://dl.acm.org/citation.cfm?id=1597148.1597222
44. Hammond, T., Davis, R.: Debugging shape definitions for use in sketch recognition. In: MIT Lab Abstract. MIT, Cambridge (2004)
45. Hammond, T., Davis, R.: Shady: a shape description debugger for use in sketch recognition. In: AAAI Fall Symposium on Making Pen-Based Interaction Intelligent and Natural (AAAI). AAAI, Arlington (2004)
46. Hammond, T., Davis, R.: Ladder, a sketching language for user interface developers. Comput. Graph. **29**(4), 518–532 (2005)
47. Hammond, T., Davis, R.: Interactive learning of structural shape descriptions from automatically generated near-miss examples. In: Proceedings of the 11th International Conference on Intelligent User Interfaces, IUI'06, pp. 210–217. ACM, New York (2006). doi:10.1145/1111449.1111495. http://doi.acm.org/10.1145/1111449.1111495
48. Hammond, T., Davis, R.: Creating the perception-based ladder sketch recognition language. In: Proceedings of the 8th ACM Conference on Designing Interactive Systems, DIS'10, pp. 141–150. ACM, New York (2010). doi:10.1145/1858171.1858197. http://doi.acm.org/10.1145/1858171.1858197

49. Hammond, T., Eoff, B., Paulson, B., Wolin, A., Dahmen, K., Johnston, J., Rajan, P.: Free-sketch recognition: putting the chi in sketching. In: CHI'08 Extended Abstracts on Human Factors in Computing Systems, CHI EA'08, pp. 3027–3032. ACM, New York (2008). doi:10.1145/1358628.1358802. http://doi.acm.org/10.1145/1358628.1358802
50. Hammond, T., Gajos, K., Davis, R., Shrobe, H.: An agent-based system for capturing and indexing software design meetings. In: Proceedings of International Workshop on Agents In Design, WAID. MIT, Cambridge (2002)
51. Hammond, T., Logsdon, D., Paulson, B., Johnston, J., Peschel, J., Wolin, A., Taele, P.: A sketch recognition system for recognizing free-hand course of action diagrams. In: Innovative Applications of Artificial Intelligence, pp. 1781–1786. Atlanta (2010). http://www.aaai.org/ocs/index.php/IAAI/IAAI10/paper/view/1581
52. Hammond, T., Logsdon, D., Peschel, J., Johnston, J., Taele, P., Wolin, A., Paulson, B.: A sketch recognition interface that recognizes hundreds of shapes in course-of-action diagrams. In: CHI'10 Extended Abstracts on Human Factors in Computing Systems, CHI EA'10, pp. 4213–4218. ACM, New York (2010). doi:10.1145/1753846.1754128. http://doi.acm.org/10.1145/1753846.1754128
53. Hammond, T., O'Sullivan, B.: Recognizing free-form hand-sketched constraint network diagrams by combining geometry and context. In: Proceedings of the Eurographics Ireland, pp. 67–74. Eurographics, Dublin (2007)
54. Hammond, T., Paulson, B.: Recognizing sketched multistroke primitives. ACM Trans. Interact. Intell. Syst. **1**(1), 4:1–4:34 (2011). doi:10.1145/2030365.2030369. http://doi.acm.org/10.1145/2030365.2030369
55. Hammond, T., Prasad, M., Dixon, D.: Art 101: learning to draw through sketch recognition. In: Proceedings of the 10th International Conference on Smart Graphics, SG'10, pp. 277–280. Springer, Berlin/Heidelberg (2010). http://dl.acm.org/citation.cfm?id=1894345.1894383
56. Hammond, T., Valentine, S., Adler, A., Payton, M.: The Impact of Pen and Touch Technology on Education, 1st edn. Springer Publishing Company, Incorporated (2015)
57. Hammond, T.A., Davis, R.: Recognizing interspersed sketches quickly. In: Proceedings of Graphics Interface 2009, GI'09, pp. 157–166. Canadian Information Processing Society, Toronto (2009). http://dl.acm.org/citation.cfm?id=1555880.1555917
58. Herot, C.F.: Graphical input through machine recognition of sketches. In: Proceedings of the 3rd Annual Conference on Computer Graphics and Interactive Techniques, SIGGRAPH'76, pp. 97–102. ACM, New York (1976). doi:10.1145/563274.563294. http://doi.acm.org/10.1145/563274.563294
59. Hilton, E.C., WilliFord, B., Li, W., McTigue, E., Hammond, T., Linsey, J.: Consistently evaluating sketching ability in engineering curriculum. In: 4th ICDC, International Conference on Design and creativity. ICDC (2016)
60. Hong, Y., Vollmer Dahlke, D., Ory, M., Goldberg, D., Cargill, J., Kellstedt, D., Pulczinski, J., Hammond, T., Hernandez, E.: Development of icanfit: a mobile device application to promote physical activity and access to health information among older cancer survivors. In: 142nd APHA Annual Meeting and Exposition, 15–19 Nov 2014. APHA (2014)
61. Johnston, J., Hammond, T.: Computing confidence values for geometric constraints for use in sketch recognition. In: Proceedings of the Seventh Sketch-Based Interfaces and Modeling Symposium, SBIM'10, pp. 71–78. Eurographics Association, Aire-la-Ville (2010). http://dl.acm.org/citation.cfm?id=1923363.1923376
62. Kaster, B.L., Jacobson, E.R., Hammond, T.A.: Sssousa: automatically generating secure and searchable data collection studies. In: International Workshop on Visual Languages and Computing. VLC/DMS, Redwood City (2009)
63. Kaul, P., Rajanna, V., Hammond, T.: Exploring users' perceived activities in a sketch-based intelligent tutoring system through eye movement data. In: ACM Symposium on Applied Perception (SAP'16). SAP (2016)
64. Kebodeaux, K., Field, M., Hammond, T.: Defining precise measurements with sketched annotations. In: Proceedings of the Eighth Eurographics Symposium on Sketch-Based Interfaces and Modeling, SBIM'11, pp. 79–86. ACM, New York (2011). doi:10.1145/2021164.2021179. http://doi.acm.org/10.1145/2021164.2021179

65. Kim, H.H., Taele, P., Seo, J., Jeffrey, L., Hammond, T.: A novel sketch-based interface for improving children's fine motor skills and school readiness. In: Proceedings of the International Symposium on Sketch-Based Interfaces and Modeling, SAP, pp. 1–10, 2016
66. Kim, H.H., Taele, P., Valentine, S., Hammond, T.: Developing intelligent sketch-based applications for children's fine motor sketching skill development. In: 2014 International Conference on Intelligent User Interfaces (IUI) Workshop on Sketch: Pen and Touch Recognition. IUI, ACM, Haifa (2014)
67. Kim, H.H., Taele, P., Valentine, S., McTigue, E., Hammond, T.: Kimchi: a sketch-based developmental skill classifier to enhance pen-driven educational interfaces for children. In: Proceedings of the International Symposium on Sketch-Based Interfaces and Modeling, SBIM'13, pp. 33–42. ACM, New York (2013). doi:10.1145/2487381.2487389. http://doi.acm.org/10.1145/2487381.2487389
68. Kim, H.H., Valentine, S., Taele, P., Hammond, T.: Easysketch: a sketch-based fine motor skill recognizing educational interface for children emerging technology research strand. In: Workshop on the Impact of Pen & Touch Technology on Education (WIPTTE). WIPTTE, College Station (2014)
69. Kim, H.H., Valentine, S., Taele, P., Hammond, T.: Easysketch: a sketch-based fine motor skill recognizing educational interface for children emerging technology research strand. In: Hammond, T., Valentine, S., Adler, A., Payton, M. (eds.) The Impact of Pen and Touch Technology on Education, 1st edn., pp. 35–46. Springer Publishing Company, Incorporated (2015)
70. Lara-Garduno, R., Leslie, N., Hammond, T.: Smartstrokes: Digitizing paper-based neuropsychological tests. In: Hammond, T., Valentine, S., Adler, A. (eds.) Revolutionizing Education with Digital Ink: The Impact of Pen and Touch Technology on Education, pp. 163–175. Springer, Cham (2016). doi:10.1007/978-3-319-31193-7_11. http://dx.doi.org/10.1007/978-3-319-31193-7_11
71. Li, W., Hammond, T.: Recognizing text through sound alone. In: AAAI Conference on Artificial Intelligence (2011). http://www.aaai.org/ocs/index.php/AAAI/AAAI11/paper/view/3791
72. Li, W., Hammond, T.: Using scribble gestures to enhance editing behaviors of sketch recognition systems. In: CHI'12 Extended Abstracts on Human Factors in Computing Systems, CHI EA'12, pp. 2213–2218. ACM, New York (2012). doi:10.1145/2212776.2223778. http://doi.acm.org/10.1145/2212776.2223778
73. Lucchese, G., Field, M., Ho, J., Gutierrez-Osuna, R., Hammond, T.: Gesturecommander: continuous touch-based gesture prediction. In: CHI'12 Extended Abstracts on Human Factors in Computing Systems, CHI EA'12, pp. 1925–1930. ACM, New York (2012). doi:10.1145/2212776.2223730. http://doi.acm.org/10.1145/2212776.2223730
74. Lupfer, N., Field, M., Kerne, A., Hammond, T.: sketchy: morphing user sketches for artistic assistance. In: Proceedings of the 2011 ACM International conference on intelligent user interface. ACM, Palo Alto (2011)
75. MacKenzie, I.S., Zhang, S.X.: The immediate usability of graffiti. In: Graphics Interface, vol. 97, pp. 129–137, 1997
76. Miller, J., Hammond, T.: Wiiolin: a virtual instrument using the Wii remote. In: Proceedings of the 2010 Conference on New Interfaces for Musical Expression (NIME), pp. 497–500. Sydney (2010)
77. Nelligan, T., Polsley, S., Ray, J., Helms, M., Linsey, J., Hammond, T.: Mechanix: a sketch-based educational interface. In: Proceedings of the 20th International Conference on Intelligent User Interfaces Companion, IUI Companion'15, pp. 53–56. ACM, New York (2015). doi:10.1145/2732158.2732194. http://doi.acm.org/10.1145/2732158.2732194
78. Paulson, B., Cummings, D., Hammond, T.: Object interaction detection using hand posture cues in an office setting. Int. J. Hum. Comput. Stud. **69**(1–2), 19–29 (2011). doi:10.1016/j.ijhcs.2010.09.003. http://dx.doi.org/10.1016/j.ijhcs.2010.09.003
79. Paulson, B., Eoff, B., Wolin, A., Johnston, J., Hammond, T.: Sketch-based educational games: drawing kids away from traditional interfaces. In: Proceedings of the 7th International Conference on Interaction Design and Children, IDC'08, pp. 133–136. ACM, New York (2008). doi:10.1145/1463689.1463739. http://doi.acm.org/10.1145/1463689.1463739

80. Paulson, B., Hammond, T.: A system for recognizing and beautifying low-level sketch shapes using NDDE and DCR. In: ACM Symposium on User Interface Software and Technology (UIST). ACM, Newport (2007)
81. Paulson, B., Hammond, T.: MARQS: retrieving sketches learned from a single example using a dual-classifier. J. Multimodal User Interface **2**(1), 3–11 (2008). doi:10.1007/s12193-008-0006-0. http://dx.doi.org/10.1007/s12193-008-0006-0
82. Paulson, B., Hammond, T.: Office activity recognition using hand posture cues. In: Proceedings of the 22nd British HCI Group Annual Conference on People and Computers: Culture, Creativity, Interaction, BCS-HCI'08, vol. 2, pp. 75–78. British Computer Society, Swinton (2008). http://dl.acm.org/citation.cfm?id=1531826.1531845
83. Paulson, B., Hammond, T.: Paleosketch: accurate primitive sketch recognition and beautification. In: Proceedings of the 13th International Conference on Intelligent User Interfaces, IUI'08, pp. 1–10. ACM, New York (2008). doi:10.1145/1378773.1378775. http://doi.acm.org/10.1145/1378773.1378775
84. Paulson, B., Hammond, T.: Towards a framework for truly natural low-level sketch recognition. In: Proceedings of the Workshop on Sketch Recognition at the 14th International Conference of Intelligent User Interfaces (IUI). ACM, Sanibel (2009)
85. Paulson, B., Rajan, P., Davalos, P., Gutierrez-Osuna, R., Hammond, T.: What!?! no rubine features? using geometric-based features to produce normalized confidence values for sketch recognition. In: HCC Workshop: Sketch Tools for Diagramming (VL/HCC), pp. 57–63. VL/HCC, Herrsching am Ammersee (2008)
86. Paulson, B., Wolin, A., Johnston, J., Hammond, T.: SOUSA: sketch-based online user study applet. In: Proceedings of the Fifth Eurographics Conference on Sketch-Based Interfaces and Modeling, SBM'08, pp. 81–88. Eurographics Association, Aire-la-Ville (2008). doi:10.2312/SBM/SBM08/081-088. http://dx.doi.org/10.2312/SBM/SBM08/081-088
87. Peschel, J.M., Hammond, T.A.: Strat: a sketched-truss recognition and analysis tool. In: 2008 International Workshop on Visual Languages and Computing (VLC) at the 14th International Conference on distributed Multimedia Systems (DMS), pp. 282–287. Knowledge Systems Insistute, Boston (2008)
88. Peschel, J.M., Paulson, B., Hammond, T.: A surfaceless pen-based interface. In: Proceedings of the Seventh ACM Conference on Creativity and Cognition, C&C'09, pp. 433–434. ACM, New York (2009). doi:10.1145/1640233.1640338. http://doi.acm.org/10.1145/1640233.1640338
89. Polsley, S., Ray, J., Nelligan, T., Helms, M., Linsey, J., Hammond, T.: Leveraging trends in student interaction to enhance the effectiveness of sketch-based educational software. In: Hammond, T., Valentine, S., Adler, A. (eds.) Revolutionizing Education with Digital Ink: The Impact of Pen and Touch Technology on Education, pp. 103–114. Springer, Cham (2016). doi:10.1007/978-3-319-31193-7_7. http://dx.doi.org/10.1007/978-3-319-31193-7_7
90. Prasad, M., Hammond, T.: Observational study on teaching artifacts created using tablet pc. In: CHI'12 Extended Abstracts on Human Factors in Computing Systems, CHI EA'12, pp. 301–316. ACM, New York (2012). doi:10.1145/2212776.2212809. http://doi.acm.org/10.1145/2212776.2212809
91. Prasad, M., Russell, M., Hammond, T.A.: Designing vibrotactile codes to communicate verb phrases. ACM Trans. Multimed. Comput. Commun. Appl. **11**(1s), 11:1–11:21 (2014). doi:10.1145/2637289. http://doi.acm.org/10.1145/2637289
92. Prasad, M., Russell, M.I., Hammond, T.A.: A user centric model to design tactile codes with shapes and waveforms. In: Haptics Symposium (HAPTICS), pp. 597–602. IEEE (2014). doi:10.1109/HAPTICS.2014.6775523
93. Prasad, M., Taele, P., Goldberg, D., Hammond, T.A.: Haptimoto: turn-by-turn haptic route guidance interface for motorcyclists. In: Proceedings of the 32Nd Annual ACM Conference on Human Factors in Computing Systems, CHI'14, pp. 3597–3606. ACM, New York (2014). doi:10.1145/2556288.2557404. http://doi.acm.org/10.1145/2556288.2557404

94. Prasad, M., Taele, P., Olubeko, A., Hammond, T.: Haptigo: a navigational tap on the shoulder. In: Haptics Symposium (HAPTICS), pp. 339–345. IEEE (2014). doi:10.1109/HAPTICS.2014.6775478
95. Rajan, P., Hammond, T.: From paper to machine: extracting strokes from images for use in sketch recognition. In: Proceedings of the Fifth Eurographics Conference on Sketch-Based Interfaces and Modeling, SBM'08, pp. 41–48. Eurographics Association, Aire-la-Ville (2008). doi:10.2312/SBM/SBM08/041-048. http://dx.doi.org/10.2312/SBM/SBM08/041-048
96. Rajan, P., Hammond, T.: Applying online sketch recognition algorithms to a scanned-in sketch. In: Proceedings of the Workshop on Sketch Recognition at the 14th International Conference of Intelligent User Interfaces Posters (IUI). ACM, Sanibel (2009)
97. Rajan, P., Taele, P., Hammond, T.: Evaluation of paper-pen based sketching interface. In: Proceedings of the 16th International Conference on Distributed Multimedia Systems (DMS), pp. 321–326 (2010)
98. Rajanna, V., Alamudun, F., Goldberg, D., Hammond, T.: Let me relax: toward automated sedentary state recognition and ubiquitous mental wellness solutions. In: Proceedings of the 5th EAI International Conference on Wireless Mobile Communication and Healthcare, MOBIHEALTH'15, pp. 28–33. ICST (Institute for Computer Sciences, Social-Informatics and Telecommunications Engineering), Brussels (2015). doi:10.4108/eai.14-10-2015.2261900. http://dx.doi.org/10.4108/eai.14-10-2015.2261900
99. Rajanna, V., Lara-Garduno, R., Behera, D.J., Madanagopal, K., Goldberg, D., Hammond, T.: Step up life: a context aware health assistant. In: Proceedings of the Third ACM SIGSPATIAL International Workshop on the Use of GIS in Public Health, HealthGIS'14, pp. 21–30. ACM, New York (2014). doi:10.1145/2676629.2676636. http://doi.acm.org/10.1145/2676629.2676636
100. Rajanna, V., Vo, P., Barth, J., Mjelde, M., Grey, T., Oduola, C., Hammond, T.: Kinohaptics: an automated, wearable, haptic assisted, physio-therapeutic system for post-surgery rehabilitation and self-care. J. Med. Syst. **40**(3), 1–12 (2015). doi:10.1007/s10916-015-0391-3. http://dx.doi.org/10.1007/s10916-015-0391-3
101. Raymond, D., Liew, J., Hammond, T.A.: A vision for education: transforming how formal systems are taught within mass lectures by using pen technology to create a personalized learning environment. In: Hammond, T., Valentine, S., Adler, A., Payton, M. (eds.) The Impact of Pen and Touch Technology on Education, pp. 355–363. Springer, Cham (2015). doi:10.1007/978-3-319-15594-4
102. Sezgin, T.M., Stahovich, T., Davis, R.: Sketch based interfaces: early processing for sketch understanding. In: ACM SIGGRAPH 2006 Courses. ACM (2006)
103. Shahzad, N., Paulson, B., Hammond, T.: Urdu qaeda: recognition system for isolated urdu characters. In: Proceedings of the Workshop on Sketch Recognition at the 14th International Conference of Intelligent User Interfaces (IUI). ACM, Sanibel (2009)
104. Taele, P., Barreto, L., Hammond, T.: Maestoso: an intelligent educational sketching tool for learning music theory. In: Proceedings of the Twenty-Ninth AAAI Conference on Artificial Intelligence, AAAI'15, pp. 3999–4005. AAAI Press (2015). http://dl.acm.org/citation.cfm?id=2888116.2888271
105. Taele, P., Dixon, D., Hammond, T.: Telling the user, "no"İ: Sketch recognition for improving sketch technique. In: SkCHI: Designing Sketch Recognition Interfaces, A CHI 2010 Workshop. Atlanta (2010)
106. Taele, P., Hammond, T.: Chinese characters as sketch diagrams using a geometric-based approach. In: Proceedings of the 2008 IEEE Symposium on Visual Languages and Human-Centric Computing (VL/HCC) Workshop on Sketch Tools for Diagramming, pp. 74–82. VL/HCC, Herrsching am Ammersee (2008)
107. Taele, P., Hammond, T.: Using a geometric-based sketch recognition approach to sketch chinese radicals. In: Proceedings of the 23rd National Conference on Artificial Intelligence, AAAI'08, vol. 3, pp. 1832–1833. AAAI Press (2008). http://dl.acm.org/citation.cfm?id=1620270.1620376

108. Taele, P., Hammond, T.: Hashigo: a next-generation sketch interactive system for Japanese kanji. In: Proceedings of the Twenty-First Innovative Applications of Artificial Intelligence Conference (IAAI), pp. 153–158. AAAI, Pasadena (2009)
109. Taele, P., Hammond, T.: Lamps: a sketch recognition-based teaching tool for mandarin phonetic symbols I. J. Vis. Lang. Comput. **21**(2), 109–120 (2010). doi:10.1016/j.jvlc.2009.12.004. http://dx.doi.org/10.1016/j.jvlc.2009.12.004
110. Taele, P., Hammond, T.: Initial approaches for extending sketch recognition to beyond-surface environments. In: CHI'12 Extended Abstracts on Human Factors in Computing Systems, CHI EA'12, pp. 2039–2044. ACM, New York (2012). doi:10.1145/2212776.2223749. http://doi.acm.org/10.1145/2212776.2223749
111. Taele, P., Hammond, T.: Adapting surface sketch recognition techniques for surfaceless sketches. In: Proceedings of the Twenty-Third International Joint Conference on Artificial Intelligence, IJCAI'13, pp. 3243–3244. AAAI Press (2013). http://dl.acm.org/citation.cfm?id=2540128.2540630
112. Taele, P., Hammond, T.: Developing sketch recognition and interaction techniques for intelligent surfaceless sketching user interfaces. In: Proceedings of the Companion Publication of the 19th International Conference on Intelligent User Interfaces, IUI Companion'14, pp. 53–56. ACM, New York (2014). doi:10.1145/2559184.2559185. http://doi.acm.org/10.1145/2559184.2559185
113. Taele, P., Hammond, T.: Enhancing instruction of written east asian languages with sketch recognition-based intelligent language workbook interfaces. In: Hammond, T., Valentine, S., Adler, A., Payton, M. (eds.) The Impact of Pen and Touch Technology on Education, 1st edn., pp. 119–126. Springer Publishing Company, Incorporated (2015)
114. Taele, P., Hammond, T.: Invisishapes: a recognition system for sketched 3D primitives in continuous interaction spaces. In: Proceedings of the 2015 International Symposium on Smart Graphics. Chengdu (2015)
115. Taele, P., Hammond, T.: An intelligent sketch-based educational interface for learning complex written east asian phonetic symbols. In: Hammond, T., Valentine, S., Adler, A. (eds.) Revolutionizing Education with Digital Ink: The Impact of Pen and Touch Technology on Education, pp. 129–140. Springer, Cham (2016). doi:10.1007/978-3-319-31193-7_9. http://dx.doi.org/10.1007/978-3-319-31193-7_9
116. Taele, P., Hammond, T.A.: A geometric-based sketch recognition approach for handwritten mandarin phonetic symbols I. In: 2008 International Workshop on Visual Languages and Computing (VLC) at the 14th International Conference on distributed Multimedia Systems (DMS). Knowledge Systems Insistute, Boston (2008)
117. Taele, P., Peschel, J., Hammond, T.: A sketch interactive approach to computer-assisted biology instruction. In: Proceedings of the Workshop on Sketch Recognition at the 14th International Conference of Intelligent User Interfaces Posters (IUI). ACM, Sanibel (2009)
118. Valentine, S., Field, M., Smith, A., Hammond, T.: A shape comparison technique for use in sketch-based tutoring systems. In: Proceedings of the 2011 Intelligent User Interfaces Workshop on Sketch Recognition. ASEE Conferences, Palo Alto (2011)
119. Valentine, S., Lara-Garduno, R., Linsey, J., Hammond, T.: Mechanix: A sketch-based tutoring system that automatically corrects hand-sketched statics homework. In: Hammond, T., Valentine, S., Adler, A., Payton, M. (eds.) The Impact of Pen and Touch Technology on Education, 1st edn., pp. 91–105. Springer Publishing Company, Incorporated (2015)
120. Valentine, S., Leyva-McMurtry, A., Borgos-Rodriguez, K., Hammond, T.: The digital sash: a sketch-based badge system in a social network for children. In: Hammond, T., Valentine, S., Adler, A. (eds.) Revolutionizing Education with Digital Ink: The Impact of Pen and Touch Technology on Education, pp. 179–189. Springer, Cham (2016). doi:10.1007/978-3-319-31193-7_12. http://dx.doi.org/10.1007/978-3-319-31193-7_12
121. Valentine, S., Vides, F., Lucchese, G., Turner, D., Kim, H.h., Li, W., Linsey, J., Hammond, T.: Mechanix: a sketch-based tutoring system for statics courses. In: Proceedings of the Twenty-Fourth Innovative Applications of Artificial Intelligence Conference (IAAI), pp. 2253–2260. AAAI, Toronto (2012)

122. Valentine, S., Vides, F., Lucchese, G., Turner, D., Kim, H.H., Li, W., Linsey, J., Hammond, T.: Mechanix: a sketch-based tutoring and grading system for free-body diagrams. AI Mag. **34**(1), 55–66 (2013)
123. Vangavolu, S., Wood, H., Newman, J., Polsley, S., Hammond, T.: Frontier: a directed graph system for web navigation. In: Proceedings of the 11th Biannual Conference on Italian SIGCHI Chapter, CHItaly 2015, pp. 82–85. ACM, New York (2015). doi:10.1145/2808435.2808465. http://doi.acm.org/10.1145/2808435.2808465
124. Veselova, O., Davis, R.: Perceptually based learning of shape descriptions for sketch recognition. ACM, New York (2006)
125. Vides, F., Taele, P., Kim, H., Ho, J., Hammond, T.: Intelligent feedback for kids using sketch recognition. In: ACM SIGCHI 2012 Conference on Human Factors in Computing Systems Workshop on Educational Interfaces, Software, and Technology. ACM (2012)
126. Williford, B., Taele, P., Nelligan, T., Li, W., Linsey, J., Hammond, T.: Persketchtivity: an intelligent pen-based educational application for design sketching instruction. In: Hammond, T., Valentine, S., Adler, A. (eds.) Revolutionizing Education with Digital Ink: The Impact of Pen and Touch Technology on Education, pp. 115–127. Springer, Cham (2016). doi:10.1007/978-3-319-31193-7_8. http://dx.doi.org/10.1007/978-3-319-31193-7_8
127. Wolin, A., Eoff, B., Hammond, T.: Shortstraw: a simple and effective corner finder for polylines. In: Proceedings of the Fifth Eurographics Conference on Sketch-Based Interfaces and Modeling, SBM'08, pp. 33–40. Eurographics Association, Aire-la-Ville (2008). doi:10.2312/SBM/SBM08/033-040. http://dx.doi.org/10.2312/SBM/SBM08/033-040
128. Wolin, A., Eoff, B., Hammond, T.: Search your mobile sketch: improving the ratio of interaction to information on mobile devices. In: Proceedings of the Workshop on Sketch Recognition at the 14th International Conference of Intelligent User Interfaces (IUI). ACM, Sanibel (2009)
129. Wolin, A., Field, M., Hammond, T.: Combining corners from multiple segmenters. In: Proceedings of the Eighth Eurographics Symposium on Sketch-Based Interfaces and Modeling, SBIM'11, pp. 117–124. ACM, New York (2011). doi:10.1145/2021164.2021185. http://doi.acm.org/10.1145/2021164.2021185
130. Wolin, A., Paulson, B., Hammond, T.: Eliminating false positives during corner finding by merging similar segments. In: Proceedings of the 23rd National Conference on Artificial Intelligence, AAAI'08, vol. 3, pp. 1836–1837. AAAI Press (2008). http://dl.acm.org/citation.cfm?id=1620270.1620378
131. Wolin, A., Paulson, B., Hammond, T.: Sort, merge, repeat: an algorithm for effectively finding corners in hand-sketched strokes. In: Proceedings of the 6th Eurographics Symposium on Sketch-Based Interfaces and Modeling, SBIM'09, pp. 93–99. ACM, New York (2009). doi:10.1145/1572741.1572758. http://doi.acm.org/10.1145/1572741.1572758
132. Zhu, Y., Johnston, J., Hammond, T.: Ringedit: A control point based editing approach in sketch recognition systems. In: Proceedings of the Workshop on Sketch Recognition at the 14th International Conference of Intelligent User Interfaces Posters (IUI). ACM, Sanibel (2009)

MIX
Papier aus verantwortungsvollen Quellen
Paper from responsible sources
FSC® C105338

If you have any concerns about our products,
you can contact us on
ProductSafety@springernature.com

In case Publisher is established outside the EU,
the EU authorized representative is:
Springer Nature Customer Service Center GmbH
Europaplatz 3, 69115 Heidelberg, Germany

Printed by Libri Plureos GmbH
in Hamburg, Germany